Studies in Big Data

Volume 22

Series editor

Janusz Kacprzyk, Polish Academy of Sciences, Warsaw, Poland
e-mail: kacprzyk@ibspan.waw.pl

About this Series

The series "Studies in Big Data" (SBD) publishes new developments and advances in the various areas of Big Data-quickly and with a high quality. The intent is to cover the theory, research, development, and applications of Big Data, as embedded in the fields of engineering, computer science, physics, economics and life sciences. The books of the series refer to the analysis and understanding of large, complex, and/or distributed data sets generated from recent digital sources coming from sensors or other physical instruments as well as simulations, crowd sourcing, social networks or other internet transactions, such as emails or video click streams and other. The series contains monographs, lecture notes and edited volumes in Big Data spanning the areas of computational intelligence incl. neural networks, evolutionary computation, soft computing, fuzzy systems, as well as artificial intelligence, data mining, modern statistics and Operations research, as well as self-organizing systems. Of particular value to both the contributors and the readership are the short publication timeframe and the world-wide distribution, which enable both wide and rapid dissemination of research output.

More information about this series at http://www.springer.com/series/11970

Constandinos X. Mavromoustakis
George Mastorakis · Ciprian Dobre
Editors

Advances in Mobile Cloud Computing and Big Data in the 5G Era

 Springer

Editors
Constandinos X. Mavromoustakis
Department of Computer Science
University of Nicosia
Nicosia
Cyprus

Ciprian Dobre
Department of Computer Science
University Politehnica of Bucharest
Bucharest
Romania

George Mastorakis
Department of Commerce and Marketing
Technological Educational Institute of Crete
Crete
Greece

ISSN 2197-6503 ISSN 2197-6511 (electronic)
Studies in Big Data
ISBN 978-3-319-83222-7 ISBN 978-3-319-45145-9 (eBook)
DOI 10.1007/978-3-319-45145-9

Printed on acid-free paper

This Springer imprint is published by Springer Nature
The registered company is Springer International Publishing AG
The registered company address is: Gewerbestrasse 11, 6330 Cham, Switzerland

To my wife Afrodyte for her continuous love, patience and support.

—Constandinos X. Mavromoustakis

To my wife Athina for her constant support and love.

—George Mastorakis

Special thanks go to my family, my daughter Anamaria Raluca and my wife Iulia Raluca, for their unconditional support, understanding and patience.

—Ciprian Dobre

Contents

Part III MCC and Big Data Paradigm in Smart Ambient Systems

Part IV MCC and Big Data Control and Data Management

Part I
Introduction and Applications of MCC and Big Data Paradigm in 5G Access

Handling Big Data in the Era of Internet of Things (IoT)

Koralia Papadokostaki, George Mastorakis, Spyros Panagiotakis,
Constandinos X. Mavromoustakis, Ciprian Dobre
and Jordi Mongay Batalla

Abstract In the emerging 5G mobile computing environments, the need for cutting edge technologies regarding data transmission, storage and processing will be more critical than ever. In addition, handling of Big Data that is produced by Internet of Things (IoT) devices and extracting value out of it, is a real challenge for scientists and markets, towards providing extra profit to the society. In this context, this chapter aims to shortly present the correlation between Internet of Things and the expansion of Big Data. At first, a short reference to the evolution of IoT and Big Data is provided and their features are then introduced. In addition, the lifecycle of Big Data in IoT—from capturing to storage and analysis—is shortly described. Finally, two different approaches for the implementation of Big Data are presented, as well as issues in privacy and security are addressed.

Keywords Internet of things (IoT) · Big data · Data storage · Privacy · Cloud computing · Wearables · RFIDs · WSNs

K. Papadokostaki (✉) · G. Mastorakis · S. Panagiotakis
Department of Informatics Engineering, Technological Educational Institute of Crete,
Heraklion, Crete, Greece
e-mail: mtp130@edu.teicrete.gr

G. Mastorakis
e-mail: gmastorakis@staff.teicrete.gr

S. Panagiotakis
e-mail: spanag@teicrete.gr

C.X. Mavromoustakis
Department of Computer Science, University of Nicosia, Nicosia, Cyprus
e-mail: mavromoustakis.c@unic.ac.cy

C. Dobre
Faculty of Automatic Control and Computers, Department of Computer Science,
University Politehnica of Bucharest, Bucharest, Romania
e-mail: ciprian.dobre@cs.pub.ro

J.M. Batalla
Warsaw University of Technology, Nowowiejska Str. 15/19, Warsaw, Poland
e-mail: jordim@interfree.it

© Springer International Publishing Switzerland 2017
C.X. Mavromoustakis et al. (eds.), *Advances in Mobile Cloud Computing and Big Data in the 5G Era*, Studies in Big Data 22,
DOI 10.1007/978-3-319-45145-9_1

1 Introduction

Internet of things (IoT) is an emerging paradigm in the science of computers and technology in general. In the last years it has invaded our lives and is gaining ground as one of the most promising technologies. According to the European Commission, IoT involves "Things having identities and virtual personalities operating in smart spaces using intelligent interfaces to connect and communicate within social, environmental, and user contexts" [1, 2]. IoT has been in the centre of focus for quite some years now, but in the last years it has actually become reality allowing "people and things to be connected Anytime, Anyplace, with Anything and Anyone, ideally using Any path/network and Any service" [3].

Internet of Things has a big impact in communities. Smart cities making use of sensors will be able to monitor air pollution or traffic, better manage waste and provide smart agriculture [4–6]. In the case of health, sensors on patients will monitor their health condition and initiate an alarm in critical cases [7, 8]. IoT will also change retail, logistics and transportation [6] whereas it can have a great impact on energy management [2, 5, 6]. As IoT gradually becomes reality, novel appliances appear in the market, promising to make our homes smarter and our lives easier: e.g. a smart fork helping its owner monitor and track his eating habits [9] or a smart lighting system adjusting automatically to the outside light intensity, the presence of people in the house and their preferences [1]. But not only are homes made smart, but also our accessories have changed: small wearable devices monitor our everyday movements and are capable of calculating the steps we made and the calories we consumed, or can even store details about our sleeping habits [10]. Moreover, smart watches with e-SIM technology can substitute our mobile phones and jewellery has turned into powerful gadgets (aka smart jewellery) [11]. This rapid evolution of non-mobile gadgets and mobile wearables has become the beginning of a new era! Imagine small, wireless devices on cars, in homes, even on clothes or food items [1]. The tracking of merchandise will be automatically monitored, the conditions at which food supply is stored will also be recorded and our clothes will be separated automatically—based on colour and textile—for the smart washing machines. But then, the production of IoT data will be enormous! So, the prevalence of Internet in our lives and especially of Internet of the Things will cause an explosion of data. According to a report from International Data Corporation (IDC) the overall created and copied data volume of the world was 1.8ZB ($\approx 10^{21}$ B) in 2011 which increased by 9 times within five years [12]. On top of that, Internet companies handle each day huge volumes of data, e.g. Google processes hundreds of PB and generates log data of over 10 PB per month.

This massive amount of data is referred to as Big Data and its capturing, storage and analysis is a challenge in the era of IoT. Discovering patterns provided by Big Data analytics can cause breakthroughs in science, e.g. medicine and cause an impact on worldwide finance. In public sector making data widely available can reduce search and processing time. In industry, the results of R&D can improve quality. When it comes to business, Big Data allows organizations to customize products and services precisely to meet the needs of the clients. Handling data and being able to extract conclusions out of it can also be important in decision making, e.g. discover trends in pricing or sales. Moreover, it can be the root of innovation; new services and products can be created; for example casualty insurance can be based on the driving habits of the driver [13]. It is indisputable that the integration of IoT with Big Data will be in the focus of research and markets in the future.

This chapter focuses on the vast amounts of data produced by IoT which are expected to grow immensely in the years to come. In Sect. 2, Big Data and its features are defined, and sectors where Big Data has its biggest impact are shortly presented. Section 3 approaches the Big Data from the perspective of IoT, examining their characteristics and requirements to be met and Sect. 4 describes two technologies developed for Big Data. Cloud technologies used in Big Data are then discussed in Sect. 5 and this chapter ends with an analysis of the security and privacy risks concerning Big Data from IoT.

2 Big Data: Definition, Value and Features

Defining Big Data is not easy. Up to 2011, Big Data was described by 3 V's (Volume, Variety and Velocity) [12] but with a report by IDC [14] where "big data technologies describe a new generation of technologies and architectures, designed to economically extract value from very large volumes of a wide variety of data, by enabling the high-velocity capture, discovery, and/or analysis.", a new dimension has been added to the Big Data characteristics. Currently four characteristics—also known as 4 V's—can be identified as the main features of Big Data [12, 15]:

a. Volume refers to the constantly growing volume of data generated from different sources and which traditional databases cannot handle [16].
b. Variety represents various data collected via sensors, smartphones and social networks. The data collected may be text, image, audio, video, in structured, unstructured or semi-structured format.
c. Velocity indicates the acquisition speed of the data and additionally the speed at which the data should be processed and analyzed.
d. Value of Big Data implies the extraction of knowledge or patterns out of the raw data.

More recent approaches add an extra V to the features of Big Data; that of Veracity which refers to the possible untrustworthiness and noise that may be hidden in the data [17, 18].

The Value of Big data can prove invaluable. In [13] it is estimated that Europe's potential value to Europe's public sector administration would reach 250 billion of Euros if Big Data was exploited. And not only public sector but also retail and sales is an area where Big Data analysis produces great value and revenues. According to

[13], if retailers made use of Big Data they would gain an extra profit of 60 %, while analyzing data on in-store behavior in a supermarket can influence the layout of the store, product mix, and shelf positioning of the product. It can even affect the price of products; for example in rural areas, where rice and butter are of higher buying priority, the prices are not elastic. On the contrary, cereals and candy are rated much higher among the priorities of urban consumers [13].

Another highly important area where value of Big Data can be extracted is Health care. A patient's electronic medical file will not only be used for his own health monitoring. It will be compared and analyzed with thousands of others, and may lead to prediction of specific threats or discovery of patterns that emerge during the comparison [13, 19]. A doctor will be able to assess the possible result of a treatment, based upon data concerning other patients with the same condition, genetic factors and lifestyle. Data analysis can prove to be not only precious but life-saving [20].

As mentioned before, Big Data is moreover characterized by its big Volume and its Variety of types of data. Therefore, it is very hard for Relational DataBase Management Systems (RDBMS) to be utilized in storing and processing Big Data, since Big Data has some special features [12, 16]:

- Data representation is critical in selecting models to store big data, because it affects the data analysis.
- Redundancy reduction and data compression: Especially when it comes to data generated from sensors, the data is highly redundant and should be filtered or compressed.
- Data cycle life management: As data aggregates at unbelievable rates, it will be hard for traditional storage systems to support such an explosion of data. Moreover, the age of data is very important, in order for our conclusions to be up-to-date.
- Analytical mechanism: Masses of heterogeneous data should be processed within limited time. Therefore, hybrid architectures combining relational and non-relational databases have been proposed.
- Data confidentiality: Data transmitted and analyzed may be sensitive, e.g. credit card numbers and should be treated with care.
- Energy management: Despite the massive generation of data, it is still important to keep energy demands low during storage, processing and transmission.
- Scalability and Expendability. The models handling the data must be scalable, flexible and able to adjust to future datasets.
- Cooperation: As big data can be used in various sectors, cooperation on the analysis and exploiting of big data is significant.

It is now evident, that such data cannot be handled using SQL-queried RDBMS and new tools and technologies (e.g. Hadoop, which is an open source distributed data processing system developed by Apache or NoSQL databases) need to be deployed.

3 Big Data and Internet of Things

With the prevalence of IoT into our lives, a huge number of 'things' can join the IoT: low-cost, low-power sensors connected wirelessly produce huge amounts of information and make use of a plethora of internet addresses, thanks to the IPv6 protocol (2^128 addresses). Seagate predicts that, by 2025, more than 40 billion devices will be web-connected and the bulk of IP traffic would be driven by non-PC devices. IP traffic due to Machine-to-machine (M2M) devices is estimated to account for 64 %, with smartphones contributing 26 %, and tablets only for 5 %. What is amazing is that laptop PCs will contribute only 1 % [21]. Moreover, wearable devices seem to become part of our lives. Seagate claims that 11 million smart watches and 32 activity trackers were sold in 2015. According to predictions, sales of fitness wearables will triple to 210 million by 2020, from 70 million in 2013 [21]. As Big Data is already on its way, the two technologies depend on each other and integrate intimately: The expansion of IoT can lead to a faster development of big data, whereas the application of Big Data to IoT can accelerate the research advances in IoT.

When it comes to data acquisition and transmission, three layers comprise the architecture in IoT: (i) the sensing/perception layer: that is the technologies responsible for data acquisition. It mainly consists of sensors (RFIDs) or wireless sensor networks (WSNs) [11, 22] that generate small quantities of data (ii) the network layer: whose main role is sending the data in the sensor network or through the internet and finally (iii) the application network which implements specific applications in IoT. Below we will look into the most usual methods used for data acquisition in IoT in the sensing layer:

- Through RFID Sensing: Each sensor carries an RFID (Radio Frequency IDentification) tag. It allows another sensor (reader) to read, from a distance a unique product identification code (EPC standing for Electronic Product Code) associated with that tag. The reader then sends the data to a data collection point (server). The RFID tag need not have power supply, as it is powered by the sensor reader; therefore its life time is big. (Actually, Active RFID tags have been developed, which are energy-sufficient) [2, 12, 23].
- Wireless Sensor Nodes (WSN) is another popular solution for the interconnection of sensors, as it meets the need for energy efficiency, scalability and reliability. A WSN consists of various nodes, each one containing sensors, processors, transceivers and power supplies. The nodes of each WSN should implement light-weight protocols in order to communicate and interact with the internet. They mainly implement the IEEE 802.15.4 protocol and support peer-to-peer communication [12, 22, 23].
- Through Mobile equipment: Mobile equipment is a powerful tool in our hands and also a medium for sending various types of data: positioning systems acquire information about the location, images or videos can be captured through the camera, audio can be recorded by the microphone and text can be input through the (sometimes virtual) keyboard. Moreover, mobile phones

nowadays have a large set of embedded sensors, such as GPS, compasses, accelerometers, proximity and meteorological sensors [24] and together with the applications that have access to data stored in a smartphone, mobile devices serve as a source of growing data as well [12, 23, 25].

Data generated by IoT technology can be in structured, unstructured or semi-structured format [15]. Structured data is often managed by SQL, might be easy to input, query, store, and analyze and can be stored in RDBMS. Examples of structured data include numbers, words, and dates. Unstructured data may consist of text messages, location information, videos, and social media data and does not apply to a specific format. The need to analyze and extract value out of such data has become a challenge since the size of this type of data rises daily through the use of smart devices. In between, lays the semi-structured data, which does not follow a conventional database system, but may be in form of structured data. Additionally data generated by IoT has the following features [12]:

- Large scale: In IoT not only is the amount of data and of sensors vast, but often a need for historical data should be met (e.g. temperature in a room, or a home surveillance video). Therefore, data generated is of very large scale.
- Heterogeneity: As the sensors vary in type and input, the data provided by IoT also varies in type: for example it might be numeric (temperature), text (message by a person) or image (a photo of a room) [16].
- Time and space correlation: In applications in IoT, time and place are important and therefore, data should be also tagged with place and time information. For example the heart-rate of a person should also be combined with the time and place where it was measured (e.g. gym).
- Effective data is only a small part of the big data: often among data collected, only a small portion is usable.

As Big Data derives from various sources and is often redundant, pre-processing is necessary, in order to reduce storage needs and maximize the analysis efficiency. Pre-processing Big Data includes:

- Integration: Data Integration is the method of combining data from different sources and integrating them to appear as homogeneous. For example, it might consist of a virtual database showing data from different data sources, but actually containing no data, but rather "indexes" to the actual data. Integration is vital in traditional databases [12].
- Cleaning: Cleaning the data involves identifying inaccurate and incomplete data and modifying them in order to enhance data quality. In the case of IoT, data generated from RFIDs can be abnormal and need to be corrected. This is due to the physical design of the RFIDs and the environmental noise that may corrupt the data [12, 23].
- Redundancy elimination: As mentioned earlier in this chapter, IoT data can be quite redundant. That is, the same data can repeat without change and without surplus value; for example, when an RFID stays at the same place. In practice, it

is useful to transmit only interesting movements and activities on the item. As data redundancy wastes storage space, leads to data inconsistency and can slow down the data analysis, it is significant to reduce the redundancy. Redundancy detection, data filtering and data compression are proposed for data redundancy elimination. The method selected depends on the data collected and the requirements of the application [12, 23].

As aforementioned, the large volumes of Big Data together with the importance of extracting value from them are two basic features of big data. It is therefore vital to store Big Data assuring that the storage is reliable and that query and analysis of the data can be made quickly and with accuracy. Various proposals have been made for the storage of Big Data, such as DAS, NAS and SAN [12, 26], but are beyond the scope of this review. Moreover, distributed systems have been implemented for the best storage of big data. Below, we will shortly present Storage Mechanisms which have been introduced for Big Data. Google, a giant of Internet and Big Data, has developed a File System to provide for the low level storage of Big Data. The Google File System (GFS) was designed "as a scalable distributed file system for large distributed data-intensive applications", providing "fault tolerance while running on inexpensive commodity hardware," and delivering "high aggregate performance to a large number of clients [27, 28]. Certain deficiencies in GFS drove to the design of Colossus, the successor of GFS, which is specifically built for real-time services. As RDBMS fail to meet the needs for Big Data Storage, NoSQL (Not only SQL) databases gain popularity for Big Data [23, 26]. They can handle large volumes of data, support simple and flexible non-relational data models and offer high availability. The most basic models of NoSQL databases and a brief presentation of each are presented below:

- Key-value databases: These NoSQL databases are based on a simple data model based on key-value pairs, which reminds of a dictionary. The key is the unique identifier of the value and allows us to store and retrieve the value. This method provides a schema—free data model, which is suitable for distributed data, but cannot represent relations. Dynamo—developed by Amazon—and Voldemort—used by LinkedIn—are the most important representatives of this category [15, 29–31].
- Column—family databases: Inspired by Google BigTable, these databases store data in a column oriented way. The key concept here is that the dataset consists of several rows, each of which addressed by a primary key. Each row consists of a set of column families, but different rows may have different column families. Cassandra, HBase and HyperTable also belong to Column-family databases [15, 29–31].
- Document stores: Data usually stored in these databases are represented by documents using JSON (JavaScript Object Notation) or some format derived from it. These databases are best suitable for cases in which the input data can be represented in a document format. MongoDB, SimpleDB and CouchDB are popular examples of Document stores [15, 29–31].

- Graph databases: Graph databases are based on graph theory and use nodes and edges to represent objects and relations. They can handle highly interconnected data and therefore are very efficient in representing relationships between different entities. They are extremely helpful for social networking applications or pattern recognition [15, 29–31].

One important issue that should be taken into account about Big Data is the technology used to query the data stored in the databases. One of the most popular tools, MapReduce, was first developed by Google and offers distributed data processing on a cluster of computers [23]. However, as SQL is extremely popular in querying, it has now also been adopted in the NoSQL world. Some of the prominent NoSQL data stores like MongoDB offer an SQL-like query language or similar variants such as CQL offered by Cassandra and SparQL.

The analysis of Big Data is the next phase in the chain of Big Data. It aims to extract value out of the data in various fields and is quite complex and demanding. Some of the methods applied for analyzing Big Data are discussed in the following [12, 13, 19]:

- A/B Testing is a technology for determining how to improve target variables by comparing the tested group. Big data will require a large number of tests to be executed and analyzed; but at the same time enables large amount of tests to be executed, ensuring, this way, that groups are statistically significant to detect differences between the control and treatment groups.
- Statistical Analysis can use probability to make conclusions about data relations between variables (for example, using the "null hypothesis"). Statistics can also reduce the likelihood of Type I errors ("false positives") and Type II errors ("false negatives").
- Data mining algorithms are algorithms which extract patterns from large datasets with the use of statistics and machine learning with database management. Methods include association rule learning, cluster analysis, classification, and regression; the most popular algorithms are C4.5, k-means, SVM, Apriori, EM, Naive Bayes, and Cart.
- Cluster (Correlation) Analysis is a statistical method for classifying objects. It divides them into categories (clusters) according to common features; objects in the same cluster have high homogeneity, while different clusters will be very different from each other.
- Regression Analysis is a mathematical method used to reveal correlations between one variable and other variables. It is used for forecasting or prediction and is common in sales forecasting.
- Neural networks are computational intelligence tools that can be valuable for classification or prediction.
- Genetic algorithms mimic the natural process of natural evolutions and are used as a methodology for optimization.

Furthermore, specific tools have been developed for the data mining and analysis of Big Data. They include both commercial and open source software the most popular of which are summarized below [12]:

- R is an open-source programming language and software environment for statistical computing and visualization. It has become very popular for statistical software development and data analysis and several data giants, like Oracle have developed products to support R [32, 33].
- Rapid-I Rapidminer is open source software suitable for data mining, machine learning, and predictive analysis. The data mining flow is represented in XML and displayed through a graphic user interface (GUI). The source code is written in Java and can support the R language [33].
- WEKA/Pentaho belong to non-commercial, open-source machine learning and data mining software written in Java. Weka is very popular for data processing, feature selection, classification, regression, clustering, association rule, and visualization. Pentaho is one of the most popular open-source BI (Business Intelligence) software. Weka and Pentaho can cooperate very intimately [26, 33, 34].
- KNIME is another open-source platform used for rich data integration, data processing, data analysis, and data mining. It is user-friendly, was written in Java and can extend its functionality with additional plug-ins (including WEKA and toolkit R) [33].
- Excel provided in the Microsoft Office Suite also offers tools for statistical analysis and data processing. It is however commercial and may require the installation of additional plug-ins to provide full functionality.

Finally, the results extracted from the previous steps should be displayed visually. As knowledge should be extracted out of large-scale and complex data, visualization data is an essential tool of Business Intelligence and various techniques have been designed to meet diverse needs. Bar charts, line charts, pie charts and scatter plots are common examples of techniques used for the visualization of numerical data with only a few dimensions, whereas heat maps and scatter plot matrices are used for the display of numeric data with multiple dimensions. As data can often be text, new representations have been proposed, such as the word clouds (often seen as tag clouds) [13, 15, 16, 19]. Additionally, in the era of IoT, the data generated by wearables or mobile devices is also often geotagged and therefore Geo Maps and geo-spatial visualizations are adopted for the visualization of the location information [3, 17, 19]. Lastly, data can be hierarchically represented with tree maps or hyperbolic views, while cluster analysis can be achieved through clustergrams [17–19].

However, apart from the content that visualization tools provide, interaction is also significant as it provides the data analyst with flexibility while viewing the graphical results. Such interactive Data Visualization software, Tableau, is used by eBay employees to conduct tendency analysis based on their customers' feedback [26].

4 Implementations

4.1 The Hadoop Project

While Google was working on GFS, Yahoo! was implementing Hadoop. Hadoop is an Apache open source distributed file system widely used in industry for Big Data applications as well as a framework for the analysis and transformation of very large data sets [15, 35–37]: Yahoo claims their clusters to consist of more than 4000 nodes, while at the same time, Facebook makes use of the Hadoop Distributed File System as a source for reporting/analytics and machine learning [37]. Various large and small scale companies use the Hadoop [38]. At the moment, Hadoop has evolved into a complete system of storing and querying big data, consisting of [23, 39, 40]:

- HDFS: the Hadoop Distributed File System storing large files with replication across multiple machines.
- Map Reduce Engine: a framework for parallel processing of large data sets. Its name implies its functions: this framework gathers all sets with the common key from all records and joins them together. Additionally, it can offer indexing capabilities in the semantic web.
- HBASE provides a scalable, distributed database that supports structured data storage for large tables. It resembles Google's BigTable but is open source.
- HIVE is Hadoop's infrastructure used for data summarization and ad-hoc querying and can presently be used autonomously as well.
- Pig framework involves a high-level scripting language (Pig Latin) and offers a run-time platform that allows users to execute MapReduce on Hadoop. Moreover, its data model and transformation language resemble RDF and the SPARQL query language respectively.
- Mahout is a machine-learning and data-mining library that has four main functionalities: collective filtering, categorization, clustering, and parallel frequent pattern mining.

4.2 Semantic Sensor Web

Apart from traditional Big Data storing and analysis tools that can be used on data generated by IoT, tools specifically designed for IoT have been developed. In the case of sensors, data volume is enormous; thus the semantic annotation of the underlying data is extremely important in order to analyze the data and produce value. Semantic Sensor Web is the use of Semantic Web oriented in the data generated by sensors. Its main features are [23, 39]:

- Data are encoded with XML. Extensible Markup Language (XML) is used to represent the semantic meaning of data; making data thus understandable by machines.

- The Resource Description Framework (RDF) is used to express XML identifiers. RDF uses triples, with each triple being a subject, verb, and object of an element. Each element is defined as a Uniform Resource Identifier (URI) on the Web.
- Ontologies can express relationships between identifiers. They can represent the relationships among these sensors in order to be able to make the appropriate conversion.

In simple terms, the Semantic Sensor Web enables true semantic interoperability and integration over sensor data. Two important knowledge representation systems are here presented: RDF, standing for "Resource Description Framework", is a language to describe resources [41]. A resource can actually be anything or in the world. For instance, it could be a person, a place, a restaurant entree etc. OWL stands for Ontology Web Language [42, 43]. This was developed to overcome the challenges with RDF, which do not provide ways to represent constraints (such as domain or range constraints). While RDF and OWL can be used to represent data, the challenge in Big Data is in querying. SPARQL is an RDF query language for querying and manipulating data which are stored in the RDF format [23, 44, 45]. SPARQL allows writing queries over data as perceived as triples. Moreover, an extension of SPARQL, called SPARQLstream [23] offers querying over RDF streams. This is valuable in the context of sensor data, which is generally stream-based. In general, it seems that Semantic Web could prove as a precious tool in the hands of the researchers for the notation of data to be stored, and that research should be driven this way.

5 Big Data Over the Cloud

As Big Data handles enormous amounts of data and needs large scale resources for storing, and analyzing it, the cloud technology is ideal for its implementation. Besides, the evolution of both IoT and Big Data can be attributed to the evolvement of cloud computing as well [46–48]. At the moment, three types of cloud deployment models exist, namely, a public, private and hybrid cloud [49, 50]. The decision between these three models can be crucial.

In public cloud, there are companies offering cloud services to organizations that need it, e.g. enterprises. The infrastructure is physically common but virtually separated for the customers of the company. This model is a pay-per-use model and can offer significant advantages to the organizations that make use of it. They can increase or decrease the use of the cloud according to their demands and need not invest large amounts to implement their own cloud as in private cloud [14, 50]. Private cloud allows the organization to set up cloud technologies in its own datacenter. Companies that provide tailor-made solutions for private cloud can install the necessary infrastructure for the implementation of the cloud locally; the cloud belongs to and is managed by the organization itself. This offers greater security and full control of the data to the organization [50].

Hybrid cloud has emerged as a combination of private and public cloud; an organization can use both public and private cloud, depending on the demands of the organization. Usually, the public cloud is used for increased computer capacity demands, without having to grow the private cloud. This technology can make use of the enormous potential of the public cloud in tools, computing power and storing space, whereas it can still maintain high level of security and privacy for data. A further way of utilizing the hybrid cloud is by storing the sensitive data in the private cloud, while sending the non-sensitive large scale data to the public cloud [49, 50]. With the rapid grow of mobile users and devices, the technology of Mobile Cloud Computing (MCC) is where mobile devices, cloud and the Internet of Things converge [46, 51]. With the future applications enabled by 5G, MCC will radically change our digital lives, while various issues, such as energy efficiency [46–48, 52–56], availability, and security and privacy need to be addressed [51].

6 Privacy and Security Issues

As Internet of the Things has omnipresence in our lives, it is critical to consider privacy and security issues that may arise. Whether location information, credit card numbers, medical history or everyday habits, each user has the right to keep personal information safe and away from eavesdroppers. In a larger scale, keeping data secure and private is a great concern for companies dealing with sensitive data or with data of great value. As privacy issues may arise at the collection stage or during data management, we need to distinguish the issues of IoT according to the three layers of IoT architecture:

In the sensing/perception layer the data is collected through RFID, WSNs, and RSN (RFID Sensor Networks). The perception layer typically consists of perception nodes (usually sensors) and a network that sends data to a gateway. Privacy in RFID tags can be ensured through physical-based schemes, such as "killing" (i.e. disabling the tag), blocking (for a limited time) or use of pseudonyms [23, 57]. Another way to protect privacy is through password-based schemes, such as hash locks, anonymous ID or re-encryption [23, 57]. Furthermore, trust management can be achieved through cryptographic algorithms and protocols to implement digital signature technology not only in RFIDs, but in readers and base stations as well to avoid eavesdropping.

As data of IoT "travel" through WSNs they may be tampered with, subject to eavesdropping or maliciously routed. It is therefore critical to ensure confidentiality, authenticity, integrity, and freshness for data collected. Cryptographic algorithms, key management, secure routing and node trust can provide solutions to those key issues. Cryptographic algorithms for WSNs include symmetric encryption algorithms and asymmetric encryption algorithms; with the former being mature but not extremely secure and the latter being more promising but still under research. Key management, i.e. generation, distribution, storage, updating and destruction of the key in WSNs is also of primary importance as the key should be safely transmitted to legitimate users only. Again the protocol implementing the key management should be lightweight and energy efficient. Secure routing in WSNs should ensure that the certification is made between the nodes and therefore secure routing protocols designed for wireless sensor networks have been proposed taking into consideration the limitations of WSNs. Last but not least, node trust is crucial in WSNs as the vulnerability of a single node to malicious assaults can produce fake data and consequently lead our system to erroneous or false results [57].

The integration of RFIDs and WSNs has led to the technology of RSNs (RFID sensor networks), which faces the challenge of heterogeneous data and large number of nodes, as data comes from both RFIDs and WSNs and therefore has different format and communication protocol. In the transport layer, three sub-layers interact with each other and in each there are security and privacy risks to be addressed: The access layer, the core layer and the Local Area Network. The access layer, providing the access to the sensors and nodes in the perception layer is implemented by Wi-Fi, Ad-Hoc or 3G technology. 3G is vulnerable to DDos/Dos attacks, phishing attacks and identity attacks while Ad-hoc networks need authentication and key management to ensure legitimate users are the only nodes of the network. On the other hand, Wi-Fi currently implements TKIP and AES to ensure encryption, PPPoE and other protocols for authentication, and WPA for access control, but privacy and security remains a critical issue. Secondly, the core layer, responsible for the data transmission, mainly uses the 6LowPan, a protocol supporting IPv6, but also caring for low power requirements. Finally, establishing security in the Local Area Network (LAN) of IoT could add extra security to our IoT implementations [57].

In the Application support layer of IoT, valid data, spam and even malicious data should be recognized and filtered in time as masses of data arrive at high speed. It may include middleware, M2M, cloud computing platform or service support platform. M2 M is a prevalent technology and as data can be critical or sensitive, security requirements should be met: User authentication, data integrity and access control are key issues, while privacy protection is also extremely important. Techniques such as k-anonymity (aggregation of data), suppression (removal of sensitive data), addition of noise, data conversion and data randomization have been proposed [57, 58]. Another risk factor for IoT data is that the data is often stored in cloud computation platforms, where leaks are scarce but not improbable; the decision to use cloud platforms for sensitive e.g. medical or financial application should be carefully taken. Above the application support layer, lays the application layer where various implementations of IoT handle Big Data according to their specifications and functions. Levels of required privacy and security depend upon the application, i.e. some IoT applications may require advanced security levels, whereas others may demand high-level privacy measures. The cooperation of security solutions across the layers of IoT can lead to a more secure and privacy aware IoT.

Many IoT applications are based on aggregate and not on individual data; therefore eliminating personal data from the aggregate data can improve the privacy significantly. For example, during the monitoring of traffic in roads plate details need not be stored but only the number of cars. Moreover, diminishing the location data can be accomplished by post-processing of data that is to be shared. Again, in the example of the traffic control, if a video was captured, the plate number should be blurred before its sharing. What would, however, be ideal, is giving the users the right to specify which kinds of their data could be shared. The W3C group has defined the Platform for Privacy Preferences (P3P) [59], which provides a language for description of privacy preferences, allowing the user to set specific privacy requirements.

Similar is the approach proposed for the semantic web: the users can specify what information can be accessed in different contexts and by various sources. This need led to the development of RDF-S, which includes markup of privacy information and of OWL-S which is of similar functionality. As data on Semantic Web is dynamic and open a lot of research yet is to be done [23]. What should finally be considered by researchers is that Big Data from IoT is both heterogeneous and generated from small IoT devices [57]. The challenge hence is to deploy solutions that can be used in lightweight and energy-efficient devices and which should be adjusted to the requirements of each application [60–72].

7 Conclusions

Exploiting Big Data to expand human knowledge in science and global finance is extremely important and will create new horizons for the next generations [73–76]. Technologies to support the giant explosion of data in the years to come are therefore necessary and should be adjusted to the era of the Internet of the Things [77–82]. In this chapter, we tried to investigate the relation between Internet of Things and Big Data and proved that they are intimately correlated and of big gravity in the years to come. A brief presentation in the lifecycle of Big Data has been attempted, whereas two technologies, the Hadoop project and the Semantic Sensor Web have been shortly described. Moreover, a short reference to the public, private and hybrid cloud has been made and the chapter concluded with security and privacy issues regarding Big Data from IoT [83–86]. With 5G communication systems expected to prevail in the years to come and Web 3.0 on its way, Big Data to be produced will exponentially grow following the three As (Anybody, Anything, Anywhere) of Internet of Things [87–90]. Accordingly, a lot of research should be done on the data produced by IoT devices, how it should be handled throughout its lifecycle and most importantly how it should be utilized [91–94].

References

1. Atzori, L., Iera, A., Morabito, G.: The internet of things: a survey. Comput. Netw. **54**(15), 2787–2805 (2010)
2. Gubbi, J., Buyya, R., Marusic, S., Palaniswami, M.: Internet of Things (IoT): a vision, architectural elements, and future directions. Fut. Gen. Comput. Syst. **29**(7), 1645–1660 (2013)
3. Vermesan, O., Friess, P., Guillemin, P., Gusmeroli, S., Sundmaeker, H., Bassi, A., Jubert, I.S., Mazura, M., Harrison, M., Eisenhauer, M., Doody, P.: Internet of things strategic research roadmap. In: Internet of Things-Global Technological and Societal Trends, pp. 9–52 (2011)
4. Kitchin, R.: The real-time city? Big data and smart urbanism. GeoJournal **79**(1), 1–14 (2014)
5. Perera, C., Zaslavsky, A., Christen, P., Georgakopoulos, D.: Sensing as a service model for smart cities supported by internet of things. Trans. Emerg. Telecommun. Technol. **25**(1), 81–93 (2014)
6. Khan, R., Khan, S.U., Zaheer, R., Khan, S.: Future internet: the internet of things architecture, possible applications and key challenges. In: 2012 10th International Conference on Frontiers of Information Technology (FIT), pp. 257–260. IEEE (2012)
7. Domingo, M.C.: An overview of the Internet of Things for people with disabilities. J. Netw. Comput. Appl. **35**(2), 584–596 (2012)
8. Bandyopadhyay, D., Sen, J.: Internet of things: applications and challenges in technology and standardization. Wirel. Pers. Commun. **58**(1), 49–69 (2011)
9. Wareable: Connected cooking: the best smart kitchen devices and appliances. http://www.wareable.com/smart-home/best-smart-kitchen-devices. Accessed 20 April 2016
10. Wareable: 2016 preview: how wearables are going to take over your life. http://www.wareable.com/wearable-tech/2016-how-wearables-will-take-over-your-life-2096. Accessed 20 April 2016
11. Wareable: Semi-precious: the best smart jewellery. http://www.wareable.com/smart-jewellery/semi-precious-the-best-smart-jewelry-582
12. Chen, M., Mao, S., Liu, Y.: Big data: a survey. Mobile Netw. Appl. **19**(2), 171–209 (2014)
13. Manyika, J., Chui, M., Brown, B., Bughin, J., Dobbs, R., Roxburgh, C., Byers, A.H.: Big data: the next frontier for innovation, competition, and productivity (2011)
14. Gantz, J., Reinsel, D.: Extracting value from Chaos State of the Universe. IDC (International Data Corporation) June (2011)
15. Hashem, I.A.T., Yaqoob, I., Anuar, N.B., Mokhtar, S., Gani, A., Khan, S.U.: The rise of "big data" on cloud computing: review and open research issues. Inf. Syst. **47**, 98–115 (2015)
16. Che, D., Safran, M., Peng, Z.: From big data to big data mining: challenges, issues, and opportunities. In: Database Systems for Advanced Applications, pp. 1–15. Springer, Berlin (2013)
17. Zikopoulos, P., deRoos, D., Bienko, C., Buglio, R., Andrews, M.: Big Data Beyond the Hype: A Guide to Conversations for Today's Data Center (2015)
18. Jin, X., Wah, B.W., Cheng, X., Wang, Y.: Significance and challenges of big data research. Big Data Res. **2**(2), 59–64 (2015)
19. Zarate Santovena, A.: Big data: evolution, components, challenges and opportunities. Doctoral dissertation, Massachusetts Institute of Technology (2013)
20. Bernard, M.: How big data is changing healthcare. http://www.forbes.com/sites/bernardmarr/2015/04/21/how-big-data-is-changing-healthcare. Accessed: 20 April 2016
21. Yu, E.: IoT to generate massive data, help grow biz revenue. http://www.zdnet.com/article/iot-to-generate-massive-data-help-grow-biz-revenue/. Accessed 20 April 2016
22. Pantelaki, K., Panagiotakis, S., Vlissidis, A.: Survey of the IEEE 802.15.4 Standard's developments for wireless sensor networking. Am. J. Mobile Syst. Appl. Serv. **2**(1), 13–31 (2016)
23. Aggarwal, C.C., Ashish, N., Sheth, A.P.: The internet of things: a survey from the data-centric perspective (2013)

24. Vakintis, I., Panagiotakis, S.: Middleware Platform for Mobile Crowd-Sensing Applications using HTML5 APIs and Web Technologies, chapter contribution in the HandBook "Internet of Things (IoT) in 5G Mobile Technologies. Springer (2016)
25. Mattern, F., Floerkemeier, C.: From the internet of computers to the internet of things. In: From Active Data Management to Event-Based Systems and More, pp. 242–259. Springer, Berlin (2010)
26. Chen, C.P., Zhang, C.Y.: Data-intensive applications, challenges, techniques and technologies: a survey on Big Data. Inf. Sci. **275**, 314–347 (2014)
27. Corbett, J.C., Dean, J., Epstein, M., Fikes, A., Frost, C., Furman, J.J., Ghemawat, S., Gubarev, A., Gubarev, Heiser, C., Hochschild, P., Hsieh, W.: Spanner: Google's globally distributed database. ACM Trans. Comput. Syst. (TOCS) **31**(3), 8 (2013)
28. Ghemawat, S., Gobioff, H., Leung, S.T.: The Google file system. ACM SIGOPS Operating Systems Review, vol. 37, no. 5, pp. 29–43. ACM (2003)
29. Grolinger, K., Higashino, W.A., Tiwari, A., Capretz, M.A.: Data management in cloud environments: NoSQL and NewSQL data stores. J. Cloud Comput.: Adv Syst. Appl. **2**(1), 22 (2013)
30. Tauro, C.J., Aravindh, S., Shreeharsha, A.B.: Comparative study of the new generation, agile, scalable, high performance NOSQL databases. Int. J. Comput. Appl. **48**(20), 1–4 (2012)
31. Sun, Y., Yan, H., Zhang, J., Xia, Y., Wang, S., Bie, R., Tian, Y.: Organizing and querying the big sensing data with event-linked network in the internet of things. Int. J. Distrib. Sensor Netw. (2014)
32. Ihaka, R., Gentleman, R.: R: a language for data analysis and graphics. J. Comput. Graph. Stat. **5**(3), 299–314 (1996)
33. Graczyk, M., Lasota, T., Trawiński, B.: Comparative analysis of premises valuation models using KEEL, RapidMiner, and WEKA. In: Computational Collective Intelligence. Semantic Web, Social Networks and Multiagent Systems, pp. 800–812. Springer, Berlin (2009)
34. Hall, M., Frank, E., Holmes, G., Pfahringer, B., Reutemann, P., Witten, I.H.: The WEKA data mining software: an update. ACM SIGKDD Explor. Newsl. **11**(1), 10–18 (2009)
35. Havre, S., Hetzler, B., Nowell, L.: ThemeRiver: visualizing theme changes over time. In: IEEE Symposium on Information Visualization, 2000. InfoVis 2000, pp. 115–123. IEEE (2000)
36. Schonlau, M.: The clustergram: a graph for visualizing hierarchical and non-hierarchical cluster analyses. The Stata J. **3**, 316–327 (2002)
37. Shvachko, K.V.: Apache hadoop: the scalability update. Login: The Magazine of USENIX **36**, 7–13 (2011)
38. Perera, C., Zaslavsky, A., Christen, P., Georgakopoulos, D.: Powered by Apache Hadoop. Context aware computing for the internet of things: a survey. IEEE Commun. Surv. Tutor. **16**(1), 414–454 (2014)
39. Aly, H., Elmogy, M., Barakat, S.: Big data on internet of things: applications, architecture, technologies, techniques, and future directions
40. "hadoop.apache.org." www.hadoop.apache.org. Accessed 20 April 2016
41. https://www.w3.org/RDF/. https://www.w3.org/RDF/. Accessed 20 April 2016
42. Perera, C., Zaslavsky, A., Christen, P., Georgakopoulos, D.: Context aware computing for the internet of things: A survey. IEEE Commun. Surv. Tutor. **16**(1), 414–454 (2014)
43. https://www.w3.org/2001/sw/wiki/OWL. https://www.w3.org/2001/sw/wiki/OWL. Accessed 20 April 2016
44. Cavanillas, J.M., Curry, E., Wahlster, W.: New Horizons for a Data-Driven Economy, pp. 63–86
45. https://www.w3.org/TR/rdf-sparql-query/. https://www.w3.org/TR/rdf-sparql-query. Accessed 20 April 2016
46. Mastorakis, G. (ed.): Resource Management of Mobile Cloud Computing Networks and Environments. IGI Global (2015)

47. Batalla, J.M., Gajewski, M., Latoszek, W., Krawiec, P., Mavromoustakis, C.X., Mastorakis, G.: ID-based service-oriented communications for unified access to IoT. Comput. Electr. Eng. (2016)
48. Mavromoustakis, C.X., Kormentzas, G., Mastorakis, G., Bourdena, A., Pallis, E., Rodrigues, J.: Context-oriented opportunistic cloud offload processing for energy conservation in wireless devices. In: Globecom Workshops (GC Wkshps), 2014, pp. 24–30. IEEE (2014)
49. Ramgovind, S., Eloff, M.M., Smith, E.: The management of security in cloud computing. In Information Security for South Africa (ISSA), 2010, pp. 1–7. IEEE (2010)
50. Zhang, Q., Cheng, L., Boutaba, R.: Cloud computing: state-of-the-art and research challenges. J. Internet Serv. Appl. **1**(1), 7–18 (2010)
51. Fernando, N., Loke, S.W., Rahayu, W.: Mobile cloud computing: a survey. Futur. Gener. Comput. Syst. **29**(1), 84–106 (2013)
52. Bourdena, A., Mavromoustakis, C., Mastorakis, G., Rodrigues, J., Dobre, C.: Using socio-spatial context in mobile cloud offload process for energy conservation in wireless devices. IEEE Trans. Cloud Comput. (2015)
53. Mousicou, P., Mavromoustakis, C.X., Bourdena, A., Mastorakis, G., Pallis, E.: Performance evaluation of dynamic cloud resource migration based on temporal and capacity-aware policy for efficient resource sharing. In: Proceedings of the 2nd ACM Workshop on High Performance Mobile Opportunistic Systems, pp. 59–66. ACM (2013)
54. Skourletopoulos, G., Bahsoon, R., Mavromoustakis, C.X., Mastorakis, G., Pallis, E.: Predicting and quantifying the technical debt in cloud software engineering. In: 2014 IEEE 19th International Workshop on Computer Aided Modeling and Design of Communication Links and Networks (CAMAD), pp. 36–40. IEEE (2014)
55. Mavromoustakis, C.X., Mousicou, P., Papanikolaou, K., Mastorakis, G., Bourdena, A., Pallis, E.: Dynamic cloud resource migration for efficient 3D video processing in mobile computing environments. In: Novel 3D Media Technologies, pp. 119–134. Springer New York (2015)
56. Mavromoustakis, C.X., Mastorakis, G., Bourdena, A., Pallis, E., Stratakis, D., Perakakis, E., Kopanakis, I., Papadakis, S., Zaharis, Z.D., Skeberis, C., Xenos, T.D.: A social-oriented mobile cloud scheme for optimal energy conservation. In: Resource Management of Mobile Cloud Computing Networks and Environments, pp. 97–121 (2015)
57. Jing, Q., Vasilakos, A.V., Wan, J., Lu, J., Qiu, D.: Security of the internet of things: perspectives and challenges. Wirel. Netw. **20**(8), 2481–2501 (2014)
58. Wu, X., Zhu, X., Wu, G.Q., Ding, W.: Data mining with big data. IEEE Trans. Knowl. Data Eng. **26**(1), 97–107 (2014)
59. https://www.w3.org/P3P. https://www.w3.org/P3P. Accessed 20 April 2016
60. Kryftis, Y., Mastorakis, G., Mavromoustakis, C.X., Batalla, J.M., Rodrigues, J., Drobre, C.: Resource usage prediction models for optimal multimedia content provision. IEEE Syst. J. (2016)
61. Markakis, E., Mastorakis, G., Negru, D., Pallis, E., Mavromoustakis, C.X., Bourdena, A.: A context-aware system for efficient peer-to-peer content provision. In: Dobre, C., Xhafa, F. (eds.) Elsevier Book on Pervasive Computing: Next Generation Platforms for Intelligent Data Collection (2016)
62. Batalla, J.M., Mastorakis, G., Mavromoustakis, C.X., Zurek, J.: On cohabitating networking technologies with common wireless access for home automation systems purposes. In: The Special Issue on "Enabling Wireless Communication and Networking Technologies for the Internet of Things". IEEE Wirel. Commun. Mag. (2016)
63. Vakintis, I., Panagiotakis, S., Mastorakis, G., Mavromoustakis, C.X.: Evaluation of a Web Crowd-Sensing IoT Ecosystem providing Big Data Analysis. In: Pop, F., Kołodziej, J., di Martino, B. (eds.) Resource Management for Big Data Platforms and Applications. Studies in Big Data Springer series. Springer International Publishing (2017)
64. Batalla, J.M., Mavromoustakis, C.X., Mastorakis, G., Sienkiewicz, K.: On the track of 5G radio access network for IoT wireless spectrum sharing in device positioning applications. In: Internet of Things (IoT) in 5G Mobile Technologies, pp. 25–35. Springer International Publishing (2016)

65. Hadjioannou, V., Mavromoustakis, C.X., Mastorakis, G., Batalla, J.M., Kopanakis, I., Perakakis, E., Panagiotakis, S.: Security in smart grids and smart spaces for smooth IoT deployment in 5G. In: Internet of Things (IoT) in 5G Mobile Technologies, pp. 371–397. Springer International Publishing (2016)

66. Goleva, R., Stainov, R., Wagenknecht-Dimitrova, D., Mirtchev, S., Atamian, D., Mavromoustakis, C.X., Mastorakis, G., Dobre, C., Savov, A., Draganov, P.: Data and traffic models in 5G network. In: Internet of Things (IoT) in 5G Mobile Technologies, pp. 485–499. Springer International Publishing (2016)

67. Mavromoustakis, C.X., Dimitriou, C., Mastorakis, G., Pallis, E.: Real-Time Performance evaluation of F-BTD scheme for optimized QoS energy conservation in wireless devices. In: Proceedings of the IEEE Globecom 2013, 2nd IEEE Workshop on Quality of Experience for Multimedia Communications (QoEMC2013), Atlanta, GA, USA, 09–13 Dec 2013 (grant received by COST ESR)

68. Stratakis, D., Miaoudakis, A., Mastorakis, G., Pallis, E., Xenos, T., Yioultsis, T., Mavromoustakis, C.X.: Noise reduction for accurate power measurements of low level signals. In: International Conference on Telecommunications and Multimedia TEMU 2014. IEEE Communications Society Proceedings, 28–30 July 2014, Crete, Greece, pp. 162–166 (2014)

69. Papadakis, S., Stykas, E., Mastorakis, G., Mavromoustakis, C.X.: A hyper-box approach using relational databases for large scale machine learning. In: International Conference on Telecommunications and Multimedia TEMU 2014, IEEE Communications Society Proceedings, 28–30 July, Crete, Greece, pp. 69–73

70. Mastorakis, G., E.K. Markakis, Evangelos, P., Mavromoustakis, C.X., Skourletopoulos, G.: Virtual network functions exploitation through a prototype resource management framework. In: International Conference on Telecommunications and Multimedia TEMU 2014, IEEE Communications Society proceedings, 28–30 July, Crete, Greece, pp. 24–28

71. Papadopoulos, M., Mavromoustakis, C.X., Skourletopoulos, G., Mastorakis, G., Pallis, E.: Performance analysis of reactive routing protocols in mobile Ad hoc networks. In: International Conference on Telecommunications and Multimedia TEMU 2014, IEEE Communications Society Proceedings, 28–30 July, Crete, Greece, pp. 104–110

72. Mavromoustakis, C.X., Mastorakis, G., Papadakis, S., Andreou, A., Bourdena, A., Stratakis, D.: Energy consumption optimization through pre-scheduled opportunistic offloading in wireless devices. In: The Sixth International Conference on Emerging Network Intelligence, EMERGING 2014, 24–28 Aug 2014, Rome, Italy, pp. 22–28 (best paper invited for journal publication)

73. Mavromoustakis, C.X., Andreou, A., Mastorakis, G., Bourdena, A., Batalla, J.M., Dobre, C.: On the performance evaluation of a novel offloading-based energy conservation mechanism for wireless devices. In: Proceedings of the 6th International Conference on Mobile Networks and Management (MONAMI 2014), 22–24 Sept 2014 Wuerzburg, Germany

74. Ciobanu, N.-V., Comaneci, D.-G., Dobre, C., Mavromoustakis, C.X., Mastorakis, G.: OpenMobs: Mobile broadband internet connection sharing. In: Proceedings of the 6th International Conference on Mobile Networks and Management (MONAMI 2014), 22–24 Sept 2014 Wuerzburg, Germany

75. Ciocan, M., Dobre, C., Mavromoustakis, C.X., Mastorakis, G.: Analysis of vehicular storage and dissemination services based on floating content. In: Proceedings of International Workshop on Enhanced Living EnvironMENTs (ELEMENT 2014), 6th International Conference on Mobile Networks and Management (MONAMI 2014), Wuerzburg, Germany, September 2014

76. Papanikolaou, K., Mavromoustakis, C.X., Mastorakis, G., Bourdena, A., Dobre, C.: Energy consumption optimization using social interaction in the mobile cloud. In: Proceedings of International Workshop on Enhanced Living EnvironMENTs (ELEMENT 2014), 6th International Conference on Mobile Networks and Management (MONAMI 2014), Wuerzburg, Germany, September 2014

77. Kryftis, Y., Mavromoustakis, C.X., Batalla, J.M., Mastorakis, G., Pallis, E., Skourletopoulos, G.: Resource usage prediction for optimal and balanced provision of multimedia services. In: Proceedings of the 19th IEEE International Workshop on Computer-Aided Modeling Analysis and Design of Communication Links and Networks (IEEE CAMAD 2014), Athens, Greece, 1–3 Dec 2014
78. Mavromoustakis, C.X., Mastorakis, G., Bourdena, A., Pallis, E., Kormentzas, G., Dimitriou, C.: Joint energy and delay-aware scheme for 5G mobile cognitive radio networks. In: Proceedings of IEEE GlobeCom 2014, track Globecom 2014—Symposium on Selected Areas in Communications: GC14 SAC Green Communication Systems and Networks—GC14 SAC Green Communication Systems and Networks, Austin, TX, USA
79. Batalla, J.M., Kantor, M., Mavromoustakis, C.X., Skourletopoulos, G., Mastorakis, G.: A novel methodology for efficient throughput evaluation in virtualized routers. In: Proceedings of IEEE International Conference on Communications 2015 (IEEE ICC 2015), London, UK, 08–12 June 2015
80. Kryftis, Y., Mavromoustakis, C.X., Mastorakis, G., Pallis, E., Batalla, J.M., Rodrigues, J., Dobre, C., Kormentzas, G.: Resource usage prediction algorithms for optimal selection of multimedia content delivery methods. In: Proceedings of IEEE International Conference on Communications 2015 (IEEE ICC 2015), London, UK, 08–12 June 2015
81. Ciobanu, R.-I., Marin, R.-C., Dobre, C., Cristea, V., Mavromoustakis, C.X., Mastorakis, G.: Opportunistic dissemination using context-based data aggregation over interest spaces. In: Proceedings of IEEE International Conference on Communications 2015 (IEEE ICC 2015), London, UK, 08–12 June 2015
82. Posnakides, D., Mavromoustakis, C.X., Skourletopoulos, G., Mastorakis, G., Pallis, E., Batalla, J.M.: Performance analysis of a rate-adaptive bandwidth allocation scheme in 5G mobile networks. In: Proceedings of the 2nd IEEE International Workshop on a 5G Wireless Odyssey: 2020 in Conjunction with the ISCC 2015- The Twentieth IEEE Symposium on Computers and Communications (ISCC 2015), 6–9 July 2015
83. Mavromoustakis, C.X., Mastorakis, G., Mysirlidis, C., Dagiuklas, T., Politis, I., Dobre, C., Papanikolaou, K., Pallis, E.: On the perceived quality evaluation of opportunistic mobile P2P scalable video streaming. In: Proceedings of the IEEE IWCMC 2015 Conference, Dubrovnik, Croatia, 24–27 Aug 2015, pp. 1515–1519
84. Kryftis, Y., Mavromoustakis, C.X., Mastorakis, G., Batalla, J.M., Chatzimisios, P.: Epidemic models using resource prediction mechanism for optimal provision of multimedia services. In: 2015 IEEE 20th International Workshop on Computer Aided Modeling and Design of Communication Links and Networks (CAMAD) - IEEE CAMAD 2015, University of Surrey, Guildford, UK/General Track, 07–09 Sept 2015, pp. 91–96
85. Pop, C., Ciobanu, R., Marin, R.C., Dobre, C., Mavromoustakis, C.X., Mastorakis, G., Rodrigues, J.J.P.C.: Data dissemination in vehicular networks using context spaces. In: IEEE GLOBECOM 2015, Fourth International Workshop on Cloud Computing Systems, Networks, and Applications (CCSNA), 6–10 Dec 2015
86. Skourletopoulos, G., Mavromoustakis, C.X., Mastorakis, G., Rodrigues, J.J.P.C. Chatzimisios, P., Batalla, J.M.: A fluctuation-based modelling approach to quantification of the technical debt on mobile cloud-based service level. In: IEEE GLOBECOM 2015, Fourth International Workshop on Cloud Computing Systems, Networks, and Applications (CCSNA), 6–10 Dec 2015
87. Skourletopoulos, G., Mavromoustakis, C.X., Mastorakis, G., Pallis, E., Chatzimisios, P., Batalla, J.M.: Towards the evaluation of a big data-as-a-service model: a decision theoretic approach. In: IEEE INFOCOM session on Big Data Sciences, Technologies and Applications (BDSTA 2016)-2016 IEEE Infocom BDSTA Workshop, IEEE International Conference on Computer Communications, 10–15 April 2016, San Francisco, CA, USA
88. Gosman, C., Cornea, T., Dobre, C., Mavromoustakis, C.X., Mastorakis, G.: Secure model to share data in intelligent transportation systems. In: 18th Mediterranean Electrotechnical conference—MELECON 2016, session: Internet of Things, Cloud-Based Systems and Big Data Analytics, Limassol, Cyprus, 18–20 April 2016

89. Skourletopoulos, G., Mavromoustakis, C.X., Mastorakis, G., Pallis, E., Batalla, J.M., Kormentzas, G.: Quantifying and evaluating the technical debt on mobile cloud-based service level. In: IEEE International Conference on Communications (IEEE ICC 2016–Communication QoS, Reliability and Modeling Symposium/Main Track). IEEE ICC 2016—Communication QoS, Reliability and Modeling Symposium, main track, Kuala Lumpur, Malaysia, 23–27 May 2016

90. Hadjioannou, V., Mavromoustakis, C.X., Mastorakis, G., Pallis, E., Stratakis, D., Valavani, D.: On the performance comparison of the agent-based rate adaptivity scheme for IEEE 802.11n and ZigBee. In: International Conference on Telecommunications and Multimedia TEMU 2016, IEEE Communications Society proceedings, Heraklion, Greece, 25–27 July 2016

91. Markakis, E., Sideris, A., Alexiou, G., Bourdena, A., Pallis, E., Mastorakis, G., Mavromoustakis, C.X.: A virtual network functions brokering mechanism. In: International Conference on Telecommunications and Multimedia TEMU 2016, IEEE Communications Society Proceedings, Heraklion, Greece, 25–27 July 2016

92. Zaharis, Z.D., Yioultsis, T., Skeberis, C., Xenos, T., Lazaridis, P., Mastorakis, G., Mavromoustakis, C.X.: Implementation of antenna array beamforming by using a novel neural network structure. In: International Conference on Telecommunications and Multimedia TEMU 2016, IEEE Communications Society proceedings, Heraklion, Greece, 25–27 July 2016

93. Bormpantonakis, P., Stratakis, D., Mastorakis, G., Mavromoustakis, C.X., Skeberis, C., Bechet, P.: Exposure EMF measurements with spectrum analyzers using free and open source software. In: International Conference on Telecommunications and Multimedia TEMU 2016, IEEE Communications Society Proceedings, Heraklion, Greece, 25–27 July 2016

94. Hadjioannou, V., Mavromoustakis, C.X., Mastorakis, G., Markakis, E., Pallis, E.: Context awareness location-based android application for tracking purposes in assisted living. In: International Conference on Telecommunications and Multimedia TEMU 2016, IEEE Communications Society Proceedings, Heraklion, Greece, 25–27 July 2016

Big Data and Cloud Computing: A Survey of the State-of-the-Art and Research Challenges

Georgios Skourletopoulos, Constandinos X. Mavromoustakis,
George Mastorakis, Jordi Mongay Batalla, Ciprian Dobre,
Spyros Panagiotakis and Evangelos Pallis

Abstract The proliferation of data warehouses and the rise of multimedia, social media and the Internet of Things (IoT) generate an increasing volume of structured, semi-structured and unstructured data. Towards the investigation of these large volumes of data, big data and data analytics have become emerging research fields, attracting the attention of the academia, industry and governments. Researchers, entrepreneurs, decision makers and problem solvers view 'big data' as the tool to revolutionize various industries and sectors, such as business, healthcare, retail, research, education and public administration. In this context, this survey chapter

G. Skourletopoulos (✉) · C.X. Mavromoustakis
Mobile Systems (MoSys) Laboratory, Department of Computer Science,
University of Nicosia, Nicosia, Cyprus
e-mail: skourletopoulos.g@unic.ac.cy

C.X. Mavromoustakis
e-mail: mavromoustakis.c@unic.ac.cy

G. Mastorakis · S. Panagiotakis · E. Pallis
Department of Informatics Engineering, Technological Educational
Institute of Crete, Heraklion, Crete, Greece
e-mail: gmastorakis@staff.teicrete.gr

S. Panagiotakis
e-mail: spanag@teicrete.gr

E. Pallis
e-mail: pallis@pasiphae.eu

J.M. Batalla
Warsaw University of Technology, Nowowiejska Str. 15/19, Warsaw, Poland
e-mail: jordim@interfree.it

C. Dobre
Faculty of Automatic Control and Computers, Department of Computer Science,
University Politehnica of Bucharest, Bucharest, Romania
e-mail: ciprian.dobre@cs.pub.ro

© Springer International Publishing Switzerland 2017
C.X. Mavromoustakis et al. (eds.), *Advances in Mobile Cloud Computing
and Big Data in the 5G Era*, Studies in Big Data 22,
DOI 10.1007/978-3-319-45145-9_2

presents a review of the current big data research, exploring applications, opportunities and challenges, as well as the state-of-the-art techniques and underlying models that exploit cloud computing technologies, such as the big data-as-a-service (BDaaS) or analytics-as-a-service (AaaS).

Keywords Big data · Data analytics · Data management · Big data-as-a-service · Analytics-as-a-service · Business intelligence · Lease storage · Cloud computing · Cost-benefit analysis model

1 Introduction

The Internet penetration constantly increases, as more and more people browse the Web, use email and social network applications to communicate with each other or access wireless multimedia services, such as mobile TV [27, 43]. Additionally, several demanding mobile network services are now available, which require increased data rates for specific operations, such as device storage synchronization to cloud computing servers or high resolution video [34–36]. The access to such a global information and communication infrastructure along with the advances in digital sensors and storage have created very large amounts of data, such as Internet, sensor, streaming or mobile device data. Additionally, data analysis is the basis for investigations in many fields of knowledge, such as science, engineering or management. Unlike web-based big data, location data is an essential component of mobile big data, which are harnessed to optimize and personalize mobile services. Hence, an era where data storage and computing become utilities that are ubiquitously available is now introduced.

Furthermore, algorithms have been developed to connect datasets and enable more sophisticated analysis. Since innovations in data architecture are on our doorstep, the 'big data' paradigm refers to very large and complex data sets (i.e., petabytes and exabytes of data) that traditional data processing systems are inadequate to capture, store and analyze, seeking to glean intelligence from data and translate it into competitive advantage. As a result, big data needs more computing power and storage provided by cloud computing platforms. In this context, cloud providers, such as IBM [23], Google [17], Amazon [2] and Microsoft [38], provide network-accessible storage priced by the gigabyte-month and computing cycles priced by the CPU-hour [8].

Although big data is still in the preliminary stages, comprehensive surveys exist in the literature [1, 9–11, 20, 37, 59]. This survey article aims at providing a holistic perspective on big data and big data-as-a-service (BDaaS) concepts to the research community active on big data-related themes, including a critical revision of the current state-of-the-art techniques, definition and open researches issues. Following this introductory section, Sect. 2 presents related work approaches in the literature, including the architecture and possible impact areas. Section 3 demonstrates the business value and long-term benefits of adopting big data-as-a-service business models and attempts to communicate the findings to non-technical stakeholders, while Sect. 4 points out opportunities, challenges and open research issues in the big data domain. Finally, Sect. 5 concludes this tutorial chapter.

2 Big Data: Background and Architecture

IBM data scientists argue that the key dimensions of big data are the "4Vs": volume, velocity, variety and veracity [21]. As large and small enterprises constantly attempt to design new products to deal with big data, the open source platforms, such as Hadoop [53], give the opportunity to load, store and query a massive scale of data and execute advanced big data analytics in parallel across a distributed cluster. Batch-processing models, such as MapReduce [14], enable the data coordination, combination and processing from multiple sources. Many big data solutions in the market exploit external information from a range of sources (e.g., social networks) for modelling and sentiment analysis, such as the IBM Social Media Analytics Software as a Service solution [22]. Cloud providers have already begun to establish new data centers for hosting social networking, business, media content or scientific applications and services. In this direction, the selection of the data warehouse technology depends on several factors, such as the volume of data, the speed with which the data is needed or the kind of analysis to be performed [25]. A conceptual big data warehouse architecture is presented in Fig. 1 [24].

Another significant challenge is the delivery of big data capabilities through the cloud. The adoption of big data-as-a-service (BDaaS) business models enables the effective storage and management of very large data sets and data processing from an outside provider, as well as the exploitation of a full range of analytics capabilities (i.e., data and predictive analytics or business intelligence are provided as service-based applications in the cloud). In this context, Zheng et al. [59] critically review the service-generated big data and big data-as-a-service (see Fig. 2) towards

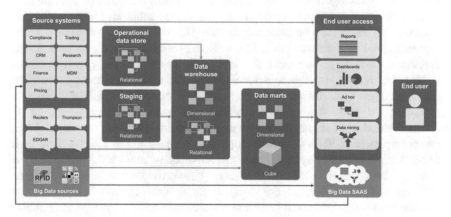

Fig. 1 A conceptual big data warehouse architecture

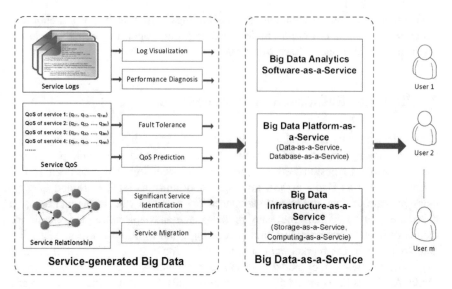

Fig. 2 Service-generated big data and big data-as-a-service as presented by Zheng et al. [59]

the proposal of an infrastructure to provide functionality for managing and analyzing different types of service-generated big data. A big data-as-a-service framework has been also employed to provide big data services and data analytics results to users, enhance efficiency and reduce cost.

The development of a cloud-supported big data mining platform, which provides statistical and data analytics functions, has been also explored [56]. In this research work, the platform's architecture is composed of four layers (i.e., infrastructure, virtualization, data set processing and services), implementing the K-means algorithm. A big data analytics-related platform was proposed by Park et al. [40], which includes a CCTV metadata analytics service and aims to manage big data and develop analytics algorithms through collaboration between data owners, scientists and developers. Since modern enterprises request new solutions for enterprise data warehousing (EDW) and business intelligence (BI), a big data provisioning solution was elaborated by Vaquero et al. [55], combining hierarchical and peer-to-peer data distribution techniques to reduce the data loading time into the virtual machines (VMs). The proposed solution includes dynamic topology and software configuration management techniques for better quality of experience (QoE) and achieves to reduce the setup time of virtual clusters for data processing in the cloud. A cloud-based big data analytics service provisioning platform, named CLAaaS, has been presented in the literature along with a taxonomy to identify significant features of the workflow systems, such as multi-tenancy for a wide range of analytic tools and back-end data sources, user group customization and web collaboration [60]. An overview of the analytics workflow for big data is shown in Fig. 4 [3]. On the other hand, an admission control and resource scheduling algorithm is examined in another work [58], which manages to satisfy the quality of service requirements

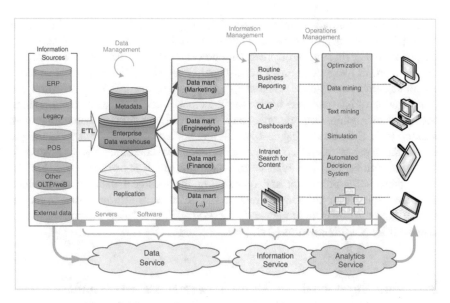

Fig. 3 A conceptual architecture of service-oriented decision support systems as presented by Demirkan and Delen [15]

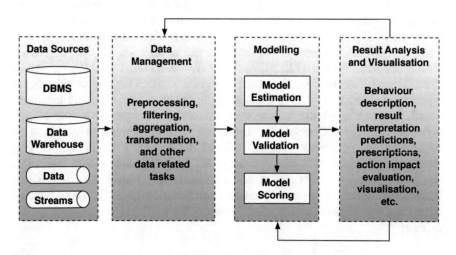

Fig. 4 Analytics workflow for big data as presented by Assunção et al. [3]

of requests, adhering to the Service Level Agreements (SLAs) guarantees, and improve the Analytics-as-a-Service (AaaS) providers' competitiveness and profitability. A framework for service-oriented decision support systems (DSS) in the cloud has been also investigated, focusing on the product-oriented decision support systems environment and exploring engineering-related issues [15]. A conceptual architecture of service-oriented decision support systems is shown in Fig. 3.

The growth of cloud computing, big data and analytics [52] compels businesses to turn into big data-as-a-service solutions in order to overcome common challenges, such as data storage or processing power. Although there is related work in the literature in the general area of cost-benefit analysis in cloud and mobile cloud computing environments, a research gap is observed towards the evaluation and classification of big data-as-a-service business models. Several research efforts have been devoted comparing the monetary cost-benefits of cloud computing with desktop grids [26], examining cost-benefit approaches of using cloud computing to extend the capacity of clusters [13] or calculating the cloud total cost of ownership and utilization cost [30] to evaluate the economic efficiency of the cloud. Finally, novel metrics for predicting and quantifying the technical debt on cloud-based software engineering and cloud-based service level were also proposed in the literature from the cost-benefit viewpoint [44, 45] and extended evaluation results are discussed by Skourletopoulos et al. [46].

3 Cloud-Supported Big Data: Towards a Cost-Benefit Analysis Model in Big Data-as-a-Service (BDaaS)

In previous research works, the cloud was considered as a marketplace [7], where the storage and computing capabilities of the cloud-based system architectures can be leased off [47, 49, 50]. Likewise, the rise of large data centers has created new business models, where businesses lease storage in a pay-as-you-go service-oriented manner [32, 57]. In this direction, the big data-as-a-service (BDaaS) model was introduced in order to provide common big data services, boost efficiency and reduce cost [51]. Communicating the business value and long-term benefits of adopting big data-as-a-service business models against the conventional high-performance data warehouse appliances to non-technical stakeholders is imperative. In this book chapter, a brief survey of a novel quantitative, cloud-inspired cost-benefit analysis metric in big data-as-a-service is presented, based on previous research studies in the literature [48]. Hence, the cost analysis (CA) modelling from the conventional data warehouse appliance (DWH) viewpoint takes the following form and the descriptions of the exploited variables are shown in Table 1:

$$CA_i = 12 * (C_{s/m} * S_{max}), \quad i \geq 1 \text{ and } S_{curr} \leq S_{max} \tag{1}$$

where,

$$C_{s/m} = C_{s/m(max)} = C_{\alpha/m(max)} + C_{\gamma/m(max)} + C_{\eta/m(max)} + C_{\theta/m(max)} + C_{\kappa/m(max)} + C_{\lambda/m(max)} + C_{\mu/m(max)} + C_{\sigma/m(max)}$$

As the benefits of cloud computing (i.e., scalability) do not stand in data warehouse appliances, the cost analysis approach adopted in this study does not consider the storage capacity currently used (S_{curr}). Therefore, the cost variations,

Table 1 Notations and variable descriptions

Symbol	Variable description
CA	The cost analysis calculation results, which are expressed in monetary units
C_D	The benefits calculation results, which are expressed in monetary units
i	The index of the year
$C_{s/m}$	The initial monthly cost for leasing cloud storage, which is expressed in monetary units
S_{max}	The maximum storage capacity
S_{curr}	The storage currently used
Δ_0	The cost formation for leasing cloud storage regarding the second year of the period of l-years, once the corresponding variation in the monthly cost is applied, which is expressed in monetary units
δ_1	The total variation regarding the cost for leasing cloud storage for the second year of the period of l-years, which is represented by a percentage value
Δ_i	The cost formation for leasing cloud storage from the third year and onwards, once the corresponding variation in the monthly cost is applied, which is expressed in monetary units
δ_i	The total variation regarding the cost for leasing cloud storage from the third year and onwards, which is represented by a percentage value
B_0	The storage used during the second year, once the corresponding variation in the demand is applied
β_1	The variation in the demand for storage capacity regarding the second year, which is represented by a percentage value
B_i	The storage used from the third year and onwards, once the corresponding variation in the demand is applied
β_i	The variation in the demand for storage capacity from the third year and onwards, which is represented by a percentage value
C_α	The data storage cost
C_γ	The document storage cost
C_η	The maintenance services cost
C_θ	The network cost
C_κ	The on-demand I/O cost
C_λ	The operations cost
C_μ	The server cost
C_σ	The technical support cost
$\alpha_i\%$	The variation in the monthly data storage cost, which is represented by a percentage value
$\gamma_i\%$	The variation in the monthly document storage cost, which is represented by a percentage value
$\eta_i\%$	The variation in the monthly maintenance services cost, which is represented by a percentage value
$\theta_i\%$	The variation in the monthly network cost, which is represented by a percentage value
$\kappa_i\%$	The variation in the monthly on-demand I/O cost, which is represented by a percentage value

(continued)

Table 1 (continued)

Symbol	Variable description
$\lambda_i\%$	The variation in the monthly operations cost, which is represented by a percentage value
$\mu_i\%$	The variation in the monthly server cost, which is represented by a percentage value
$\sigma_i\%$	The variation in the monthly technical support cost, which is represented by a percentage value

due to the fluctuations in the demand for storage capacity, do not apply as long as $S_{curr} \leq S_{max}$ and the true benefits are always zero ($C_D = 0$) over the years. In case of such an increase in the demand for storage capacity that $S_{curr} > S_{max}$, incremental capacity should be added to the storage systems with overhead and downtime. On the contrary, the cost-benefit analysis modelling from the big data-as-a-service point of view takes the following form during the first year (i.e., Eqs. 2 and 4) and from the second year and onwards (i.e., Eqs. 3 and 5):

$$CA_1 = 12 * (C_{s/m} * S_{curr}) \tag{2}$$

$$CA_i = 12 * (\Delta_{i-2} * B_{i-2}), i \geq 2 \tag{3}$$

$$C_{D_1} = 12 * [C_{s/m} * (S_{max} - S_{curr})] \tag{4}$$

$$C_{D_i} = 12 * [\Delta_{i-2} * (S_{max} - B_{i-2})], i \geq 2 \tag{5}$$

where,

$$C_{s/m} = C_{s/m(curr)} = C_{\alpha/m(curr)} + C_{\gamma/m(curr)} + C_{\eta/m(curr)} + C_{\theta/m(curr)} + C_{\kappa/m(curr)} + C_{\lambda/m(curr)} + C_{\mu/m(curr)} + C_{\sigma/m(curr)}$$

$$\Delta_0 = (1 + \delta_1\%) * C_{s/m}$$

$$\Delta_i = (1 + \delta_{i+1}\%) * \Delta_{i-1}, i \geq 1$$

$$\delta_i\% = a_i\% + \gamma_i\% + \eta_i\% + \theta_i\% + \kappa_i\% + \lambda_i\% + \mu_i\% + \sigma_i\%, i \geq 1$$

$$B_0 = (1 + \beta_1\%) * S_{curr}$$

$$B_i = (1 + \beta_{i+1}\%) * B_{i-1}, i \geq 1$$

The amount of profit not earned due to the underutilization of the storage capacity is measured, under the assumption that fluctuations in the demand for cloud storage occur. The possible upgradation of the storage and the risk of entering into new and accumulated costs in the future are also examined. The cloud storage capacity to be leased off is evaluated with respect to the following assumptions:

- The cloud storage is subscription-based and the billing vary over the period of *l*-years due to the fluctuations in the demand for storage capacity (i.e., gigabyte per month).
- The total network cost consists of bandwidth usage, egress and data transfer costs between regional and multi-region locations. As the cloud-based, always-on mobile services are usually sensitive to network bandwidth and latency [42], the additional network cost is expected to satisfy the outbound network traffic demands in order to avoid delays.
- Since the content retrieval from a bucket should be faster than the default, the additional on-demand I/O cost enables to increase the throughput [4, 39].
- The additional server cost stems from the additional CPU cores and the amount of memory required for processing.

Two possible types of benefits calculation results are encountered, when leasing cloud storage:

- Positive calculation results, which point out the underutilization of the storage capacity.
- Negative calculation results, which reveal the immediate need for upgradation. This need stimulates additional costs; however, the total amount of accumulated cost in conventional data warehouse appliances is not comparable, as the earnings by adopting a big data-as-a-service business model can be reinvested on the additional storage required, maximizing the return on investment.

Towards the evaluation of big data-as-a-service business models and the increase in the return on investment, the way the benefits overcome the costs is of significant importance [12, 28, 32, 41]. An illustrative example emphasizes on the need to consolidate data from different sources. Cost analysis and benefits comparisons are performed during a 5-year period of time ($l = 5$) prior to adoption of either a conventional data warehouse or a big data-as-a-service business model. The predicted variations in the demand for cloud storage with respect to two case scenarios are shown in Table 2.

In this framework, the first case scenario reveals that adopting a big data-as-a-service business model is more cost-effective than a conventional data warehouse, as the cost analysis results for the big data-as-a-service model have the least positive values throughout the 5-year period. The benefits calculation results are positive in big data-as-a-service business models, while the benefits results are always zero in conventional data warehouse business models (Figs. 5 and 6).

Term	Case scenario 1	Case scenario 2
Year 1–2	$\beta_{1\%} = 5\%$	$\beta_{1\%} = 10\%$
Year 2–3	$\beta_{2\%} = 15\%$	$\beta_{2\%} = 22\%$
Year 3–4	$\beta_{3\%} = 20\%$	$\beta_{3\%} = 35\%$
Year 4–5	$\beta_{4\%} = 23\%$	$\beta_{4\%} = 40\%$

Table 2 Variations in the demand for cloud storage regarding two case scenarios

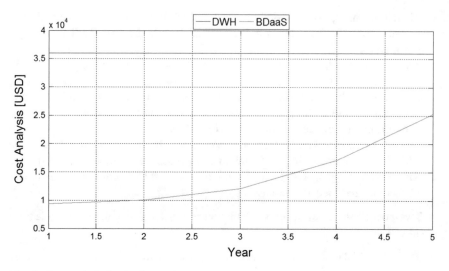

Fig. 5 Cost analysis for the first case scenario

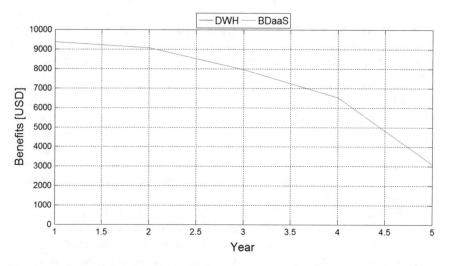

Fig. 6 Benefits analysis for the first case scenario

On the other hand, the second case scenario points out the cost-effectiveness and the benefits gained by adopting the big data-as-a-service model during the first four years. However, the benefits calculations results become negative during the fifth year, indicating the need for immediate upgradation to meet the demand requirements. The necessity for upgradation is also witnessed at the increased costs compared to those in the traditional data warehouse approach. In this direction, the earnings gained throughout the period, due to the selection of the dig data-as-a-service business model, will be reinvested on the additional storage required, maximizing the return on investment (ROI) (Figs. 7 and 8).

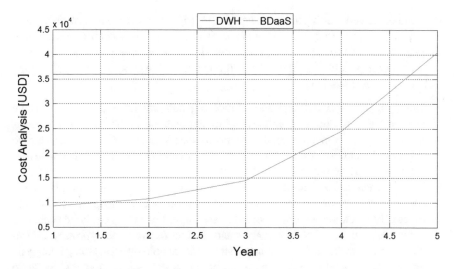

Fig. 7 Cost analysis for the second case scenario

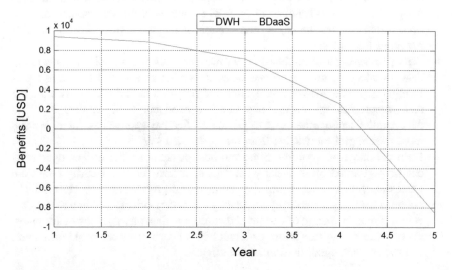

Fig. 8 Benefits analysis for the second case scenario

4 Challenges and Open Research Issues

The rise and development of social networks, multimedia, electronic commerce (e-Commerce) and cloud computing have increased considerably the data. Additionally, since the needs of enterprise analytics are constantly growing, the conventional hub-and-spoke architectures cannot satisfy the demands and, therefore, new and enhanced architectures are necessary [15]. In this context, new challenges and open research issues are encountered, including storage, capture, processing, filtering, analysis, curation, search, sharing, visualization, querying and privacy of the very large volumes of data. The aforementioned issues are categorized and elaborated as follows [11]:

- **Data storage and management**: Since big data are dependent on extensive storage capacity and data volumes grow exponentially, the current data management systems cannot satisfy the needs of big data due to limited storage capacity. In addition, the existing algorithms are not able to store data effectively because of the heterogeneity of big data.
- **Data transmission and curation**: Since network bandwidth capacity is the major drawback in the cloud, data transmission is a challenge to overcome, especially when the volume of data is very large. For managing large-scale and structured datasets, data warehouses and data marts are good approaches. Data warehouses are relational database systems that enable the data storage, analysis and reporting, while the data marts are based on data warehouses and facilitate the analysis of them. In this context, NoSQL databases [19] were introduced as a potential technology for large and distributed data management and database design. The major advantage of NoSQL databases is the schema-free orientation, which enables the quick modification of the structure of data and avoids rewriting the tables.
- **Data processing and analysis**: Query response time is a significant issue in big data, as adequate time is needed when traversing data in a database and performing real-time analytics. A flexible and reconfigured grid along with the big data preprocessing enhancement and consolidation of application- and data-parallelization schemes can be more effective approaches for extracting more meaningful knowledge from the given data sets.
- **Data privacy and security**: Since the host of data or other critical operations can be performed by third-party services or infrastructures, security issues are witnessed with respect to big data storage and processing. The current technologies used in data security are mainly static data-oriented, although big data entails dynamic change of current and additional data or variations in attributes. Privacy-preserving data mining without exposing sensitive personal information is another challenging field to be investigated.

5 Summary and Conclusion

Since networking is ubiquitous and vast amounts of data are now available, big data is envisioned to be the tool for productivity growth, innovation and consumer surplus. Huge opportunities related to advanced big data analytics and business intelligence are at the forefront of research, focusing on the investigation of innovative business-centric methodologies that can transform various sectors and industries, such as e-commerce, market intelligence, e-government, healthcare and security [4, 29, 31, 54]. To this end, this tutorial paper discusses the current big data research and points out the research challenges and opportunities in this field by exploiting cloud computing technologies and building new models [5, 6, 16, 18, 33–36]. A cost-benefit analysis is also performed towards measuring the long-term benefits of adopting big data-as-a-service business models in order to support data-driven decision making and communicate the findings to non-technical stakeholders.

Acknowledgements The authors would like to thank the anonymous reviewers for their constructive comments and insights on the manuscript. Their suggestions have contributed greatly to the high quality and improvement of this article. The authors would also like to acknowledge networking support by the ICT COST Action IC1303: Algorithms, Architectures and Platforms for Enhanced Living Environments (AAPELE) and the ICT COST Action IC1406 "High-Performance Modelling and Simulation for Big Data Applications" (cHiPSet).

References

1. Agrawal, D., Das, S., El Abbadi, A.: Big data and cloud computing: current state and future opportunities. In: Proceedings of the 14th International Conference on Extending Database Technology (EDBT/ICDT'11), pp. 530–533 (2011)
2. Amazon Web Services, Inc.: Elastic Compute Cloud (EC2). http://aws.amazon.com/ec2 (2015). Accessed 18 Oct 2015
3. Assunção, M.D., Calheiros, R.N., Bianchi, S., Netto, M.A.S., Buyya, R.: Big data computing and clouds: trends and future directions. J. Parallel Distrib. Comput. **79–80**, 3–15 (2015)
4. Batalla, J.M., Kantor, M., Mavromoustakis, C.X., Skourletopoulos, G., Mastorakis, G.: A novel methodology for efficient throughput evaluation in virtualized routers. In: Proceedings of the IEEE International Conference on Communications (ICC 2015)—Communications Software, Services and Multimedia Applications Symposium (CSSMA), London, UK, pp. 6899–6905 (2015)
5. Batalla, J.M., Mavromoustakis, C.X., Mastorakis, G., Sienkiewicz, K.: On the track of 5G radio access network for IoT wireless spectrum sharing in device positioning applications. In: Internet of Things (IoT) in 5G Mobile Technologies, pp. 25–35. Springer International Publishing (2016)
6. Batalla, J.M., Mastorakis, G., Mavromoustakis, C.X., Zurek, J.: On cohabitating networking technologies with common wireless access for Home Automation Systems purposes. Special Issue on "Enabling Wireless Communication and Networking Technologies for the Internet of Things". IEEE Wirel. Commun. Mag. (2016)
7. Buyya, R., Yeo, C.S., Venugopal, S., Broberg, J., Brandic, I.: Cloud computing and emerging IT platforms: vision, hype, and reality for delivering computing as the 5th utility. Future Gener. Comput. Syst. **25**, 599–616 (2009)
8. Buyya, R., Ranjan, R., Calheiros, R.N.: Intercloud: utility-oriented federation of cloud computing environments for scaling of application services. Algorithms Arch. Parallel Process. **6081**, 13–31 (2010)
9. Chen, C.L.P., Zhang, C.Y.: Data-intensive applications, challenges, techniques and technologies: a survey on big data. Inf. Sci. **275**, 314–347 (2014)

10. Chen, H., Chiang, R.H.L., Storey, V.C.: Business intelligence and analytics: from big data to big impact. MIS Q. **36**, 1165–1188 (2012)
11. Chen, M., Mao, S., Liu, Y.: Big data: a survey. Mobile Netw. Appl. **19**, 171–209 (2014)
12. Ciobanu, R.-I., Marin, R.-C., Dobre, C., Cristea, V., Mavromoustakis, C.X., Mastorakis, G.: Opportunistic dissemination using context-based data aggregation over interest spaces. In: Proceedings of IEEE International Conference on Communications 2015 (IEEE ICC 2015), London, UK, 08–12 June 2015
13. De Assunção, M.D., Di Costanzo, A., Buyya, R.: A cost-benefit analysis of using cloud computing to extend the capacity of clusters. Clust. Comput. **13**, 335–347 (2010)
14. Dean, J., Ghemawat, S.: Mapreduce: simplified data processing on large clusters. Commun. ACM **51**, 107–113 (2008)
15. Demirkan, H., Delen, D.: Leveraging the capabilities of service-oriented decision support systems: putting analytics and big data in cloud. Decis. Support Syst. **55**, 412–421 (2013)
16. Goleva, R., Stainov, R., Wagenknecht-Dimitrova, D., Mirtchev, S., Atamian, D., Mavromoustakis, C.X., Mastorakis, G., Dobre, C., Savov, A., Draganov, P.: Data and traffic models in 5G network. In: Internet of Things (IoT) in 5G Mobile Technologies, pp. 485–499. Springer International Publishing (2016)
17. Google, Inc.: App engine—platform as a service. https://cloud.google.com/appengine (2015). Accessed 18 Oct 2015
18. Hadjioannou, V., Mavromoustakis, C.X., Mastorakis, G., Batalla, J.M., Kopanakis, I., Perakakis, E., Panagiotakis, S.: Security in smart grids and smart spaces for smooth IoT deployment in 5G. In: Internet of Things (IoT) in 5G Mobile Technologies, pp. 371–397. Springer International Publishing (2016)
19. Han, J., Haihong, E., Le, G., Du, J.: Survey on NoSQL database. In: Proceedings of the 2011 6th International Conference on Pervasive Computing and Applications (ICPCA), Port Elizabeth, pp. 363–366 (2011)
20. Hashem, I.A.T., Yaqoob, I., Anuar, N.B., Mokhtar, S., Gani, A., Khan, S.U.: The rise of "big data" on cloud computing: review and open research issues. Inf. Syst. **47**, 98–115 (2015)
21. IBM Corporation: IBM big data & analytics hub: the four V's of big data. http://www.ibmbigdatahub.com/infographic/four-vs-big-data (2014). Accessed 18 Oct 2015
22. IBM Corporation: IBM social media analytics software as a service. http://www-03.ibm.com/software/products/en/social-media-analytics-saas (2015a). Accessed 18 October 2015
23. IBM Corporation: IBM bluemix. http://www.ibm.com/cloud-computing/bluemix (2015b). Accessed 18 Oct 2015
24. Informatica Corporation: Making sense of big data. https://marketplace.informatica.com/solutions/making_sense_of_big_data (2015). Accessed 18 Oct 2015
25. Inmon, W.H.: Building the Data Warehouse. Wiley (2005)
26. Kondo, D., Javadi, B., Malecot, P., Cappello, F., Anderson, D.P.: Cost-benefit analysis of cloud computing versus desktop grids. In: Proceedings of the 2009 IEEE International Symposium on Parallel and Distributed Processing (IPDPS 2009), Rome, pp. 1–12 (2009)
27. Kryftis, Y., Mavromoustakis, C.X., Batalla, J.M., Mastorakis, G., Pallis, E., Skourletopoulos, G.: Resource usage prediction for optimal and balanced provision of multimedia services. Proceedings of the 2014 IEEE 19th International Workshop on Computer Aided Modeling and Design of Communication Links and Networks (CAMAD 2014), pp. 255–259. Greece, Athens (2014)
28. Kryftis, Y., Mavromoustakis, C.X., Mastorakis, G., Pallis, E., Batalla, J.M., Rodrigues, J., Dobre, C., Kormentzas, G.: Resource usage prediction algorithms for optimal selection of multimedia content delivery methods. In: Proceedings of IEEE International Conference on Communications 2015 (IEEE ICC 2015), London, UK, 08–12 June 2015
29. Kryftis, Y., Mastorakis, G., Mavromoustakis, C.X., Batalla, J.M., Rodrigues, J., Drobre, C.: Resource usage prediction models for optimal multimedia content provision. IEEE Syst. J. (2016)

30. Li, X., Li, Y., Liu, T., Qiu, J., Wang, F.: The method and tool of cost analysis for cloud computing. In: 2009 IEEE International Conference on Cloud Computing (CLOUD'09), Bangalore, pp. 93–100 (2009)
31. Markakis, E., Mastorakis, G., Negru, D., Pallis, E., Mavromoustakis, C.X., Bourdena, A.: A context-aware system for efficient peer-to-peer content provision. In: Dobre, C., Xhafa, F. (eds.) Pervasive Computing: Next Generation Platforms for Intelligent Data Collection. Elsevier (2016)
32. Mastorakis, G., Markakis, E., Pallis, E., Mavromoustakis, C.X., Skourletopoulos, G.: Virtual network functions exploitation through a prototype resource management framework. Proceedings of the 2014 IEEE 6th International Conference on Telecommunications and Multimedia (TEMU 2014), pp. 24–28. Heraklion, Crete, Greece (2014)
33. Mavromoustakis, C.X., Dimitriou, C., Mastorakis, G., Pallis, E.: Real-time performance evaluation of F-BTD scheme for optimized QoS energy conservation in wireless devices. In: Proceedings of IEEE Globecom 2013, 2nd IEEE Workshop on Quality of Experience for Multimedia Communications (QoEMC2013), Atlanta, GA, USA, 09–13 Dec 2013
34. Mavromoustakis, C.X., Mastorakis, G., Bourdena, A., Pallis, E., Kormentzas, G., Rodrigues, J.J.P.C.: Context-oriented opportunistic cloud offload processing for energy conservation in wireless devices. Proceedings of the IEEE Global Communications Conference (GLOBECOM 2014)—The Second International Workshop on Cloud Computing Systems, Networks, and Applications (CCSNA), pp. 24–30. Austin, Texas, USA (2014a)
35. Mavromoustakis, C.X., Andreou, A., Mastorakis, G., Bourdena, A., Batalla, J.M., Dobre, C.: On the performance evaluation of a novel offloading-based energy conservation mechanism for wireless devices. In: Proceedings of the 6th International Conference on Mobile Networks and Management (MONAMI 2014), 22–24 Sept 2014. Wuerzburg, Germany (2014b)
36. Mavromoustakis, C.X., Mastorakis, G., Bourdena, A., Pallis, E., Kormentzas, G., Dimitriou, C.: Joint energy and delay-aware scheme for 5G mobile cognitive radio networks. In: Proceedings of IEEE GlobeCom 2014, Symposium on Selected Areas in Communications: GC14 SAC Green Communication Systems and Networks, Austin, TX, USA (2014c)
37. McAfee, A., Brynjolfsson, E., Davenport, T.H., Patil, D.J., Barton, D., Court, D.: Big data: the management revolution. Harv. Bus. Rev. 59–68 (2012)
38. Microsoft Corporation: Microsoft azure: cloud computing platform and services. https://azure.microsoft.com (2015). Accessed 18 Oct 2015
39. Papadopoulos, M., Mavromoustakis, C.X., Skourletopoulos, G., Mastorakis, G., Pallis, E.: Performance analysis of reactive routing protocols in mobile ad hoc networks. Proceedings of the 2014 IEEE 6th International Conference on Telecommunications and Multimedia (TEMU 2014), pp. 104–110. Heraklion, Crete, Greece (2014)
40. Park, K., Nguyen, M.C., Won, H.: Web-based collaborative big data analytics on big data as a service platform. In: Proceedings of the 2015 17th International Conference on Advanced Communication Technology (ICACT), Seoul, pp. 564–567 (2015)
41. Pop, C., Ciobanu, R., Marin, R.C., Dobre, C., Mavromoustakis, C.X., Mastorakis, G., Rodrigues, J.J.P.C.: Data dissemination in vehicular networks using context spaces. In: IEEE GLOBECOM 2015, Fourth International Workshop on Cloud Computing Systems, Networks, and Applications (CCSNA), 6–10 Dec 2015
42. Posnakides, D., Mavromoustakis, C.X., Skourletopoulos, G., Mastorakis, G., Pallis, E., Batalla, J.M.: Performance analysis of a rate-adaptive bandwidth allocation scheme in 5G mobile networks. Proceedings of the 20th IEEE Symposium on Computers and Communications (ISCC 2015)—The 2nd IEEE International Workshop on A 5G Wireless Odyssey:2020, pp. 955–961. Larnaca, Cyprus (2015)
43. Skourletopoulos, G., Xanthoudakis, A.: Developing a business plan for new technologies: application and implementation opportunities of the interactive digital (iDTV) and internet protocol (IPTV) television as an advertising tool. Bachelor's Degree Dissertation, Technological Educational Institute of Crete, Greece (2012)

44. Skourletopoulos, G.: Researching and quantifying the technical debt in cloud software engineering. Master's Degree Dissertation, University of Birmingham, UK (2013)
45. Skourletopoulos, G., Bahsoon, R., Mavromoustakis, C.X., Mastorakis, G., Pallis, E.: Predicting and quantifying the technical debt in cloud software engineering. Proceedings of the 2014 IEEE 19th International Workshop on Computer Aided Modeling and Design of Communication Links and Networks (CAMAD 2014), pp. 36–40. Greece, Athens (2014)
46. Skourletopoulos, G., Bahsoon, R., Mavromoustakis, C.X., Mastorakis, G.: The technical debt in cloud software engineering: a prediction-based and quantification approach. In: Mastorakis, G., Mavromoustakis, C.X., Pallis, E. (eds.) Resource Management of Mobile Cloud Computing Networks and Environments, 1st edn, pp. 24–42. IGI Global, Hershey, PA (2015)
47. Skourletopoulos, G., Mavromoustakis, C.X., Mastorakis, G., Rodrigues, J.J.P.C., Chatzimisios, P., Batalla, J.M.: A fluctuation-based modelling approach to quantification of the technical debt on mobile cloud-based service level. In: Proceedings of the IEEE Global Communications Conference (GLOBECOM 2015)—The Fourth IEEE International Workshop on Cloud Computing Systems, Networks, and Applications (CCSNA), San Diego, California, USA (2015b)
48. Skourletopoulos, G., Mavromoustakis, C.X., Mastorakis, G., Pallis, E., Chatzimisios, P., Batalla, J.M.: Towards the evaluation of a big data-as-a-service model: a decision theoretic approach. In: Proceedings of the IEEE International Conference on Computer Communications (INFOCOM 2016)—First IEEE International Workshop on Big Data Sciences, Technologies and Applications (BDSTA), San Francisco, California, USA (2016a)
49. Skourletopoulos, G., Mavromoustakis, C.X., Mastorakis, G., Batalla, J.M., Sahalos, J.N.: An evaluation of cloud-based mobile services with limited capacity: a linear approach. Soft Comput. (2016). doi:10.1007/s00500-016-2083-4
50. Skourletopoulos, G., Mavromoustakis, C.X., Mastorakis, G., Pallis, E., Batalla, J.M., Kormentzas, G.: Quantifying and evaluating the technical debt on mobile cloud-based service level. In: Proceedings of the IEEE International Conference on Communications (ICC 2016)—Communication QoS, Reliability and Modeling (CQRM) Symposium, Kuala Lumpur, Malaysia (2016c)
51. Sun, X., Gao, B., Fan, L., An, W.: A cost-effective approach to delivering analytics as a service. Proceedings of the 2012 IEEE 19th International Conference on Web Services (ICWS), pp. 512–519. Honolulu, HI (2012)
52. Talia, D.: Clouds for scalable big data analytics. IEEE Comput. Sci. 98–101 (2013)
53. The Apache Software Foundation: Apache hadoop. http://hadoop.apache.org (2014). Accessed 18 Oct 2015
54. Vakintis, I., Panagiotakis, S., Mastorakis, G., Mavromoustakis, C.X.: Evaluation of a Web Crowd-sensing IoT ecosystem providing big data analysis. In: Pop, F., Kołodziej, J., di Martino, B. (eds.) Resource Management for Big Data Platforms and Applications. Studies in Big Data Springer International Publishing, 2017
55. Vaquero, L.M., Celorio, A., Cuadrado, F., Cuevas, R.: Deploying large-scale datasets on-demand in the cloud: treats and tricks on data distribution. IEEE Trans. Cloud Comput. 3, 132–144 (2015)
56. Ye, F., Wang, Z., Zhou, F., Wang, Y., Zhou, Y.: Cloud-based big data mining and analyzing services platform integrating R. In: Proceedings of the 2013 International Conference on Advanced Cloud and Big Data (CBD), Nanjing, pp. 147–151 (2013)
57. Yeo, C.S., Venugopal, S., Chu, X., Buyya, R.: Autonomic metered pricing for a utility computing service. Future Gener. Comput. Syst. 26, 1368–1380 (2010)
58. Zhao, Y., Calheiros, R.N., Gange, G., Ramamohanarao, K., Buyya, R.: SLA-based resource scheduling for big data analytics as a service in cloud computing environments. In: Proceedings of the 2015 44th International Conference on Parallel Processing (ICPP), Beijing, pp. 510–519 (2015)

59. Zheng, Z., Zhu, J., Lyu, M.R.: Service-generated big data and big data-as-a-service: an overview. Proceedings of the 2013 IEEE International Congress on Big Data (BigData Congress), pp. 403–410. Santa Clara, California (2013)
60. Zulkernine, F., Martin, P., Zou, Y., Bauer, M., Gwadry-Shridhar, F., Aboulnaga, A.: Towards cloud-based analytics-as-a-service (CLAaaS) for big data analytics in the cloud. Proceedings of the 2013 IEEE International Congress on Big Data (BigData Congress), pp. 62–69. Santa Clara, California (2013)

Author Biographies

Georgios Skourletopoulos is currently a Doctoral Researcher in the Mobile Systems (MoSys) Laboratory, Department of Computer Science at the University of Nicosia, Cyprus. He also works as a Junior Business Analytics and Strategy Consultant in the Business Analytics and Strategy (BA&S) Service Line within IBM's Global Business Services (GBS) Business Unit at IBM Hellas S.A., Greece and he is a member of the IBM Big Data and Business Analytics Center of Competence in Greece. He obtained his M.Sc. in Computer Science from the University of Birmingham, UK in 2013 and his B.Sc. in Commerce and Marketing (with major in e-Commerce and Digital Marketing) from the Technological Educational Institute of Crete, Greece in 2012 (including a semester as an Erasmus exchange student at the Czech University of Life Sciences Prague, Czech Republic). In the past, he has worked as an e-Banking Platforms Use Case and Quality Assurance Analyst at Scientia Consulting S.A., Greece and as a Junior Research Analyst at Hellastat S.A., Greece. Mr. Georgios Skourletopoulos has publications at various international journals, conference proceedings, workshops and book chapters. His research interests lie in the general areas of mobile cloud computing, cloud computing, cloud-based software engineering, communications and networks.

Constandinos X. Mavromoustakis is currently an Associate Professor in the Department of Computer Science at the University of Nicosia, Cyprus. He received a five-year dipl.Eng in Electronic and Computer Engineering from Technical University of Crete, Greece, his M.Sc. in Telecommunications from University College of London, UK and his Ph.D. from the Department of Informatics at Aristotle University of Thessaloniki, Greece. He serves as the Chair of C16 Computer Society chapter of the Cyprus IEEE section, whereas he is the main recipient of various grants including the ESR-EU. His research interests are in the areas of spatial and temporal scheduling, energy-aware self-scheduling and adaptive behaviour in wireless and multimedia systems.

George Mastorakis received his B.Eng. in Electronic Engineering from UMIST, UK in 2000, his M.Sc. in Telecommunications from UCL, UK in 2001 and his Ph.D. in Telecommunications from the University of the Aegean, Greece in 2008. He is serving as an Assistant Professor at the Technological Educational Institute of Crete and as a Research Associate in Research & Development of Telecommunications Systems Laboratory at the Centre for Technological Research of Crete, Greece. His research interests include cognitive radio networks, networking traffic analysis, radio resource management and energy efficient networks. He has more than 80 publications at various international conferences proceedings, workshops, scientific journals and book chapters.

Jordi Mongay Batalla received his M.Sc. degree from Universitat Politecnica de Valencia in 2000 and his Ph.D. degree from Warsaw University of Technology in 2009, where he still works as Assistant Professor. In the past, he has worked in Telcordia Poland (Ericsson R&D Co.) and later in the National Institute of Telecommunications, Warsaw, where he is the Head of Internet

Architectures and Applications Department from 2010. He took part (coordination and/or participation) in more than 10 national and international ICT research projects, four of them inside the EU ICT Framework Programmes. His research interests focus mainly on Quality of Service (Diffserv, NGN) in both IPv4 and IPv6 infrastructures, Future Internet architectures (Content Aware Networks, Information Centric Networks) as well as applications for Future Internet (Internet of Things, Smart Cities, IPTV). He is author or co-author of more than 100 papers published in books, international and national journals and conference proceedings.

Ciprian Dobre completed his Ph.D. at the Computer Science Department, University Politehnica of Bucharest, Romania, where he is currently working as a full-time Associate Professor. His main research interests are in the areas of modeling and simulation, monitoring and control of large scale distributed systems, vehicular ad hoc networks, context-aware mobile wireless applications and pervasive services. He has participated as a team member in more than 10 national projects the last four years and he was member of the project teams for 5 international projects. He is currently involved in various organizing committees or committees for scientific conferences. He has developed MONARC 2, a simulator for LSDS used to evaluate the computational models of the LHC experiments at CERN and other complex experiments. He collaborated with Caltech in developing MonALISA, a monitoring framework used in production in more than 300 Grids and network infrastructures around the world. He is the developer of LISA, a lightweight monitoring framework that is used for the controlling part in projects like EVO or the establishment of world-wide records for data transferring at SuperComputing Bandwidth Challenge events (from 2006 to 2010). His research activities were awarded with the Innovations in Networking Award for Experimental Applications in 2008 by the Corporation for Education Network Initiatives (CENIC) and a Ph.D. scholarship by Oracle between 2006 and 2008.

Spyros Panagiotakis is an Assistant Professor at the Department of Informatics Engineering of the Technological Educational Institute of Crete and leader of the Laboratory for Sensor Networks, Telematics and Industrial Information Systems. He received his Ph.D. in Communication Networks from the Department of Informatics and Telecommunications of the University of Athens, Greece in 2007. He also received a four-year fellowship for postgraduate studies from the National Center for Scientific Research "Demokritos". He has participated in several European projects (IST projects MOBIVAS, POLOS, ANWIRE, LIAISON), as well as in several national projects. He is the author of over than 40 publications in international refereed books, journals and conferences. He also serves as peer-reviewer in international conferences and journals, as well as member of technical programme committees for international conferences and workshops. Finally, he serves as evaluator of research projects for the "Competitiveness, Entrepreneurship and Innovation" Operational Programme of Greece. His research interests include mobile multimedia technologies, communications and networking, Internet of Things, pervasive computing, sensor networks, web engineering, mobile applications, automation systems, location and context awareness and informatics in education.

Evangelos Pallis is an Associate Professor in the Applied Informatics and Multimedia Department, School of Applied Technology, Technological Educational Institute of Crete, and co-director of the "PASIPHAE" Research and Development of Telecommunication Systems Laboratory. He received his B.Sc. in Electronic Engineering from the Technological Educational Institute of Crete in 1994 and his M.Sc. in Telecommunications from University of East London in 1997. He received his Ph.D. in Telecommunications from the University of East London in 2002. His research interests are in the fields of wireless networks, mobile communication systems, digital broadcasting technologies and interactive television systems, cognitive radio and dynamic access technologies, QoS/QoE techniques and network management techniques, as well as in radio-resource trading and optimisation techniques. He has participated in a number of national and European funded R&D projects, including the AC215 "CRABS", IST-2000-26298 "MAMBO", IST-2000-28521 "SOQUET", IST-2001-34692 "REPOSIT", IST-2002-FP6-507637

"ENTHRONE", "IMOSAN", and as technical/scientific coordinator for the IST-2002-FP6-507312 "ATHENA" project. Currently, he is involved in the FP7-214751 "ADAMANTIUM", FP7-ICT-224287 "VITAL++" and FP7-ICT-248652 "ALICANTE" projects. He has more than 12 publications in international referred journals, 58 conference papers and 9 book chapters in the above scientific areas. He is the general chairman of the international conference on Telecommunications and Multimedia (TEMU), member of IET/IEE and active contributor to the IETF interconnection of content distribution networks (CDNi).

Towards Mobile Cloud Computing in 5G Mobile Networks: Applications, Big Data Services and Future Opportunities

Georgios Skourletopoulos, Constandinos X. Mavromoustakis,
George Mastorakis, Jordi Mongay Batalla, Ciprian Dobre,
Spyros Panagiotakis and Evangelos Pallis

Abstract The highly computationally capable mobile devices and the continuously increasing demand for high data rates and mobility, which are required by several mobile network services, enabled the research on fifth-generation (5G) mobile networks that are expected to be deployed beyond the year 2020 in order to support services and applications with more than one thousand times of today's network traffic. On the other hand, the huge and complex location-aware datasets exceed the capability of spatial computing technologies. In this direction, the mobile cloud computing (MCC) technology was introduced as the combination of cloud computing and mobile computing, enabling the end-users to access the cloud-supported services through mobile devices (e.g., smartphones, tablets, portable computers or wearable devices). The mobile applications exploit cloud technologies for data

G. Skourletopoulos (✉) · C.X. Mavromoustakis
Mobile Systems (MoSys) Laboratory, Department of Computer Science,
University of Nicosia, Nicosia, Cyprus
e-mail: skourletopoulos.g@unic.ac.cy

C.X. Mavromoustakis
e-mail: mavromoustakis.c@unic.ac.cy

G. Mastorakis · S. Panagiotakis · E. Pallis
Department of Informatics Engineering, Technological Educational Institute of Crete,
Heraklion, Crete, Greece
e-mail: gmastorakis@staff.teicrete.gr

S. Panagiotakis
e-mail: spanag@teicrete.gr

E. Pallis
e-mail: pallis@pasiphae.eu

J.M. Batalla
National Institute of Telecommunications, Szachowa Str. 1, Warsaw, Poland
e-mail: jordim@interfree.it

C. Dobre
Faculty of Automatic Control and Computers, Department of Computer Science,
University Politehnica of Bucharest, Bucharest, Romania
e-mail: ciprian.dobre@cs.pub.ro

© Springer International Publishing Switzerland 2017   43
C.X. Mavromoustakis et al. (eds.), *Advances in Mobile Cloud Computing
and Big Data in the 5G Era*, Studies in Big Data 22,
DOI 10.1007/978-3-319-45145-9_3

processing, storage and other intensive operations, as they are executed on resource providers external to the devices. This tutorial article is a comprehensive review of the current state-of-the-art and the latest developments on mobile cloud computing under the 5G era, which helps early-stage researchers to have an overview of the existing solutions, techniques and applications and investigate open research issues and future challenges in this domain.

Keywords Mobile cloud computing · Mobile cloud-based service level · 5G mobile networks · Big data · Data-driven · Modelling · Capacity · Lease cloud-based mobile services

1   Introduction

Since the data burst, the growth of mobile communication networks and the unprecedented increase in mobile data and multimedia traffic every year will occur in 5G mobile networks, the need for increasing the wireless network capacity is motivated in this fully interconnected information society [1–4]. Location information is an intrinsic part of different mobile devices and current mobile services are developed having a context-aware orientation and enclosing location-aware features. In addition, spatial big data refers to location-aware datasets (e.g., trajectories of mobile phones and GPS devices or detailed road maps), envisaged to transform the society via next-generation routing services [5]. Since the computationally capable mobile terminals can support more complex functionalities, 5G systems are expected to satisfy the quality of service (QoS) and the rate requirements set by applications and services, such as wireless broadband access, multimedia messaging services, mobile TV and services that utilize bandwidth [6, 7]. More specifically, each node should behave as a router (i.e., there is no central-control device) forwarding data and establishing an improved scalable network. Furthermore, the need for a cloud-based framework for sensor networks is motivated to enable the sharing or trading of feeds, as big data collection, management and analytics acquire new dimensions in Internet of Things (IoT) environments [8].

Despite the positive characteristics of the 5G mobile networks [9, 10], the challenges faced have to do with the improvement of network throughput, bandwidth utilization, mobility and fading [11–13]. In this direction, the mobile cloud computing (MCC) technology [14] was introduced as a solution for mobile services to improve the computational capabilities of resource-constrained mobile devices, exploiting the elastic resources provided by cloud computing, which can be a virtual server, a service or an application platform, in pay-as-you-use manner [15–19]. The mobile cloud computing paradigm indicates an infrastructure where both the data storage and data processing happen outside of the mobile device [20, 21]. As a result, the consumer and enterprise markets are increasingly adopting mobile cloud migration approaches to provide better services to their customers, boost their profits by reducing the development cost of mobile applications and gain competitive advantage.

Since mobile cloud computing is still in the preliminary stages, comprehensive surveys exist in the literature [14, 22, 23]. This tutorial paper focuses on the current mobile cloud computing research that discusses issues, applications and services and explores future opportunities and challenges in the 5G era. Following this introductory section, the structure of this book chapter is organized as follows: Sect. 2 presents an overview of the mobile cloud computing paradigm, including the definition, architecture and possible applications. Section 3 demonstrates the business value and long-term benefits of adopting mobile cloud migration approaches from the technical debt perspective and attempts to communicate the findings to non-technical stakeholders. Finally, Sect. 4 points out future opportunities and challenges in this domain, while Sect. 5 concludes this survey chapter.

2 Mobile Cloud Computing: Concept and Core Ideas

Mobile cloud computing (MCC) is constantly attracting the attention of entepreneurs and researchers as a cost-effective concept that can significantly improve the user experience (UX). Beyond the usual advantages provided by the cloud (e.g., resources, storage and applications always available on demand, scalability, dynamic provisioning), mobile cloud computing has also other advantages from the computational offloading point of view. More specifically, the battery lifetime is prolonged, as the energy-consuming tasks are executed in the cloud, and higher data storage and processing power capabilities are provided. In addition, the reliability is improved as the data is stored and backed up in a number of different devices [1]. In this context, this section gives an overview of the definition, architecture and applications of the mobile cloud computing technology.

2.1 Mobile Cloud Computing Definition and Architecture

The mobile cloud computing paradigm was introduced as a technology that can remedy the issues and challenges of mobile computing, such as the lack of resources in mobile devices, by exploiting the resources in the cloud platforms. A conceptual mobile cloud architecture is shown in Fig. 1 [24].

As mobile cloud computing emerged not long after the cloud computing and is basically based on that concept, it is important to give an overview of the cloud computing architecture, which is divided into three layers: Infrastructure-as-a-Service (IaaS), Platform-as-a-Service (PaaS) and Software-as-a-Service (SaaS). The three service layers of cloud computing along with brief descriptions for each layer are shown thoroughly in Fig. 2 [25]. The delivery objectives of each layer are described below [26]:

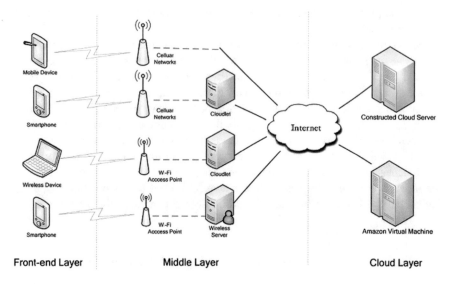

Fig. 1 A conceptual mobile cloud architecture

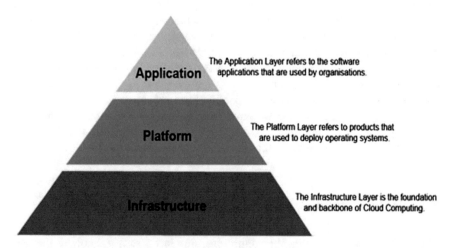

Fig. 2 The three service layers of cloud computing

- Infrastructure-as-a-Service (IaaS) offers on-demand components for building IT infrastructure, such as storage, bandwidth and virtual servers.
- Platform-as-a-Service (PaaS) offers development and runtime environments for applications that are hosted on the cloud and a scalable middleware for the management of application execution and dynamic resource provisioning.
- Software-as-a-Service (SaaS) offers multi-tenant applications and services on-demand.

The above layered architecture enabled the development of additional sophisti-cated models, such as the Big Data-as-a-Service (BDaaS) [27] or the Business Process-as-a-Service (BPaaS), where the customers are given the opportunity to consume business outcomes (e.g., payroll processing, HR) by accessing business services via web-centric interfaces [28]. In this context, the existing work on cloud computing points out three types of deployment models; the private (on-premises), public (off-premises) and hybrid.

However, the challenge of establishing a mobile cloud, using mobile devices and accessing the services provided, still exists. Satyanarayanan et al. [29] proposed an architecture, where the virtual machine-based (VM-based) cloudlets are exploited to rapidly instantiate customized service software and then use the service over a wireless LAN. Such an architecture can meet the need for quick real-time inter-active response by low latency, one-hop, high-bandwidth wireless access for multiple users. The differences between the conventional mobile cloud computing and the mobile-cloudlet-cloud computing architectures are demonstrated in Fig. 3 [30]. For the latter, the mobile devices interact via the cloudlet, using dynamic partitioning to achieve their quality of service goals. Finally, CloneCloud [31], another VM migration approach, aims to offload part of the mobile application workload to a resourceful server through either 3G or a wireless connection without the need of annotating methods, as witnessed in MAUI [32]. A cost-benefit analysis is also performed, calculating the cost involved in migration and execution on the cloud and comparing it against a monolithic execution.

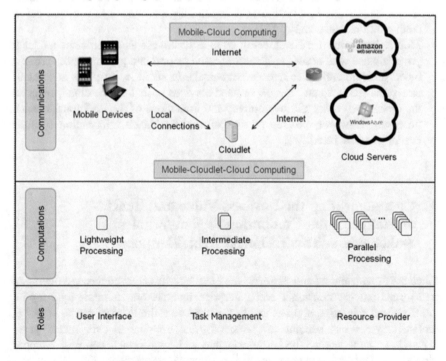

Fig. 3 Mobile cloud computing architecture versus mobile-cloudlet-cloud computing architecture

2.2 Applications of Mobile Cloud Computing and Impact Areas

Mobile devices, such as smartphones, tablets, portable computers, wearable devices or portable digital assistants (PDAs), have become tools that are used on a daily basis. A number of empirical studies [33] indicate that the number of today's connected mobile devices already exceeds five billion, while several demanding mobile network services, such as data sharing and cloud-based applications [34–38], require increased data rates. There are also numerous recent researches depicting several issues within the context of resource and cost management [39–49]. In this direction, some applications of mobile cloud computing are listed below:

- Mobile banking (m-Banking): Mobile banking is among the newest delivery channels used for accessing banking services securely and conducting transactions, payments and any kind of money transfers electronically. The nature of selling and buying financial services has now changed by improving the user experience and reducing the operational costs [50].
- Mobile commerce (m-Commerce): Mobile commerce, a subcategory of electronic commerce (e-Commerce), is a business model for commerce using mobile devices, which enables transactions, payments, messaging or even ticketing over the Internet on the move. However, there are still issues to be addressed in the future, such as the low network bandwidth, high complexity of mobile device configurations and security [22].
- Mobile healthcare (m-Healthcare): Mobile healthcare is an efficient model to provide users with access to resources, such as past and present health records, blood pressure and pulse rate measurements. In addition, hospitals and healthcare organizations can offer on-demand services hosted on the cloud, reducing the operational costs and overcoming the limitations of the traditional medical treatment. However, privacy of the personal healthcare information is still a challenge to be faced [22].

3 Communicating the Business Value and Benefits of Mobile Cloud Computing to Non-technical Stakeholders: The Technical Debt Viewpoint

Since the organizational and performance excellence is achieved through the digital transformation, the consumer and enterprise markets are increasingly adopting mobile cloud migration approaches. In this direction, the lease of cloud-supported mobile services has become an issue of great importance over recent years. Towards communicating the business value and long-term benefits of adopting mobile cloud migration approaches to non-technical stakeholders, the evaluation of such services is performed from the return on investment (ROI) point of view [51]. Investigating the technical debt on mobile cloud-based service level is a valid measure to reveal those cost-effective web services to lease off and inform effective decision making.

Considering such challenges, a number of research efforts have been devoted comparing the monetary cost-benefits of cloud computing with desktop grids [52] and investigating cost-benefit approaches of using cloud computing to extend the capacity of clusters [53]. The calculations of the cloud total cost of ownership (TCO) and utilization cost have been also examined in the literature [54], evaluating the economic efficiency of the cloud and providing indications for cost optimization. On the contrary, the technical debt term was initially introduced by Ward Cunningham [55] as a metaphor to communicate technical elements to non-technical product stakeholders. Currently, there is little work in the area of the technical debt quantification in mobile cloud computing and a research gap for introducing linear and non-linear metrics is observed. In this context, the mobile cloud can be considered as a marketplace [56], where the web services of the mobile cloud-based service-oriented architectures can be leased off [57]. Such services are differentiated on the subject of the non-functional requirements and the maximum number of users that can be supported. Two novel quantitative models for predicting and quantifying the technical debt on cloud-based software engineering and cloud-based service level were proposed in other research works [58, 59]. The quantification modelling is based on a cost-benefit analysis and extended experimental results are discussed by Skourletopoulos et al. [60]. Linear and fluctuation-based modelling approaches to estimate and measure the technical debt on mobile cloud-based service level have been also studied by Skourletopoulos et al. [61, 62]. In this book chapter, a brief survey of linear and non-linear approaches to quantification of the technical debt is presented, towards introducing the technical debt research problem in mobile cloud computing environments.

In this framework, a linear algorithm, which encompasses the mathematical formula that calculates the technical debt on mobile cloud-based service level, is shown in Table 1 [62]. The lease of cloud-supported mobile services is investigated from the capacity viewpoint. The notations and descriptions for the variables used are observed in Table 2.

On the contrary, the non-linear modelling for quantifying the technical debt on mobile cloud-based service level takes the following form, considering fluctuations in the number of users [61]:

$$
\begin{aligned}
TD_1 &= 12 * [ppm * (U_{max} - U_{curr}) - C_{u/m} * (U_{max} - U_{curr})] \\
&= 12 * (U_{max} - U_{curr}) * (ppm - C_{u/m})
\end{aligned}
\tag{1}
$$

$$
\begin{aligned}
TD_i &= 12 * \{K_{i-2} * [U_{max} - L_{i-2}] - M_{i-2} * [U_{max} - L_{i-2}]\} \\
&= 12 * (U_{max} - L_{i-2}) * (K_{i-2} - M_{i-2}), \quad i \geq 2
\end{aligned}
\tag{2}
$$

Table 1 Linear approach: the proposed algorithm encompassing the novel quantitative model

Algorithm 1 Pseudocode Implementation of Technical Debt Estimates

// This algorithm aims to calculate the technical debt in mobile cloud
// computing environments, once a linear growth in the number of users
// occurs.

1: procedure CALCULATION (U_{max}, $\beta\%$, i, U_{curr}, $\Delta\%$, λ, ppm, $\alpha\%$,

   $\gamma\%$, $\theta\%$, $\mu\%$, $\sigma\%$, $\eta\%$, $Cu_{/m}$)

2: sequential input (U_{max}, $\beta\%$, i, U_{curr}, $\Delta\%$, λ, ppm, $\alpha\%$, $\gamma\%$, $\theta\%$,

   $\mu\%$, $\sigma\%$, $\eta\%$, $Cu_{/m}$)

3: for $i=1$ to λ do // *Increasing the index of the year to get the*
   Output element with respect to the year

$$TD[i] \leftarrow 12 * \left[U_{max} - (1 + \beta\%)^{i-1} * U_{curr}\right] * \left[\left(1 + \frac{\Delta\%}{\lambda}\right)^{i-1} * \right.$$

$$\left. * ppm - \left(1 + \frac{\alpha\% + \gamma\% + \theta\% + \mu\% + \sigma\% + \eta\%}{\lambda}\right)^{i-1} * Cu_{/m}\right]$$

4: end for

5: return TD[i]

6: end procedure

where, $$K_0 = (1 + \Delta_1\%) * ppm$$

$$K_i = K_{i-1} * (1 + \Delta_{i+1}\%), \quad i \geq 1$$

$$L_0 = (1 + \beta_1\%) * U_{curr}$$

$$L_i = L_{i-1} * (1 + \beta_{i+1}\%), \quad i \geq 1$$

$$M_0 = (1 + VoC_1\%) * C_{u/m}$$

$$M_i = M_{i-1} * (1 + VoC_{i+1}\%), \quad i \geq 1$$

Table 2 Notations and variable descriptions

Symbol	Variable description
λ	The period of years
i	The index of the year
U_{max}	The maximum number of users that can be supported
U_{curr}	The initial number of users
$\beta\%$	The estimated average annual increase in the number of users, which is represented by a percentage value
ppm	The initial monthly subscription price for the end-users, which is expressed in monetary units
$\Delta\%$	The estimated average increase in the monthly subscription price over the period of λ-years, which is represented by a percentage value
$C_{u/m}$	The estimated initial monthly cost for servicing an end-user in the mobile cloud, which is expressed in monetary units
$\alpha\%$	The estimated average increase in the monthly document storage cost over the period of λ-years, which is represented by a percentage value
$\gamma\%$	The estimated average increase in the monthly data storage cost over the period of λ-years, which is represented by a percentage value
$\theta\%$	The estimated average increase in the monthly technical support cost over the period of λ-years, which is represented by a percentage value
$\mu\%$	The estimated average increase in the monthly maintenance services cost over the period of λ-years, which is represented by a percentage value
$\sigma\%$	The estimated average increase in the monthly network bandwidth cost over the period of λ-years, which is represented by a percentage value
$\eta\%$	The estimated average increase in the monthly server cost over the period of λ-years, which is represented by a percentage value
TD	The technical debt calculation result, which is expressed in monetary units

$$VoC_i\% = a_i\% + \gamma_i\% + \theta_i\% + \mu_i\% + \sigma_i\% + \eta_i\%, i \geq 1$$

The proposed quantitative models and the algorithm formulation achieve to measure the amount of profit not earned due to the underutilization of a service, examining the probability of overutilization of a service that would lead to accumulated technical debt in the long run. Any candidate cloud-based mobile service to be leased off is evaluated with repsect to the following assumptions:

- The period of λ-years is examined, aiming to provide insights about the return on investment. The optimal condition is the zero point (zero monetary units).
- Any candidate service is subscription-based for the end-users, while the SLA contract between the involved parties (i.e. provider and client, which is an organization) includes charges for servicing an end-user in the mobile cloud. The pricing and billing schemes vary over the period of λ-years due to the scalability provided by the cloud.
- Linear and non-linear approaches with respect to the demand are predicted accordingly over the period of λ-years.

- The cloud-supported mobile services are sensitive to network bandwidth and latency. The additional network bandwidth cost is expected to satisfy the outbound network traffic demands, while the additional server cost includes those costs that arise from the additional CPU cores and the amount of memory required for processing.

In this direction, two possible types of technical debt are encountered, once cloud-supported mobile services are leased off:

- Positive technical debt results, which declare the underutilization of the service and the probability to satisfy a possible future increase in the number of users.
- Negative technical debt results, which point out the overutilization of the selected service and possible SLA violations.

Towards a better understanding of the technical debt on mobile cloud computing, an indicative example emphasizes on the need to lease a cloud-based and always-on mobile service. A what-if analysis on different case scenarios and lease options is performed, exploring high, medium and low-capacity services. Furthermore, the technical debt numerical results will reveal the probability of overutilization of any of the services examined. Since linear and non-linear approaches are examined, the case scenarios with respect to the predicted demand are outlined below:

- Linear approach (i.e., linear growth in the number of users): The first case scenario forecasts a low-density annual increase in the demand ($\beta \% = 10 \%$). On the other hand, a high-density annual increase in the number of users is foreseen ($\beta \% = 80 \%$) for the second case scenario.
- Non-linear approach (i.e., fluctuations in the number of users): The first case scenario predicts an increase in the demand by 12 % during the second year, growth by 40 % during the third year, decrease by 18 % during the fourth year and increase by 35 % during the fifth year. On the contrary, regarding the second case scenario, an increase in the number of users by 20 % is forecasted during the second year, a decrease by 10 % during the third year, an increase in the demand by 60 % during the fourth year and, finally, increase in the number of users by 65 % during the fifth year.

The flow and comparisons of the technical debt estimates over the years are observed in Figs. 4 and 5 (for the linear approach) and Figs. 6 and 7 (for the non-linear point of view), respectively.

Towards the examination of the linear approach, the first case scenario indicates that the services A, B and C are always underutilized over the 4-year period due to the positive technical debt results. The technical debt is gradually paid off, as the amount of monetary units is constantly decreasing for the three services, due to the mild linear increase in the number of users per year (10 %). The lease of service C is the most cost-effective option for that case scenario, because the market needs are met over the 4-year period and the risk of entering into a new accumulated technical debt in the future does not lurk. On the other hand, the second case scenario

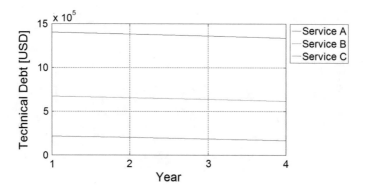

Fig. 4 Linearity: case scenario 1 with 10 % increase in the demand

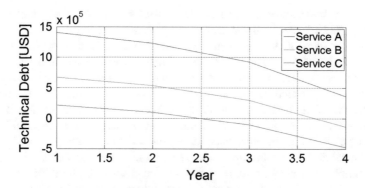

Fig. 5 Linearity: case scenario 2 with 80 % increase in the demand

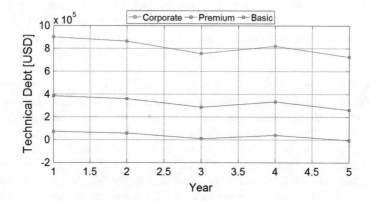

Fig. 6 Non-linearity: case scenario 1

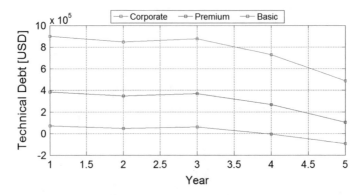

Fig. 7 Non-linearity: case scenario 2

demonstrates that service A is underutilized over the years and the technical debt is gradually paid off, despite the sheer annual increase in the number of users (80 %). The interpretation of the technical debt results with respect to service B shows underutilization of the service from the first until the end of the third year. During the fourth year, the technical debt becomes zero (i.e., optimal condition), revealing that it is totally cleared out. However, the technical debt results become negative until the end of the period as a result of the linear growth in the number of users, indicating that the service is overutilized. Likewise, service C is underutilized the first two years. During the third year, the technical debt is totally paid off; however, the technical debt results become negative until the end of the period. The negative results imply the overutilization of the service. Hence, a sheer increase in the demand (80 %) will motivate the need for abandoning/terminating either service B or C and switching to a more flexible capacity service. The risk of entering into a new and accumulated technical debt in the future is high, as these options would create additional costs. The analysis of the results for this case scenario points out that the most cost-effective cloud-supported mobile service to lease off is A, as the problem of overutilization does not lurk over the 4-year period, the technical debt is gradually paid off and the risk of entering into a new technical debt in the future is low.

Towards the evaluation of the non-liner approach, the first case scenario indicates that the Corporate and Premium services are always underutilized over the 5-year period of time due to the positive technical debt results. Despite the fluctuations in the demand, a gradual payoff of the technical debt is observed as the amount of monetary units for both services is less in the end of the period than in the beginning. The interpretation of the technical debt results concerning the Basic service reveals the underutilization of the service for the first four years. During the fifth year, the technical debt becomes zero, which constitutes the optimal condition, as it is totally cleared out. However, the technical debt result becomes negative until the end of the 5-year period of time, pointing out that the service is overutilized due to the variation in the number of users. Hence, the need for abandoning/terminating the existing service will be faced in the future in order to meet the evolving market needs. Having explained that any positive technical debt to be further incurred can be hardly managed, the Premium service is the most cost-effective lease option for that case scenario, maximizing the ROI and paying off the technical debt.

Finally, the second case scenario indicates that the Corporate and Premium services are always underutilized due to the positive technical debt results. Although the Basic service is underutilized the first three years, the technical debt calculation results become negative during the last two years. The technical debt is totally cleared out during the fourth year, achieving the optimal condition; however, the variations in the demand affect the flexibility and adaptability of the service, as it is overutilized until the end of the 5-year period of time. The need for abandoning/terminating the existing service and switching to a more flexible capacity one will be faced again in the future. Hence, the lease of the Premium service constitutes the most cost-effective investment decision for that case scenario, as the calculation results have the minimum positive values. It is significant to mention that in case of all three services were overutilized, the options of abandoning and switching would be inevitable and the need to examine other more flexible capacity services in the market from the same or different mobile cloud services providers would be motivated in order to meet the demand requirements and avoid entering into a new and accumulated technical debt in the future.

4 Future Opportunities and Challenges

Despite the advantages that mobile cloud computing brings in the 5G era, there are still issues to be addressed in this field on potentially better practices in providing better services and enforcing the standards. These challenges are strongly associated with the performance, control, security and cost. A critical investigation to each of the aforementioned challenges is performed below:

- **Performance**: Connection to clouds at a long distance is a major issue and results in high latency and delays. In addition, underperformance is witnessed when running transactions-oriented or data-intensive mobile applications, not to mention the need for enhancing the efficiency in data access.
- **Control**: Although large corporations and enterprises are willing to adopt mobile cloud migration approaches and strengthen their digital presence, the full control of the platforms, which the cloud computing providers have, is a major concern.
- **Security**: Securing the data in cloud is still a major issue in terms of integrity. Privacy concerns are also witnessed among end-users, especially vulnerability to attacks when information and critical IT resources are outside the firewall.
- **Cost**: Although adopting a mobile cloud migration approach can be very cost-effective (i.e., pay-as-you-go model), high network bandwidth charges may occur when it comes to data-intensive internet-based applications.

5 Concluding Remarks

Mobile cloud computing is recognized as the dominant model for mobile applications in the long run, providing rich functionality to the users. In this direction, this tutorial paper critically reviews the current mobile cloud computing research and indicates the motivation, research challenges and opportunities in this domain. Furthermore, the technical debt concept is being tackled analytically as a valid measure to understand and quantify the business value and the long-term benefits of adopting mobile cloud migration approaches and communicate these findings to non-technical stakeholders.

Acknowledgements The authors would like to thank the anonymous reviewers for their constructive comments and feedback on the manuscript. Their suggestions have contributed significantly to the high quality and improvement of this survey chapter. The research is supported by the ICT COST Action IC1406 "High-Performance Modelling and Simulation for Big Data Applications" (cHiPSet).

References

1. Barbarossa, S., Sardellitti, S., Di Lorenzo, P.: Communicating while computing: distributed mobile cloud computing over 5G heterogeneous networks. IEEE Signal Process. Mag. **31**, 45–55 (2014)
2. Kryftis, Y., Mavromoustakis, C.X., Batalla, J.M., Mastorakis, G., Pallis, E., Skourletopoulos, G.: Resource usage prediction for optimal and balanced provision of multimedia services. In: Proceedings of the 2014 IEEE 19th International Workshop on Computer Aided Modeling and Design of Communication Links and Networks (CAMAD 2014), Athens, Greece, pp. 255–259 (2014)
3. Kryftis, Y., Mavromoustakis, C.X., Mastorakis, G., Pallis, E., Batalla, J.M., Rodrigues, J.J.P. C., Dobre, C., Kormentzas, G.: Resource usage prediction algorithms for optimal selection of multimedia content delivery methods. In: Proceedings of the IEEE International Conference on Communications (ICC 2015), London, UK, pp. 5903–5909 (2015)
4. Mavromoustakis, C.X., Mastorakis, G., Bourdena, A., Pallis, E., Kormentzas, G., Rodrigues, J.J.P.C.: Context-oriented opportunistic cloud offload processing for energy conservation in wireless devices. In: Proceedings of the IEEE Global Communications Conference (GLOBECOM 2014)—The Second International Workshop on Cloud Computing Systems, Networks, and Applications (CCSNA), Austin, Texas, USA, pp. 24–30 (2014a)
5. Shekhar, S., Gunturi, V., Evans, M.R., Yang, K.S.: Spatial big-data challenges intersecting mobility and cloud computing. In: Proceedings of the Eleventh ACM International Workshop on Data Engineering for Wireless and Mobile Access (MobiDE'12), pp. 1–6 (2012)
6. Posnakides, D., Mavromoustakis, C.X., Skourletopoulos, G., Mastorakis, G., Pallis, E., Batalla, J.M.: Performance analysis of a rate-adaptive bandwidth allocation scheme in 5G mobile networks. In: Proceedings of the 20th IEEE Symposium on Computers and Communications (ISCC 2015)—The 2nd IEEE International Workshop on A 5G Wireless Odyssey: 2020, Larnaca, Cyprus, pp. 955–961 (2015)
7. Skourletopoulos, G., Xanthoudakis, A.: Developing a business plan for new technologies: application and implementation opportunities of the interactive digital (iDTV) and internet protocol (IPTV) television as an advertising tool. Bachelor's Degree Dissertation, Technological Educational Institute of Crete, Greece (2012)

8. Ramaswamy, L., Lawson, V., Gogineni, S.V.: Towards a quality-centric big data architecture for federated sensor services. In: Proceedings of the 2013 IEEE International Congress on Big Data (BigData Congress), Santa Clara, California, pp. 86–93 (2013)
9. Mastorakis, G., Mavromoustakis, C.X., Pallis, E.: Resource Management of Mobile Cloud Computing Networks and Environments. IGI Global, Hershey, Pennsylvania (2015)
10. Mavromoustakis, C.X., Mastorakis, G., Bourdena, A., Pallis, E., Kormentzas, G., Dimitriou, C.D.: Joint energy and delay-aware scheme for 5G mobile cognitive radio networks. In: Proceedings of the IEEE Global Communications Conference (GLOBECOM 2014)—Symposium on Selected Areas in Communications: Green Communication Systems and Networks, Austin, Texas, USA, pp. 2624–2630 (2014b)
11. Batalla, J.M., Kantor, M., Mavromoustakis, C.X., Skourletopoulos, G., Mastorakis, G.: A novel methodology for efficient throughput evaluation in virtualized routers. In: Proceedings of the 2015 IEEE International Conference on Communications (ICC 2015)—Communications Software, Services and Multimedia Applications Symposium (CSSMA), London, UK, pp. 6899–6905 (2015)
12. Mavromoustakis, C.X., Bourdena, A., Mastorakis, G., Pallis, E., Kormentzas, G.: An energy-aware scheme for efficient spectrum utilization in a 5G mobile cognitive radio network architecture. Telecommun. Syst. **59**, 63–75 (2015)
13. Papadopoulos, M., Mavromoustakis, C.X., Skourletopoulos, G., Mastorakis, G., Pallis, E.: Performance analysis of reactive routing protocols in mobile ad hoc networks. In: Proceedings of the 2014 IEEE 6th International Conference on Telecommunications and Multimedia (TEMU 2014), Heraklion, Crete, Greece, pp. 104–110 (2014)
14. Fernando, N., Loke, S.W., Rahayu, W.: Mobile cloud computing: a survey. Futur. Gener. Comput. Syst. **29**, 84–106 (2013)
15. Markakis, E., Sideris, A., Alexiou, G., Bourdena, A., Pallis, E., Mastorakis, G., Mavromoustakis, C.: A virtual network functions brokering mechanism. International Conference on Telecommunications and Multimedia (TEMU 2016), IEEE Communications Society Proceedings, Heraklion, Greece, 25–27 Jul 2016
16. Zaharis, Z., Yioultsis, T., Skeberis, C., Xenos, T., Lazaridis, P., Mastorakis, G., Mavromoustakis, C.: Implementation of antenna array beamforming by using a novel neural network structure. In: International Conference on Telecommunications and Multimedia (TEMU 2016), IEEE Communications Society proceedings, Heraklion, Greece, 25–27 Jul 2016
17. Bormpantonakis, P., Stratakis, D., Mastorakis, G., Mavromoustakis, C., Skeberis, C., Bechet, P.: Exposure EMF measurements with spectrum analyzers using free and open source software. In: International Conference on Telecommunications and Multimedia (TEMU 2016), IEEE Communications Society proceedings, Heraklion, Greece, 25–27 Jul 2016
18. Hadjioannou, V., Mavromoustakis, C., Mastorakis, G., Pallis, E., Markakis, E.: Context awareness location-based android application for tracking purposes in assisted living. In: International Conference on Telecommunications and Multimedia (TEMU 2016), IEEE Communications Society proceedings, Heraklion, Greece, 25–27 Jul 2016
19. Mastorakis, G., Markakis, E., Pallis, E., Mavromoustakis, C.X., Skourletopoulos, G.: Virtual network functions exploitation through a prototype resource management framework. In: Proceedings of the 2014 IEEE 6th International Conference on Telecommunications and Multimedia (TEMU 2014), Heraklion, Crete, Greece, pp. 24–28 (2014)
20. Bourdena, A., Mavromoustakis, C.X., Mastorakis, G., Rodrigues, J.J.P.C., Dobre, C.: Using socio-spatial context in mobile cloud process offloading for energy conservation in wireless devices. IEEE Trans. Cloud Comput. 1 (2015)
21. Chilipirea, C., Petre, A.C., Dobre, C., Pop, F.: Enabling mobile cloud wide spread through an evolutionary market-based approach. IEEE Syst. J. 1–8 (2015)
22. Dinh, H.T., Lee, C., Niyato, D., Wang, P.: A survey of mobile cloud computing: architecture, applications, and approaches. Wirel. Commun. Mob. Comput. **13**, 1587–1611 (2013)
23. Khan, A.N., Kiah, M.L.M., Khan, S.U., Madani, S.A.: Towards secure mobile cloud computing: a survey. Futur. Gener. Comput. Syst. **29**, 1278–1299 (2013)

24. Mobile Cloud Computing Lab, The Hong Kong Polytechnic University. Mobile cloud computing (2013). http://www4.comp.polyu.edu.hk/~csbxiao/MCCLab/MCCLab_background.html. Accessed 09 Oct 2015
25. IT Associates. What is cloud computing (2015). http://www.itassociates.com.au/products-services/cloud-computing.php. Accessed 09 Oct 2015
26. Calheiros, R.N., Vecchiola, C., Karunamoorthy, D., Buyya, R.: The aneka platform and QoS-driven resource provisioning for elastic applications on hybrid clouds. Futur. Gener. Comput. Syst. **28**, 861–870 (2012)
27. Skourletopoulos, G., Mavromoustakis, C.X., Mastorakis, G., Pallis, E., Chatzimisios, P., Batalla, J.M.: Towards the evaluation of a big data-as-a-service model: a decision theoretic approach. In: Proceedings of the IEEE International Conference on Computer Communications (INFOCOM 2016)—First IEEE International Workshop on Big Data Sciences, Technologies and Applications (BDSTA), San Francisco, California, USA (2016c)
28. IBM Corporation. IBM Business Process as a Service (2015). http://www-935.ibm.com/services/us/business-consulting/bpo/business-process-as-a-service.html. Accessed 09 Oct 2015
29. Satyanarayanan, M., Bahl, P., Cáceres, R., Davies, N.: The case for vm-based cloudlets in mobile computing. IEEE Pervasive Comput. **8**, 14–23 (2009)
30. Soyata, T., Ba, H., Heinzelman, W., Kwon, M., Shi, J.: Accelerating mobile-cloud computing: a survey. In: Mouftah, H.T., Kantarci, B. (eds.) Communication Infrastructures for Cloud Computing, pp. 175–197. IGI Global, Hershey, PA (2014)
31. Chun, B.G., Ihm, S., Maniatis, P., Naik, M., Patti, A.: Clonecloud: elastic execution between mobile device and cloud. In: Proceedings of the 6th Conference on Computer Systems (EuroSys'11), New York, NY, USA, pp. 301–314 (2011)
32. Cuervo, E., Balasubramanian, A., Cho, D.K., Wolman, A., Saroiu, S., Chandra, R., Bahl, P.: Maui: making smartphones last longer with code offload. In: Proceedings of the 8th International Conference on Mobile Systems, Applications, and Services (MobiSys'10), New York, NY, USA, pp. 49–62 (2010)
33. Cisco Systems, Inc. Cisco visual networking index: global mobile data traffic forecast update, 2014–2019 (2015). http://www.cisco.com/c/en/us/solutions/collateral/service-provider/visual-networking-index-vni/white_paper_c11-520862.pdf. Accessed 09 Oct 2015
34. Mavromoustakis, C., Mastorakis, G., Papadakis, G., Andreou, A., Bourdena, A., Stratakis, D.: Energy consumption optimization through pre-scheduled opportunistic offloading in wireless devices. In: The Sixth International Conference on Emerging Network Intelligence, EMERGING 2014, Rome, Italy, pp. 22–28, 24–28 Aug 2014
35. Mastorakis, G., Mavromoustakis, C., Bourdena, A., Kormentzas, G., Pallis, E.: Maximizing energy conservation in a centralized cognitive radio network architecture. In: Proceedings of the 18th IEEE International Workshop on Computer Aided Modeling Analysis and Design of Communication Links and Networks (CAMAD), Berlin, Germany, 25–27 Sept 2013, pp. 190–194
36. Mastorakis, G., Mavromoustakis, C., Bourdena, A., Pallis, E., Sismanidis, G.: Optimizing radio resource management in energy-efficient cognitive radio networks. In: Proceedings of The 16th ACM International Conference on Modeling, Analysis and Simulation of Wireless and Mobile Systems, 3–8 Nov 2013, Barcelona, Spain, pp. 75–82
37. Mousicou, P., Mavromoustakis, C., Bourdena, A., Mastorakis, G., Pallis, E.: Performance evaluation of dynamic cloud resource migration based on temporal and capacity-aware policy for efficient resource sharing. In: Proceedings of The 16th ACM International Conference on Modeling, Analysis and Simulation of Wireless and Mobile Systems, 3–8 Nov 2013, Barcelona, Spain, pp. 59–66
38. Papanikolaou, K., Mavromoustakis, C., Mastorakis, G., Bourdena, A., Dobre, C.: Energy consumption optimization using social interaction in the mobile cloud. In: Proceedings of International Workshop on Enhanced Living EnvironMENTs (ELEMENT 2014), 6th International Conference on Mobile Networks and Management (MONAMI 2014), Wuerzburg, Germany, September 2014

39. Vakintis, I., Panagiotakis, S., Mastorakis, G., Mavromoustakis, C.: Evaluation of a Web crowd-sensing IoT ecosystem providing Big data analysis. In: Pop, F., Kołodziej, J., di Martino, B. (eds.) Resource Management for Big Data Platforms and Applications. Studies in Big Data. Springer International Publishing (2016)

40. Hadjioannou, V., Mavromoustakis, C., Mastorakis, G., Batalla J.M., Kopanakis, I., Perakakis, E., Panagiotakis, S.: Security in smart grids and smart spaces for smooth IoT deployment in 5G. In: Internet of Things (IoT) in 5G Mobile Technologies. Modeling and Optimization in Science and Technologies, vol. 8, pp. 371–397, April 2016

41. Goleva, R., et al.: Data and traffic models in 5G network. In: Internet of Things (IoT) in 5G Mobile Technologies. Springer International Publishing, pp. 485–499 (2016)

42. Batalla, J.M., Mavromoustakis, C., Mastorakis, G., Sienkiewicz, K.: On the track of 5G radio access network for IoT wireless spectrum sharing in device positioning applications. In: Internet of Things (IoT) in 5G Mobile Technologies. Modeling and Optimization in Science and Technologies, vol. 8, pp. 25–35, April 2016

43. Markakis, E., Mastorakis, G., Negru, D., Pallis, E., Mavromoustakis, C.: A context-aware system for efficient peer-to-peer content provision. In: Xhafa, F. (ed.) Pervasive Computing: Next Generation Platforms for Intelligent Data Collection. Intelligent Data-Centric Systems. Morgan Kaufmann/Elsevier

44. Zaharis, Z., Yioultsis, T., Skeberis, C., Lazaridis, P., Stratakis, D., Mastorakis, G., Mavromoustakis, C., Pallis, E., Xenos, T.: Design and optimization of wideband log-periodic dipole arrays under requirements for high gain, high front-to-back ratio, optimal gain flatness and low side lobe level: the application of Invasive Weed optimization. In: Matyjas, J.D., Hu, F., Kumarto, S. (eds.) Wireless Network Performance Enhancement via Directional Antennas: Models, Protocols, and Systems. Taylor & Francis LLC, CRC Press, December 2015

45. Karolewicz, K., Beben, A., Batalla J.M., Mastorakis, G., Mavromoustakis, C.: On efficient data storage service for IoT. Int. J. Netw. Manag. Wiley, May 2016

46. Batalla, J.M., Mavromoustakis, C., Mastorakis, G., Négru, D., Borcoci, E.: Evolutionary Multiobjective optimization algorithm for multimedia delivery in critical applications through Content Aware Networks. J. Supercomput. (SUPE) (2016). Springer International Publishing

47. Batalla, J.M., Mastorakis, G., Mavromoustakis, C., Żurek, J.: On cohabitating networking technologies with common wireless access for Home Automation Systems purposes, Special Issue on Enabling Wireless Communication and Networking Technologies for the Internet of Things, IEEE Wireless Communication Magazine (2016)

48. Batalla, J.M., Gajewski, M., Latoszek, W., Krawiec, P., Mavromoustakis, C., Mastorakis, G.: ID-based service-oriented communications for unified access in IoT. Comput. Electr. Eng. J. 2016

49. Bourdena, A., Mavromoustakis, C., Kormentzas, G., Pallis, E., Mastorakis, G., Yassein, M. B.: A resource intensive traffic-aware scheme using energy-efficient routing in cognitive radio networks. Futur. Gener. Comput. Syst. J. 39, 16–28 (2014)

50. Suoranta, M., Mattila, M.: Mobile banking and consumer behaviour: new insights into the diffusion pattern. J. Financ. Serv. Mark. 8, 354–366 (2004)

51. Skourletopoulos, G., Mavromoustakis, C.X., Mastorakis, G., Batalla, J.M., Sahalos, J.N.: An evaluation of cloud-based mobile services with limited capacity: a linear approach. Soft. Comput. (2016). doi:10.1007/s00500-016-2083-4

52. Kondo, D., Javadi, B., Malecot, P., Cappello, F., Anderson, D.P.: Cost-benefit analysis of cloud computing versus desktop grids. In: Proceedings of the 2009 IEEE International Symposium on Parallel & Distributed Processing (IPDPS 2009), Rome, pp. 1–12 (2009)

53. De Assunção, M.D., Di Costanzo, A., Buyya, R.: A cost-benefit analysis of using cloud computing to extend the capacity of clusters. Clust. Comput. 13, 335–347 (2010)

54. Li, X., Li, Y., Liu, T., Qiu, J., Wang, F.: The method and tool of cost analysis for cloud computing. In: 2009 IEEE International Conference on Cloud Computing (CLOUD'09), Bangalore, pp. 93–100 (2009)

55. Cunningham, W.: The WyCash portfolio management system. In: Proceedings on Object-oriented Programming Systems, Languages, and Applications (OOPSLA), Vancouver, British Columbia, Canada, pp. 29–30 (1992)
56. Buyya, R., Yeo, C.S., Venugopal, S., Broberg, J., Brandic, I.: Cloud computing and emerging IT platforms: vision, hype, and reality for delivering computing as the 5th utility. Futur. Gener. Comput. Syst. **25**, 599–616 (2009)
57. Nallur, V., Bahsoon, R.: A decentralized self-adaptation mechanism for service-based applications in the cloud. IEEE Trans. Softw. Eng. **39**, 591–612 (2012)
58. Skourletopoulos G (2013) Researching and quantifying the technical debt in cloud software engineering. Master's Degree Dissertation, University of Birmingham, UK
59. Skourletopoulos, G., Bahsoon, R., Mavromoustakis, C.X., Mastorakis, G., Pallis, E.: Predicting and quantifying the technical debt in cloud software engineering. In: Proceedings of the 2014 IEEE 19th International Workshop on Computer Aided Modeling and Design of Communication Links and Networks (CAMAD 2014), Athens, Greece, pp. 36–40 (2014)
60. Skourletopoulos, G., Bahsoon, R., Mavromoustakis, C.X., Mastorakis, G.: The technical debt in cloud software engineering: a prediction-based and quantification approach. In: Mastorakis, G., Mavromoustakis, C.X., Pallis, E. (eds.) Resource Management of Mobile Cloud Computing Networks and Environments, 1st edn, pp. 24–42. IGI Global, Hershey, PA (2015)
61. Skourletopoulos, G., Mavromoustakis, C.X., Mastorakis, G., Rodrigues, J.J.P.C., Chatzimisios, P., Batalla, J.M.: A fluctuation-based modelling approach to quantification of the technical debt on mobile cloud-based service level. In: Proceedings of the IEEE Global Communications Conference (GLOBECOM 2015)—The Fourth IEEE International Workshop on Cloud Computing Systems, Networks, and Applications (CCSNA), San Diego, California, USA (2015b)
62. Skourletopoulos, G., Mavromoustakis, C.X., Mastorakis, G., Pallis, E., Batalla, J.M., Kormentzas, G.: Quantifying and evaluating the technical debt on mobile cloud-based service level. In: Proceedings of the IEEE International Conference on Communications (ICC 2016)—Communication QoS, Reliability and Modeling (CQRM) Symposium, Kuala Lumpur, Malaysia (2016b)

Author Biographies

Georgios Skourletopoulos is currently a Doctoral Researcher in the Mobile Systems (MoSys) Laboratory, Department of Computer Science at the University of Nicosia, Cyprus. He also works as a Junior Business Analytics and Strategy Consultant in the Business Analytics and Strategy (BA&S) Service Line within IBM's Global Business Services (GBS) Business Unit at IBM Hellas S.A., Greece and he is a member of the IBM Big Data and Business Analytics Center of Competence in Greece. He obtained his M.Sc. in Computer Science from the University of Birmingham, UK in 2013 and his B.Sc. in Commerce and Marketing (with major in e-Commerce and Digital Marketing) from the Technological Educational Institute of Crete, Greece in 2012 (including a semester as an Erasmus exchange student at the Czech University of Life Sciences Prague, Czech Republic). In the past, he has worked as an e-Banking Platforms Use Case and Quality Assurance Analyst at Scientia Consulting S.A., Greece and as a Junior Research Analyst at Hellastat S.A., Greece. Mr. Georgios Skourletopoulos has publications at various international journals, conference proceedings, workshops and book chapters. His research interests lie in the general areas of mobile cloud computing, cloud computing, cloud-based software engineering, communications and networks.

Constandinos X. Mavromoustakis is currently an Associate Professor in the Department of Computer Science at the University of Nicosia, Cyprus. He received a five-year dipl.Eng in Electronic and Computer Engineering from Technical University of Crete, Greece, his M.Sc. in Telecommunications from University College of London, UK and his Ph.D. from the Department of Informatics at Aristotle University of Thessaloniki, Greece. He serves as the Chair of C16 Computer Society chapter of the Cyprus IEEE section, whereas he is the main recipient of various grants including the ESR-EU. His research interests are in the areas of spatial and temporal scheduling, energy-aware self-scheduling and adaptive behaviour in wireless and multimedia systems.

George Mastorakis received his B.Eng. in Electronic Engineering from UMIST, UK in 2000, his M.Sc. in Telecommunications from UCL, UK in 2001 and his Ph.D. in Telecommunications from the University of the Aegean, Greece in 2008. He is serving as an Assistant Professor at the Technological Educational Institute of Crete and as a Research Associate in Research & Development of Telecommunications Systems Laboratory at the Centre for Technological Research of Crete, Greece. His research interests include cognitive radio networks, networking traffic analysis, radio resource management and energy efficient networks. He has more than 80 publications at various international conferences proceedings, workshops, scientific journals and book chapters.

Jordi Mongay Batalla received his M.Sc. degree from Universitat Politecnica de Valencia in 2000 and his Ph.D. degree from Warsaw University of Technology in 2009, where he still works as Assistant Professor. In the past, he has worked in Telcordia Poland (Ericsson R&D Co.) and later in the National Institute of Telecommunications, Warsaw, where he is the Head of Internet Architectures and Applications Department from 2010. He took part (coordination and/or participation) in more than 10 national and international ICT research projects, four of them inside the EU ICT Framework Programmes. His research interests focus mainly on Quality of Service (Diffserv, NGN) in both IPv4 and IPv6 infrastructures, Future Internet architectures (Content Aware Networks, Information Centric Networks) as well as applications for Future Internet (Internet of Things, Smart Cities, IPTV). He is author or co-author of more than 100 papers published in books, international and national journals and conference proceedings.

Ciprian Dobre completed his Ph.D. at the Computer Science Department, University Politehnica of Bucharest, Romania, where he is currently working as a full-time Associate Professor. His main research interests are in the areas of modeling and simulation, monitoring and control of large scale distributed systems, vehicular ad-hoc networks, context-aware mobile wireless applications and pervasive services. He has participated as a team member in more than 10 national projects the last four years and he was member of the project teams for 5 international projects. He is currently involved in various organizing committees or committees for scientific conferences. He has developed MONARC 2, a simulator for LSDS used to evaluate the computational models of the LHC experiments at CERN and other complex experiments. He collaborated with Caltech in developing MonALISA, a monitoring framework used in production in more than 300 Grids and network infrastructures around the world. He is the developer of LISA, a lightweight monitoring framework that is used for the controlling part in projects like EVO or the establishment of world-wide records for data transferring at SuperComputing Bandwidth Challenge events (from 2006 to 2010). His research activities were awarded with the Innovations in Networking Award for Experimental Applications in 2008 by the Corporation for Education Network Initiatives (CENIC) and a Ph.D. scholarship by Oracle between 2006 and 2008.

Spyros Panagiotakis is an Assistant Professor at the Department of Informatics Engineering of the Technological Educational Institute of Crete and leader of the Laboratory for Sensor Networks, Telematics and Industrial Information Systems. He received his Ph.D. in Communication Networks from the Department of Informatics and Telecommunications of the University of

Athens, Greece in 2007. He also received a four-year fellowship for postgraduate studies from the National Center for Scientific Research "Demokritos". He has participated in several European projects (IST projects MOBIVAS, POLOS, ANWIRE, LIAISON), as well as in several national projects. He is the author of over than 40 publications in international refereed books, journals and conferences. He also serves as peer-reviewer in international conferences and journals, as well as as member of technical programme committees for international conferences and workshops. Finally, he serves as evaluator of research projects for the "Competitiveness, Entrepreneurship and Innovation" Operational Programme of Greece. His research interests include mobile multimedia technologies, communications and networking, Internet of Things, pervasive computing, sensor networks, web engineering, mobile applications, automation systems, location and context awareness and informatics in education.

Evangelos Pallis is an Associate Professor in the Applied Informatics and Multimedia Department, School of Applied Technology, Technological Educational Institute of Crete, and co-director of the "PASIPHAE" Research and Development of Telecommunication Systems Laboratory. He received his B.Sc. in Electronic Engineering from the Technological Educational Institute of Crete in 1994 and his M.Sc. in Telecommunications from University of East London in 1997. He received his Ph.D. in Telecommunications from the University of East London in 2002. His research interests are in the fields of wireless networks, mobile communication systems, digital broadcasting technologies and interactive television systems, cognitive radio and dynamic access technologies, QoS/QoE techniques and network management techniques, as well as in radio-resource trading and optimisation techniques. He has participated in a number of national and European funded R&D projects, including the AC215 "CRABS", IST-2000-26298 "MAMBO", IST-2000-28521 "SOQUET", IST-2001-34692 "REPOSIT", IST-2002-FP6-507637 "ENTHRONE", "IMOSAN", and as technical/scientific coordinator for the IST-2002-FP6-507312 "ATHENA" project. Currently, he is involved in the FP7- 214751 "ADAMANTIUM", FP7-ICT-224287 "VITAL++" and FP7-ICT-248652 "ALICANTE" projects. He has more than 12 publications in international referred journals, 58 conference papers and 9 book chapters in the above scientific areas. He is the general chairman of the international conference on Telecommunications and Multimedia (TEMU), member of IET/IEE and active contributor to the IETF interconnection of content distribution networks (CDNi).

Part II
Architectures of MCC
and Big Data Paradigm

Heterogeneous Data Access Control Based on Trust and Reputation in Mobile Cloud Computing

Zheng Yan, Xueyun Li and Raimo Kantola

Abstract Cloud computing, as an emerging computing paradigm for service provision on-demand, has blossomed into both academia and industry. One typical cloud service is cloud data storage and processing. By stepping into the era of 5G, Mobile Cloud Computing (MCC) becomes a perfect match to further leverage the limitations of resources and locations for advanced networking services. However, adopting mobile cloud implies placing critical applications and sensitive data in Cloud Service Providers (CSPs) and accessing them with mobile devices. This leads to serious security and usability issues, particularly on cloud data access control. In this chapter, we explore the objectives of data access control in MCC. Based on a literature review, we summarize open research issues in this area. In order to achieve secure and usable cloud data access, we adopt trust and reputation management and investigate three schemes for securing cloud data access based on trust and reputation that can be flexibly applied in different cloud data access scenarios. Furthermore, we evaluate the performance of the schemes through the analysis on security, complexity, scalability and flexibility, as well as scheme implementation. Particularly, important issues and challenges are also discussed in order to propose future research trends.

Keywords Mobile Cloud Computing · Access control · Security · Privacy · Trust · Reputation · 5G

Z. Yan (✉)
The State Key Laboratory of ISN, School of Cyber Engineering,
Xidian University, Xi'an, China
e-mail: zhengyan.pz@gmail.com; zyan@xidian.edu.cn

Z. Yan · X. Li · R. Kantola
Department of Communications and Networking, Aalto University, Espoo, Finland
e-mail: xueyun.li@aalto.fi

R. Kantola
e-mail: raimo.kantola@aalto.fi

© Springer International Publishing Switzerland 2017
C.X. Mavromoustakis et al. (eds.), *Advances in Mobile Cloud Computing
and Big Data in the 5G Era*, Studies in Big Data 22,
DOI 10.1007/978-3-319-45145-9_4

1 Introduction

Cloud computing, as an emerging computing paradigm, is blooming in both academia and industry. It is defined by the U.S. National Institute of Standards and Technology (NIST) as a model for enabling convenient, on-demand network access to a shared pool of configurable computing resources that can be rapidly provisioned and released with minimal management efforts [1]. Cloud computing services can be generally segmented into three categories: Infrastructure-as-a-Service (IaaS), Platform-as-a-Service (PaaS), and Software-as-a-Service (SaaS). These cloud services provide ubiquitous communications and on-demand data access. Moreover, cloud computing created a huge market in IT industry, by providing cost-effective solutions, minimum to zero investments, and low management of computing resources. In Cisco's forecast, Cloud Service Providers (CSPs) will process more than three quarters (78 %) of workloads by 2018 [2]. There are many cloud services that can be used in our everyday life, such as Google Drive, iCloud and Dropbox, which have freed us from hardware restraints and ensured that the data we need are available at any time and in any place.

Mobile Cloud Computing (MCC) is a converged technology of mobile computing, cloud computing and wireless technologies. By stepping into the era of 5G, Mobile Cloud Computing (MCC) becomes a perfect match to further leverage the limitations of resources and locations for advanced networking services. According to Cisco's global mobile data traffic forecast, the global mobile devices and connections in 2014 grew to 7.4 billion, as many as the population on our planet [3]. Moreover, the improved wireless technologies of 5G will provide higher data rate, bandwidth, and ubiquitous service coverage to enable anticipated performance of MCC. Currently, many MCC services and business models have been proposed, such as mobile commerce (m-commerce), mobile learning and mobile healthcare. Doukas proposed a mobile Healthcare information management system, called @HealthCloud, to provide pervasive healthcare data storage, update and retrieval [4]. In @HealthCloud, the mobile clients run on Google's Android mobile operating system to consume such services as information management and processing handled by the Amazon simple Storage Service (S3). One Hour Translation [5] is an online translation service running in the cloud of Amazon Web Services. It can be used by mobile client devices and translates the information in the own languages of mobile client users.

However, while embracing the benefits brought by various mobile cloud services, we are still facing a number of problems such as data disclosure, privacy leaks and malicious attacks. For example, user's personal data were leaked from Google and Apple applications. And four times of data breaches took place at Oregon Health and Science University due to inappropriate handling and storage of unencrypted patient medical records in the cloud. Currently, access control in the cloud environment is typically provided by techniques such as Virtual Local Area Networks (VLANs) and firewalls, which are originally designed to support enterprise IT systems [6]. These techniques are not suitable for the multi-tenant and dynamic nature of cloud, thus they could lead to major data breaches. In a cloud environment, there is a large volume of users, which makes it difficult for a centralized agent to check the user's eligibility and trace their IDs. Moreover, due to the purposes of resource conserving, cost cut, and maintaining efficiency of cloud computing, many CSPs support multi-tenant services in a cloud environment. In the public cloud, many users store their data, and conduct data processing or run Virtual Machines (VMs) on the same server provided by a CSP. This service model poses security threats that a customer shares the same physical hardware with its adversaries, and its resources (e.g., data and VMs) placed in the cloud servers can be attacked by the adversaries. Insider attack is another security problem that may lead to the leakage of business sensitive and user confidential data. The Cloud Security Alliance has recently reported that insider attacks have become the third biggest threat in cloud computing. Since the user's data are managed at the cloud servers, the CSPs must ensure that all their employees who have access to the servers in the data center are trusted and do not disclose any information or conduct attacks.

In MCC, the security and privacy problem becomes more serious due to the insecure nature of wireless communications. For example, in a Wi-Fi environment, it is easy for an attacker to play a "man-in-the-middle" attack, where the attacker can intercept the communications between a user and an Access Point (AP). Therefore, the attacker can sniff communication traffic, or force the user to drop its connection and reconnect to the attacker's soft AP. The mobility of wireless networks and mobile users also introduces many security problems. In a cellular network, security threats could happen during channel setup. When a user moves and registers with a cell, a base station needs to manage and update the user's information in order to allow messages to be routed to the user correctly. This process can expose the user's information to all the entities that can access to the user's routing information. The mobility also brings the trust issues between the users and mobile service providers, because the mobile users can use resources at various locations, provided by various service providers. This fact increases the difficulty of verifying the user's identity and preventing the users from connecting to a malicious cell. However, the security problems of wireless transmission are being well studied, since they happened from the beginning of wireless communications. In the era of 5G, these problems will be further researched and addressed by more rigorous protocols and more advanced wireless technologies.

One of the most important issues in MCC is data security. Private data of users are stored in the data center of CSP to release the storage and computing burden of personal devices. Typical data examples are social security records and health statistical data monitored by wearable sensors. However, the CSP could be curious on personal privacy [7]. Thus, critical personal data stored in CSP are generally encrypted and their access is controlled. Obviously, these personal data could be accessed by other entities in order to fulfill a cloud service. How to control personal data access at CSP is a practical issue.

Rationally, the data owner should control own data access. But in many situations, the data owner is not available or has no idea how to do the control, when, for example, personal health records should be accessed by several medical experts in order to figure out a treatment solution or by a foreign physician when the data owner is traveling and falling into a health problem abroad. Therefore, an access control agent is expected in this situation in order to reduce the risk of the data owner in personal data management. Considering both above demands in practice, a heterogeneous scheme that can flexibly control data access in cloud computing is expected.

A number of solutions have been proposed for protecting data access in the cloud. Access Control List (ACL) based solutions suffer from the drawback that computation complexity grows linearly with the number of data-groups [8] or the number of users in the ACL [9]. Role Based Access Control (RBAC) cannot flexibly support various data access demands that rely on trust [10]. In recent years, access control schemes based on Attribute-Based Encryption (ABE) were proposed for controlling cloud data access based on attributes in order to enhance flexibility [11–14, 15, 16]. However, the computation cost of these solutions is generally high due to the complexity of attribute structure. The time spent on data encryption, decryption and key management is more than symmetric key or asymmetric key encryptions. Critically, most of existing schemes cannot support controlling cloud data access by either the data owner or access control agents or both. This fact greatly influences the practical deployment of existing schemes. Current research is still at the stage of academic study. Notably, trust and reputation play a decisive role in data access control. But little work was proposed to control the access of data stored at the CSP based on the trust evaluated by the data owner or the reputation generated by a Trusted Third Party (TTP) (e.g., a reputation center) or both in a uniform design.

Reputation and trust relationships in different contexts can be assessed based on mobile social networking activities, behaviors and experiences, as well as performance evaluation. With the rapid growth of mobile communications and the wide usage of mobile devices, people nowadays perform various social activities with their mobile devices, e.g., calling, chatting, sending short messages, and conducting instant social activities via various wireless network connections. Social trust relationships can be established and assessed in a digital world, as illustrated in [17]. Reputation can be assessed based on feedback and Quality of Service (QoS) of an entity [18]. Trust and reputation play a decisive role in cyber security, especially for cloud data access control.

In this book chapter, we will explore the technologies of data access control in MCC. Based on a literature review, we summarize open research issues in this area. In order to achieve secure and efficient cloud data access, we introduce trust and reputation management systems, and investigate three schemes for securing cloud data access based on trust and reputation. Furthermore, we implement the schemes and evaluate the performance of the schemes through analysis on security, computation complexity, system scalability and flexibility. Particularly, important issues and challenges for practical deployment are also discussed in order to propose future research trends.

The rest of the chapter is organized as follows. Section 2 briefly introduces the MCC system model, and the objectives of data access control. Section 3 reviews the current literature about cloud data security and privacy in MCC. We also introduce some technologies and algorithms proposed for enhancing data security, such as

Attribute Based Encryption (ABE) and trust management systems. Section 4 presents three data access control schemes for MCC, and their technical preliminaries. Section 5 reports the results of performance evaluation, analysis and comparison. Usage scenarios of the three schemes and challenges for practical deployment are further discussed in Sect. 6. Section 7 summarizes the whole chapter.

2 Data Access Control in MCC

2.1 MCC System Model

Mobile Cloud Computing is a cutting-edge computing paradigm, and it has gained great attention in both academia and industry. The Mobile Cloud Computing Forum defines MCC as [19]:

"Mobile Cloud Computing at its simplest, refers to an infrastructure where both the data storage and the data processing happen outside of mobile devices. Mobile cloud applications move the computing power and data storage away from mobile devices and into the cloud, bringing applications and mobile computing to not just smartphone users but a much broader range of mobile subscribers"

Sanaei et al. [20] defined MCC as a "rich mobile computing technology" to provide mobile devices anywhere, anytime based on a pay-as-you-use principle. MCC leverages and unifies resources of various clouds and network technologies, and enables unrestricted functionalities. Khan et al. [21] specified MCC as the outcome of interdisciplinary approaches comprising mobile computing and cloud computing. The cloud computing technology enables the mobile devices to be more resourceful in terms of computational power, memory, storage, energy, and context awareness.

Figure 1 presents the general architecture of MCC. It consists of mobile networks, the Internet, CSPs, mobile users, all producing/consuming applications or services. In MCC, the mobile users request cloud services by firstly connecting to mobile networks through a base station, a wireless access point, or other air interfaces. The requests and the mobile users' information will be transmitted to the central processor in the mobile network, where the mobile network operators can verify the users' identities, process data and provide mobile services. After that, the user's request will be forwarded from the mobile network to the cloud through the Internet, where the cloud service providers can provide the mobile user's request with corresponding cloud services.

The cloud architecture can also be presented according to different cloud services, shown in Fig. 2. Data Centers provides the hardware facilities, such as servers, for cloud services. The *Data Centers* handles data storage and processing, and are usually stable and disaster-tolerant. *IaaS* provides users the infrastructure, including networking, servers and operating systems. The infrastructure and hardware can be provided on demand, and shared by multiple users through

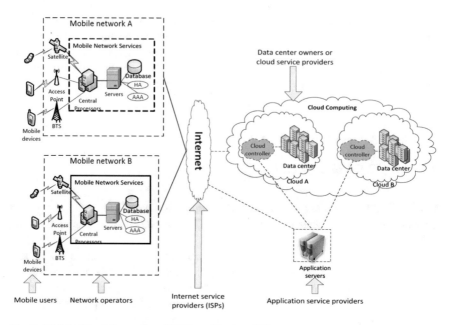

Fig. 1 Mobile Cloud Computing (MCC) architecture [83]

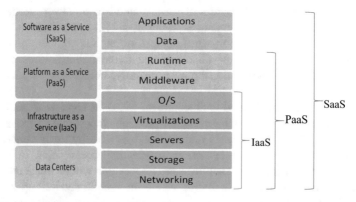

Fig. 2 Cloud service architecture

virtualization technologies. The user only needs to pay for the resources they really use. An example of IaaS is Amazon S3.

PaaS offers an integrated environment to help the users build, test, deploying customized applications. Examples of PaaS are Google AppEngine and Microsoft Azure. *SaaS* is familiared by most people. It supports software application distribution in a cloud environment. The users can access various software and information via the Internet at any time and in any place. Since SaaS supports pay-as-you-use, the users only need to pay for what they consume.

2.2 Access Control on Mobile Cloud Data

Although cloud services, especially cloud data services, are gaining a special attention from mobile users and business enterprises, data security is still a major hurdle of cloud service adoption. In particular, data access control in MCC concerns two aspects: access control in mobile network and access control in cloud.

The Mobility Management Entity (MME) and Home Subscriber Server (HSS) handle the access control in mobile networks in 4G networks and Authentication Center (AuC) in 3G networks. A Subscriber Identification Module (SIM) is used to identify and authenticate subscribers in mobile networks. A SIM card securely stores the International Mobile Subscriber Identity (IMSI) and its related keys for identity verification.

The data access control in the cloud is often carried out based on different access control schemes by applying various theories and technologies, such as Attribute Based Encryption (ABE), Trust Management, and predefined Service Level Agreement (SLA). For example, Google Cloud Storage service applies Access Control Lists (ACLs) as the control scheme to manage data access. The ACL is a list of permissions attached to an object. It specifies the users or entities that are granted access to an object and defines the operations that are allowed regarding the object. In this book chapter, we will focus on the data access control in the cloud and introduce a number of schemes based on trust and reputation for cloud data access control.

3 The State of Arts

3.1 Overview of Data Security and Privacy in MCC

While embracing dynamic service provision, ubiquitous data access, optimized resource allocation and low-cost infrastructure, mobile cloud users are still facing many security problems in terms of data confidentiality and privacy leakage.

3.1.1 Risks of Data Security in MCC

Outsourced Data Security
Cloud computing brings additional security risks since the data owners outsource their sensitive data for sharing in cloud servers. Apart from the security risks in cloud computing, mobile cloud users also face the security problems of wireless communications. The data could be intercepted or altered during wireless transmission due to the insecure nature of wireless communications. Several works [22–25] proposed to encrypt data before outsourcing and use methods such as re-encryption and TTP to protect data confidentiality. Practical security problems still exist. For example, many solutions of cloud security incur a large

computational cost for encrypting all the data stored in the cloud and spend lots of efforts for user revocation and key management. On the other hand, it is difficult for data owners to know if the data has been tampered after outsourcing, especially when there are multiple data owners [26]. This fact threats the integrity of data sharing and outsourcing in the cloud environment.

Malicious Rating

Comparing to encryption schemes with high computational costs, trust and reputation management provide a more computationally efficient way of protecting data security in cloud. However, there could exist malicious nodes or adversaries that provide unfair ratings to affect the accuracy of trust and reputation evaluation. Although a number of approaches have been proposed to detect unfair ratings and malicious nodes [27–29], either of them requires a large amount of prior knowledge or usage experiences in order to assure detection accuracy and evaluation precision. Otherwise, false positive results could be lead and eligible users could be excluded.

Multi-Tenancy Security

Virtualization technology is now widely used to provide infrastructure sharing and resource optimization in cloud services. CSPs normally support "Multi-Tenancy", that is multiple service users can store and process their data based on the same physical hardware [30]. This service model poses security threats that a user shares the same physical hardware with its adversaries, and the resources (e.g., data, virtual machines, etc.) placed in the cloud servers could be easily attacked by the adversaries. Ristenpart et al. showed how malicious behaviors are carried out towards a user when multiple users share a same cloud server [30]. Factor et al. proposed a scheme for Secure Logical Isolation for multi-tenancy cloud storage (SLIM) by predefining the principles to isolate the resources of tenants [31]. Yang and Guo integrated RBAC with an attribute check mechanism to determine the resources that a user can access [32]. Li et al. also proposed a RBAC based scheme for multi-tenancy cloud services by embedding predefined security duty separations [33]. Tang, Li and Sandhu proposed a RBAC based control scheme and integrated trust management to set trust relations among tenants [34].

3.1.2   Privacy Preservation in Cloud Computing

Privacy is a critical issue that influences user's adoption of cloud services. The privacy violation of some famous cloud services such as Facebook and iCloud had leaded to increased concerns in terms of either personal or commercial information leakage. Many existing works [35–41] proposed to solve the problems of privacy violation. The proposed schemes can be divided into two categories: non-cryptographic and cryptographic schemes. The main non-cryptographic scheme is data perturbation. Wang proposed a privacy-preserving scheme based on data anonymity [35]. It hides a part of the user information to deduct sensitive information from disclosed data. Yang et al. proposed a retrievable data perturbation method for privacy preservation based on random noise perturbation [36].

It adds random noise to perturb data values, but remains the data covariance unchanged so that users can retrieve the original values. Haas et al. proposed a privacy preserving system for electronic health records by applying a TTP to control data access, and prevent service providers of data storage from accessing and disclosing data [37]. Regarding cryptographic schemes, Wang et al. proposed a privacy-preserving scheme based on a public key based homomorphic authenticator with a third party auditor applied [38]. He also proposed a secure and efficient method for ranked key word searching to preserve privacy for outsourced data [39]. This scheme enables relevance ranking instead of sending undifferentiated results. It develops a one-to-many order-preserving mapping technique to protect sensitive information. Leng et al. proposed a privacy-preserving scheme for personal health records based on predefined policies. They applied Proxy Re-Encryption (PRE) for flexible encryption and access right delegation [40]. Narayan, Gagne and Safavi-Naini proposed a privacy-preserving Electronic Health Record (EHR) system based on attribute-based cryptography [41].

Although there are many studies on protecting user data stored in cloud, few of them investigated preserving privacy for data usage, such as data usage behavioral information, recently visited information, and usage historical statistics that can also be tracked and studied to violate user privacy [42].

3.2 Data Access Control in MCC

MCC demands a trustworthy system architecture and data protection solutions to provide qualified mobile cloud services in order to support data storage and process can be handled by CSPs, instead of the resource-constrained mobile devices. One typical solution is to protect sensitive data using encryption theories.

3.2.1 Role-Based Access Control in Cloud Computing

Ferraiolo and Kuhn firstly proposed RBAC [43]. The basic principle of RBAC is separating users and permissions by defining different roles. Users are assigned to some roles based on their job functions and responsibilities. Each role has its corresponding operational permissions. Users can only obtain the permissions after activating their roles. RBAC simplifies the permission assignment and authorization management by grouping permissions according to the roles, as well as separating job duties. By relating permissions only to roles instead of users, it is scalable and easy to manage permissions in cloud computing where a large amount of users are presented and it is difficult to track each of their identities. However, RBAC manages permissions statistically according to predefined roles, without considering some dynamic aspects such as time and context. Bertino, Bonatti, and Ferrari proposed a Temporal Role-Based Access Control (TRBAC) to address the periodic permission problems by enabling roles at different time intervals [44]. Joshi et al.

proposed a General Temporal RBAC (GTRBAC) that allows a wider range of temporal constraints of role assignments and permission assignments by clearly defining a protocol of constraints, including active duration constrains, maximum number of activations of a role, and enabling/disabling roles [45]. Besides permission time intervals and dynamic service contexts are also considered to improve RBAC. Yu proposed a Role and Task based Access Control Model (RTABC) to refine permission assignment by adding a task layer between permission sets and roles [46]. In RTABC, operating permissions are not directly assigned to roles, but tasks. A task is a minimum functional unit of an operation, and an operation process can contain a group of tasks. Users can obtain permissions of certain tasks by their allocated roles. The permissions are constrained by task state, task weight and time. Barati et al. designed a new semantic role-based access control model for cloud computing [47]. The model enables recommendations of tasks, of which a user can possibly require the permissions.

In RBAC [43], there are role hierarchies where senior roles can have permissions of their junior roles. However, sometimes it is necessary for a junior role to have a permission of a senior role in order to perform a senior role's operations. Tang et al. proposed a new RBAC model for cloud computing, in which there are two additional roles namely User Role (UR) and Owner Role (OR) [48]. Users can get permissions from owners to access some resources in the cloud. Na and Cheon also proposed a role delegation RBAC that allows junior roles to be granted with their senior roles' permissions [49]. Delegation server and delegation protocols are employed in this method, in which a delegation server is responsible for the delegation, while the delegation protocol describes the delegation process. Lin, Bie, and Lei proposed a scheme for cross-domain access control system in cloud computing, which integrates the RBAC with trust management. It establishes a set of associations with regard of roles between a local domain and other domains [50]. In order to perform various permission constraints and role delegations, Gitanjali, Sehra, and Singh proposed several policy specifications for RBAC [51]. The policies consist of permission delegation, role delegation, and also delegation of access rights to CSPs. Moreover, they designed backup and restoration policies to handle the cases of service crashes and data loss.

3.2.2   Attribute-Based Encryption for Cloud Computing

ABE has been recently well studied for data access control in cloud computing. It does not require either the identities of the data requesters or their public keys, but only the attributes the requesters own [52]. Sahai and Waters initially proposed ABE that evolves from Identity-Based Encryption [53]. In this initial ABE scheme, an identity consists of a set of descriptive attributes. A user can use a private key for an identity ω to decrypt a ciphertext encrypted with an identity ω', if and only if ω matches ω' over a certain threshold that makes it error-tolerant. The initial ABE was further extended to two varieties, Key Policy-Attribute Based Encryption (KP-ABE) and Ciphertext Policy-Attribute Based Encryption (CP-ABE).

Goyal et al. proposed KP-ABE in 2006 [54]. It is a public key cryptographic scheme built upon bilinear map and Linear Secret-Sharing Scheme (LSSS) [55]. In KP-ABE, a ciphertext is associated with a set of attributes during encryption, while a secret key is generated based on an access policy, which is an access structure over a set of data attributes. To decrypt the ciphertext, the data attributes must satisfy the access structure. Chase proposed a multi-authority attribute based encryption scheme, which allows multiple authorities to control the data access at the same time [56]. Each of the authorities maintains a set of attributes, and a user can decrypt the data only if the number of attributes he/she possesses is beyond the threshold of each authority. Wang and Luo proposed a KP-ABE scheme with constant ciphertext size no matter how many attributes are embedded [57]. Yan, Li, and Li designed a secure personal health record system using KP-ABE in a cloud computing environment [58]. Although KP-ABE is believed secure for data access control in cloud computing, it comes with high computational cost, especially if there are a large number of attributes. Lv et al. proposed an efficient and secure KP-ABE scheme for mobile cloud storage by outsourcing the KP-ABE key generation and decryption process to a trusted attribute authority [59]. Yu et al. combined the KP-ABE and Proxy Re-Encryption (PRE) to achieve fine-grained data access control [15]. Moreover, the scheme allows data owners to delegate most of computation tasks to untrusted cloud servers without disclosing the underlying data contents.

Bethencourt, Sahai, and Waters proposed the CP-ABE after KP-ABE [60]. Different from KP-ABE, a ciphertext in CP-ABE is built upon an access policy, while a private key is associated with a set of attributes. A user can decrypt the data only if its attributes embedded in the private key match the access policy. Lewko and Waters proposed a decentralized CP-ABE scheme that allows multiple authorities to issue access right and private keys with a part of the attribute set [61]. It solved the user collusion problem by applying the unique global identity of a user. Horvath further extended the decentralized CP-ABE scheme to reduce the computational burden of user revocation by removing the computations of revocation at CSPs and distributing them over service users [62]. Xu et al. proposed a fine-grained document sharing system based on CP-ABE, which refines users into different classes according to their privileges to access files [63]. A document is divided into multiple segments and encrypted by hierarchical keys. A user with a higher security class key can derive all its lower level keys. Although this scheme achieves fine-grained data access control, it is very difficult to manage both CP-ABE keys and hierarchical keys, especially where there are a large number of users in cloud. Wan, Liu, and Deng proposed a hierarchical attribute-based scheme in cloud computing by organizing user attributes into different sets, where the same attribute can be assigned different values [64]. They further applied expiration time as an access constraint for user revocation, which made the scheme efficient.

3.3   Reputation and Trust Management in Cloud Computing

Trust and reputation are studied in the recent literature. They usually have no unique definitions because of their appearance in different research disciplines. For instance, Hussain and Chang provided an overview on trust and reputation definitions. They indicated that the current notions of trust and reputation should be further formally defined [65]. Dasgupta and Prat defined trust as "the expectation of one person about the actions of others that affects the first person's choice, when an action must be taken before the actions of others are known" [66]. Abdul-Rahman and Hailes defined reputation as "an expectation about an agent's behavior based on information or observations of its past behavior" [67]. As the research of trust and reputation grows fast in both theoretical foundations and real-world deployment, a number of e-commerce and cloud service companies apply them in ranking their products and suppliers, as well as building their recommendation systems. In cloud computing, trust and reputation are capitalized on designing mechanisms for data access control or recommendation systems. They are regarded as effective incentives to form a healthy and trustworthy network with participants who may have no prior knowledge of each other.

3.3.1   Trust Management

Trust Management (TM) plays a critical role in increasing reliability, usability and security in cloud computing. Sato, Kanai, and Tanimoto proposed a trust management model by dividing the model into two layers, namely the internal trust layer and contracted trust layer [68]. The internal trust layer is based on the Trusted Platform Module (TPM), which is an international standard for a secure crypto-processor. It handles strict security operations, such as key generation, key management and private data modification. The contracted trust layer is defined as the trust determined by certain contracts between service providers and users. A contract defines the trust and security level of a service provider, and the reliance between the provider and its users. Although there can be different trust levels in the negotiation on a contract, this model does not give specific factors for evaluating the trust levels. It requires resigning the contract whenever there is a service demand or a security requirement update. Yao et al. proposed an accountability-based system to achieve Trustworthy Service Oriented Architecture (TSOA) [69]. The trust and accountability management also depends on predefined and mutually agreed policies, called Service Level Agreement (SLA). It monitors each participant's behavior, and determines which participant is accountable according to the SLA. Pawar et al. proposed a trust model to support the trustworthiness verification of cloud infrastructure providers [70]. The trust levels are calculated based on a feedback model in terms of trust, distrust and uncertainty. This approach also depends on a predefined SLA.

Prajapati, Changder, and Sarkar proposed a trust management model for SaaS in cloud [71]. In this model, trust is evaluated through Trust that indicates direct relations between participants and Recommended Trust that is based on recommendations from other participants. However, this model assumes that all the nodes are honest, and is not resilient to malicious feedback or attacks. Habib, Ries, and Muhlhauser proposed a multi-faceted TM system to help users choose trustworthy CSPs [72]. The system evaluates trust levels through comprehensive attributes of Quality of Service (QoS), such as security, compliance, availability and response time. The TM system is centralized. It registers all the CSPs in the Registration Manager (RM) and provides a Trust Manager (TMg) to allow service users to specify their requirements and feedback.

Yan, Li, and Kantola proposed a trust assessment model that enables the trust assessment between CSPs and users to help users choose trustworthy CSPs and also the trust assessment between users to allow access right delegation [17]. This trust assessment model is designed for cloud computing based on mobile social networks. It takes mobile social networking activities (e.g., mobile calls, instant messages, pervasive interactions, etc.) as a basic clue to evaluate social trust. Furthermore, this model takes weight parameters, priority levels and a punishment factor into account for assessing trust levels to issue personal data access rights.

Table 1 provides a comparison of the trust management models proposed in [68, 72, 17] according to the following properties.

Factors for trust assessment: Factors for trust assessment determine what properties and issues should be considered in trust evaluation and management. The factors depend on application context or design purposes. For example, Habib, Ries, and Muhlhauser chose QoS; Yan, Li and Kantola selected performance and social behaviors/activities in mobile social networking as trust evaluation parameters.

Enable trust assessment between CSPs and Users: Enabling trust assessment between CSPs and users is designed to help users choose trustworthy CSPs, and encourage CSPs to keep good performance.

Table 1 Comparison of trust management model

	Sato [68]	Habib [72]	Yan [17]
Factors concerned in trust assessment	Not specified	QoS	Social behaviors and activities records; Priority level; Punishment factor
Enable trust assessment between CSPs and users	√	√	√
Enable trust assessment between users	–	–	√
Trust management through policy	√	–	–
Trust management through feedback	–	√	√
Enable attack resistance	–	√	–

Enable trust assessment between users: It is designed for enabling data access right delegation among cloud service users.

Trust management through policy/feedback: This impacts the technique used for trust assessment and management. For example, Sato, Kanai, and Tanimoto managed trust levels by signing contracts [59], while Habib, Ries, and Muhlhauser managed trust levels based on feedback and recommendations [62].

Enable attack resistance: It considers if a trust management scheme is resistant to malicious attacks, such as malicious rating, self-promoting, and collusive behaviors.

3.3.2 Reputation Management

While trust management plays an important role in guaranteeing trustworthy services and interactions in cloud computing, reputation provides the basis of evaluating and quantifying trust levels. Habib, Ries, and Muhlhauser identified a set of important parameters required in reputation evaluation, which includes system performance (e.g., latency, bandwidth, availability, reliability, etc.), and security measures (e.g., physical security support, network security support, key management, etc.) [26]. Bradai, Ben-Ameur, and Afifi proposed a reputation-based architecture to help users choose trustworthy peers in cloud [73]. This architecture consists of three models, in which the first model evaluates the reputation given by the users, and the other two refine the reputation values to detect malicious ratings. Koneru et al. proposed a reputation management system based on user recommendations [74]. Muralidharan and Kumar proposed a novel reputation management system for volunteer clouds, in which computer owners can donate their computing resources [75]. This system assesses each participant's reputation according to the performance of service time, crash time, and correctness of service results. Although the three factors are important for service and reputation evaluation, the network status can easily affect different aspects of the performance. Therefore, it decreases the objectivity of a participant's reputation. Zhu et al. proposed an authenticated reputation and trust management system for cloud and sensor networks [76]. The trust and reputation are calculated according to the performance and feedback of data processing, data privacy and data transmission. However, the system still depends on a predefined SLA and a Privacy Level Agreement (PLA).

Besides different factors chosen for reputation evaluation and management in various application contexts, unfair ratings and malicious attacks are also important issues that affect the performance and efficiency of a reputation management system. Yan, Li, and Kantola proposed a centralized reputation management system for data access control in cloud computing [18]. It employs a Reputation Center (RC) as a TTP to manage and verify users' reputations. Moreover, it applies a punishment agreement to encourage good performances and honest voting. Wu et al. proposed a reputation revision method that applies a novel filter to recognize unfair ratings [27]. The reputation evaluation is based on the QoS, including response time, cost, and reliability, as well as users' ratings based on prior service experiences. The method applies a similarity theory to distinguish abnormal evaluations and calculates average ratings. Wang et al. proposed an accurate and multi-faceted reputation scheme that detects unfair ratings to improve the accuracy of reputation calculation [28]. The scheme firstly identifies suspicious users whose ratings deviate from others significantly, and then detects collusive users through similarity clustering. After removing those malicious users and unfair ratings, it calculates the overall reputation value.

3.4 Open Research Issues

One major difference of MCC from normal cloud systems is that MCC requires a more energy-efficient, flexible and usable infrastructure for access control. The mobility and unavailability of MCC users and the power/process/display-restriction of their mobile devices cause this. As a result, decreasing energy consumption becomes one of the most important targets to design an access control scheme in a MCC system. Meanwhile, A lightweight and efficient data access control scheme is highly expected in MCC, especially for MCC users. They expect a secure, flexible and usable cloud data access control mechanism no matter when and where they are located. Although there are many techniques proposed for data security in the cloud, they normally require heavy computations and complicated operations for data encryption and key management. Many CSPs providers support non-cryptographic access controls, but they may not ensure data confidentiality because the CSPs can access the user's personal data or private information. Therefore, improving energy efficiency and simplifying user client's operations are still open issues in the design of an access control scheme in MCC. More importantly, the design should be user-centric and usable, which means the designed scheme should understand the MCC user's demands and intelligently make a balance between efficiency, flexibility and data security.

Despite the problems inherited from cloud computing when designing access control schemes, MCC faces extra security challenges in terms of wireless communication channels. MCC relies on highly heterogeneous networks that consist of different wireless network interfaces. There could be many different radio access technologies used for supporting MCC in 5G networks. Frequent shift among different networks or wireless network interfaces requires dealing with different security levels and protocols. This greatly increases the risk of data leakage or tampering. How to support control data access at heterogeneous air interfaces is crucially important for MCC, but hasn't yet seriously explored in the literature.

Despite the technical aspect, it is also important to consider legal issues when deploying an access control scheme in order not to violate any data or privacy regulations. Additionally, in order to deploy a data access control scheme in practical systems or applications, it is inevitable to develop a corresponding business model that can be adopted by all involved stakeholders. The access control services should to be rational and feasible in terms of commercial business. People seldom perform their studies that really cross academic, industrial and economic domains.

4 Heterogeneous Data Access Control in MCC Based on Trust and Reputation: Three Schemes

This section introduces three data access control schemes based on trust or reputation designed for MCC. The schemes are suitable for resource-constrained mobile devices in MCC.

4.1 Scheme 1: Controlling Cloud Data Access Based on Reputation

We proposed a scheme [18] to control data access in MCC by applying PRE [77] and reputation management [16]. There are four kinds of entities in the system model: Data Owners, CSP, Reputation Center (RC), and Data Users. The RC is fully trusted and employed for reputation management. It helps the data owners check if the users meet the access policies. The CSP is responsible for data storage and data access control, which includes data re-encryption and issuing access right to eligible users. The Data Owners owns the right of data access and alteration. The Users are the system entites who request for data access.

4.1.1 Reputation Generation

The reputation model is proposed in [16]. $R_i(t_e)$ denotes the reputation of an entity i (e.g., a CSP) at time t_e. It contains two parts: the reputation contributed by the user feedback, denoted as Rf_i and the reputation contributed by performance monitoring and reporting, denoted as Rp_i. Many existing reputation mechanisms can be applied to evaluate an entity's reputation based on user's feedback. One example mechanism is described below based on the scheme in [78], but tailored for the scenario of cloud computing. Suppose the user k's vote on the entity i at time t is $V_i(k,t)$, and the user's own credibility of providing feedback at time t is $C(k,t)$, we calculate the reputation value Rf_i of the entity contributed by the user feedback as:

$$Rf_i(t_e) = \frac{\theta(K)}{O} \sum_{k=1}^{K} V_i(k,t) \times C(k,t) \times e^{-\frac{|t_e - t|^2}{\tau}},$$

where $O = \sum\limits_{k=1}^{K} C(k,t) \times e^{-\frac{|t_e - t|^2}{\tau}}$; K is the total number of votes on the entity i; t_e is the reputation evaluation time; t is the time of vote $V_i(k,t)$, $(V_i(k,t) \in [0,1])$; parameter τ is used to control time decaying since we pay more attention to the recent votes and apply $C(k,t)$ to tailor the consideration of $V_i(k,t)$ in order to overcome some potential attacks, such as bad mouthing attack. $\theta(K)$ is the Rayleigh cumulative distribution function $\theta(K) = \{1 - exp(\frac{-K^2}{2\sigma^2})\}$ that is applied to model the impact of integer number K on reputation generation, where $\sigma > 0$, is a parameter that inversely controls how fast the number K impacts the increase of $\theta(K)$. Parameter σ can be theoretically set from 0 to, ∞ to capture the characteristics of different scenarios. $C(k,t)$ $(C(k,t) \in [0,1])$ is generated based on the performance of user's feedback by considering the feedback provided by all users on the entity's service and the performance monitoring and reporting.

Concretely, $C(k,t)$ is generated at RC. If the feedback $V_i(k,t)$ provided by k doesn't match the final evaluation result, that is $\delta = \frac{1}{2} - |V_i(k,t) - R_i(t)| \leq 0, \gamma + +$; If $V_i(k,t)$ matches the fact, that is $\delta = \frac{1}{2} - |V_i(k,t) - R_i(t)| > 0$, γ is not changed; If no $V_i(k,t)$ is provided, γ is not changed. The $C(k,t)$ of k at time t is:

$$C(k,t) = \begin{cases} C(k,t) + \omega\delta(\gamma < thr) \\ C(k,t) + \omega\delta - \mu\gamma(\gamma \geq thr) \end{cases} = \begin{cases} 1(C(k,t) > 1) \\ 0(C(k,t) < 0) \end{cases},$$

where γ is a parameter to record the number of mismatching votes, μ is a parameter to control $C(k,t)$'s deduction caused by bad voting performance, parameter δ is a deviation factor to indicate the difference between a user's vote and the real reputation, and ω is a parameter to control the adjustment scale of $C(k,t)$. We have proved that the above design can effectively overcome bad-mouthing attack and on-off attack in our previous work [78, 79].

For generating the reputation based on the performance monitoring and performance records on the services provided by the entity, suppose the monitored performance or recorded performance is P_m, while the threshold performance is P_t, if P_m is better than P_t, the positive number of performance P is increased by 1, otherwise, the negative number of performance N is increased by 1. The initial values of P and N are 0. Obviously, the percentage of positive evidence indicates the probability of satisfying the expected performance. We design the reputation value Rp_i of entity i contributed by performance monitoring at time t_e as:

$$Rp_i(t_e) = \frac{P}{P+N+\xi}(\xi = 1),$$

where ξ is a parameter to control the rate of loss of uncertainty for different scenarios (we often take $\xi = 1$).

Thus the final reputation $R_i(t_e)$ of entity i at time t_e can be aggregated by fusing Rp_i and Rf_i as below:

$$R_i(t_e) = \frac{\theta(K_{Rp}) \times Rp_i(t_e)}{\theta(K_{Rp}) + \theta(K_{Rf})} + \frac{\theta(K_{Rf}) \times Rf_i(t_e)}{\theta(K_{Rp}) + \theta(K_{Rf})},$$

where K_{Rp} is the total number of performance monitoring reports from the entity i and performance records provided by i's users. K_{Rf} is the total number of feedback records. We use the ratio between $\theta(K_{Rp})$ or $\theta(K_{Rf})$ and their sum to respectively weight the contributions of $Rp_i(t_e)$ and $Rf_i(t_e)$ to $R_i(t_e)$. This is because the more evidence aggregated, the more convincing the contribution is from the statistical analysis point of view. Note that $R_i(t_e) \in [0, 1]$.

4.1.2 The Proposed Scheme 1

We adopt PRE that enables RC to issue a re-encryption key and access right to eligible entities. The procedures in the scheme can be grouped into four phases: System Setup, New Data Creation, Data Access, and User Revocation.

System Setup: Each entity in the system matins a public key and private key pair under the public key infrastructure of PRE. The global parameters for key generation are shared within the system.

New Data Creation: A data owner encrypts its data using a symmetric key *DEK*, and then encrypts the symmetric key *DEK* using the RC's public key *pk_RC*. Then the data owner stores the data along with the encrypted *DEK* to the CSP, and specifies an access policy to RC. The plaintext of data and *DEK* is hidden from CSP, thus provides data protection and privacy preservation. The CSP in this scheme functions as a proxy, which indirectly delegates an access right to an entity without learning anything about the secret data.

Data Access: A user *u* firstly sends an access request to the CSP. The CSP would forward the request to RC, who is responsible for evaluating the user's latest reputation and decides if the user meets the access policy. If the user is eligible, RC will generate a re-encryption key *rk_RC- > u* if needed. The CSP conducts the cipherkey re-encryption based on the *rk_RC- > u* received from the RC, and send the re-encrypted *DEK* to the user. The user can decrypt the data using its own private key *sk_u* and obtained *DEK*.

User Revocation: RC retains a list of revoked users that are no longer eligible to access the data and handles user revocation. And the CSP blocks the revoked user from accessing the data.

We adopt a punishment rate in an insurance agreement between an entity and RC. The punishment rate is set based on an entity's reputation level, in case of data disclosure from the entity. The higher the reputation, the lower the punishment rate in an agreement. In this way, the entities in a system are encouraged to behave honestly and perform well. In practice, an insurance business can be established by

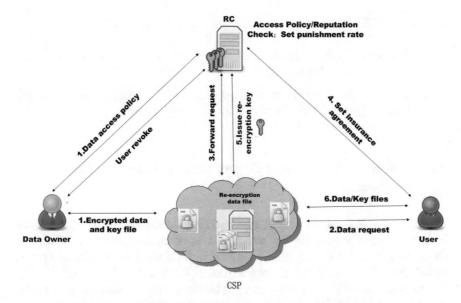

Fig. 3 System model of cloud data access control based on reputation

RCs to guarantee legal data access and compensate data owners if there is any illegal data disclosure. CSPs will pay an annual fee to RC for getting re-encryption keys and allowing it to manage reputations.

Detailed steps of access procedure in this scheme are illustrated in Fig. 3.

Step 1: The data owner firstly encrypts the data using *DEK*, denoted as *E(DEK, data)*. And it encrypts the *DEK* using RC's public key *pk_RC*, denoted as *E(pk_RC, DEK)*. Then the data owner uploads the encrypted data to the CSP, and specifies the data access policy to RC.

Step 2: The user sends a data access request to CSP for cipherkey *E(pk_RC, DEK)*, and encrypted data *E(DEK, data)*.

Step 3: CSP firstly verifies the user's ID to check if it is valid in the system. If the user's ID is valid, the CSP forwards the data request to the RC. Otherwise, the request is rejected.

Step 4: The RC evaluates the user's latest reputation and decides if the user meets the access policy. If the user is eligible, the RC sets an insurance agreement based on the user's reputation level, in case of data disclosure. Otherwise, the request is rejected.

Step 5: The RC generates a re-encryption key *rk_RC- > u = RG(pk_RC, sk_RC, pk_u)*, in which *RG(pk_RC, sk_RC, pk_u)* is the re-encryption key generation function, based on its own private and public key pair *sk_RC, pk_u* and user's public key *pk_u*.

Step 6: The CSP conducts the cipherkey re-encryption $R(rk\_RC- > u,\ E(pk\_RC,$ $DEK)) = E(pk\_u,\ DEK)$, and sends the re-encrypted $DEK\ E(pk\_u,\ DEK)$ to the user. Therefore, the user can decrypt the data using its own secret key sk_u and obtain DEK.

If later on, the user wants to access the data again but no longer eligible, the RC will inform the CSP to block the user from accessing the data.

4.2 Scheme 2: Cloud Data Access Control Based on Trust Assessment

This scheme is designed for controlling cloud data access based on trust assessment in mobile social networking by using trust assessment algorithm in [16] and CP-ABE [80] for *DEK* encryption. There are three kinds of entities in the system model: Data Owner, CSP, and Users who request the data. The CSP cannot be fully trusted. It is responsible for storing data and checking user validity. The access right is fully controlled by Data Owners, who issue the decryption keys based on the Users' trust levels evaluated based on their social networking activities, behaviors and experiences.

4.2.1 Trust Assessment

The trust assessment model is built based on the user behaviors in the mobile social network [16]. The behaviors can be classified into three categories:

- Mobile calls
- Messages
- Local instant interaction

The trust level between two persons then can be automatically assessed based on the statistics and properties (e.g. number of calls, weight, etc.) of the above three categories. Table 2 describes the notations that are used in the trust assessment, designed below:

$$
\begin{aligned}
TL(i,j) = f\{TL^{'}(i,j) + pl(i,j) \\
* [\omega 1 * \theta(N\_c(i,j) + N\_c(j,i)) + \omega 2 * \theta(N\_m(i,j) \\
+ N\_m(j,i)) + \omega 3 * \theta(N\_i(i,j) + N\_i(j,i))] - pu(i,j)\}
\end{aligned}
$$

Besides taking the behaviors as part of the parameters, the formula also considers the previous trust level, priority and punishment factor. In addition, the trust level assessment can be linked to a certain context based on keyword extraction, in which the data owner can set the data access policies according to the context.

Table 2 The notations used in trust assessment based on mobile social networking

Symbols	Description
$N\_c(i,j)$	The number of calls made by i to j
$N\_m(i,j)$	The number of messages sent by i to j
$N\_i(i,j)$	The number of interactions initiated by i to j
$TL(i,j)$	The trust level of j assessed by i; $TL'(i,j)$ is the old value of $TL(i,j)$
$\theta(I)$	The Rayleigh cumulative distribution function $\theta(I) = \left\{1 - \exp(\frac{-I^2}{2\sigma^2}\right\}$ to model the impact of integer number I, where $\sigma > 0$, is a parameter that inversely controls how fast the number I impacts the increase of $\theta(I)$. Parameter σ can be set from 0 to theoretically ∞, o capture the characteristics of different scenarios. We use $\theta(I)$ to model the influence of the number of calls, messages and interactions on social relationships
$\omega 1$	The weight parameter to show the importance of voice call, $\omega 1 + \omega 2 + \omega 3 = 1$
$\omega 2$	The weight parameter to show the importance of message
$\omega 3$	The weight parameter to show the importance of instant interaction
$pl(i,j)$	The priority level of j in i's social networks
$pu(i,j)$	The punishment factor of j in i's view
$f(x)$	The Sigmoid function $f(x) = \frac{1}{1+e^{-\alpha x}}$, which is used to normalize a value into (0, 1) in order to unify an evaluated trust level into an expected scope. Parameter α applied to control the normalization

4.2.2 The Proposed Scheme 2

This scheme can be applied into MCC. A mobile user can save its sensitive personal data at a data center offered by a CSP. For ensuring safe data access by other trustworthy users in the network, the mobile users make use of trust levels accumulated and analyzed from their individual mobile social networking records. The mobile user can issue secret keys to eligible users with sufficient trust to access the personal data at the CSP.

Trust level of a user is evaluated based on the activities, behaviors, and experiences in mobile networks. In the scheme, we divide individual trust into discrete levels. For example, TL_i represents the ith trust level, TL can be from 0 to $\bar{I}_{tl}$, where $\bar{I}_{tl}$ is the maximum level of TL.

In our scheme, plaintext is hidden from the CSPs in order to provide data protection and privacy preservation. The CSPs are responsible for data storage, verifying users' ID, and blocking eligible users from accessing the data. Trust assessment and secret key issuing is handled by the data owners to ensure safe data access by trustworthy users. A user firstly sends an access request to the CSP, which checks if the user's ID is valid. If it is the case, the CSP will forward the access request to the data owner, who will decide if the user who sends the request is eligible to access the data.

The user revocation is handled by the CSP based on the data owner's notifications about the non-eligible users. The encrypted data could be renewed by encrypting with a new access policy, or the access can be blocked by announcing the expired secret key. Although the CSP is not fully trusted, it is encouraged to perform well by applying a reputation mechanism to evaluate the performance of the CSP. The CSP's reputation is generated based on the user feedback, and will be published to all users in the system. Mobile users won't select the CSPs with bad reputations.

4.2.3 Scheme Algorithms

The scheme contains two main algorithms: trust assessment and CP-ABE based access control. Trust assessment algorithm is based on the formula in [16]. We adopt conventional CP-ABE, but integrate Individual Trust Level (TL) as an attribute in our modified CP-ABE algorithm. Concretely, the operations for our modified CP-ABE algorithm are: Setup, InitiateUser, CreateTrustPK, IssueTrustSK, Encrypt and Decrypt.

Setup: The Setup operation generates a system public key PK and a master key MK based on bilinear paring,

$$PK = \{G, G_T, e, g, P, e(g, g)^y\}, \ MK = g^y$$

where P is a random point in G, and $y \in Z_p$.

InitiateUser: The InitiateUser takes public key PK and master key MK as inputs, and generates a user's public key PK_u and secret key SK_u,

$$PK\_u = g^{mk_u}, \ SK\_u = MK * P^{mk_u} = g^y * P^{mk_u}$$

where $mk_u \in Z_p$. It also chooses a random hash function $H_{SK\_u}:\{0,1\}^* \rightarrow Z_p$ from a finite family of hash functions.

CreateTrustPK: The CreateTrustPK operation generates CP-ABE public attribute key based on trust levels. It is executed by the data owners to encrypt the data and control the access right. The CreateTrustPK checks the TL related policies for data access, and outputs the public attribute key $PK\_(TL_i, u)$ for each attribute TL_i, where $i \in [0, \bar{I}_{tl}]$ and $\bar{I}_{tl}$ is the maximum level of TL. The $PK\_(TL_i, u)$ consists of two parts:

$$PK\_(TL_i, u) = \ <PK\_(TL_i, u)' = g^{H_{SK\_u}(TL_i)},$$
$$PK\_(TL_i, u)'' = e(g, g)^{yH_{SK\_u}(TL_i)} >$$

Encrypt: The Encrypt operation takes the attribute public key $PK_-(TL_i, u)(i \in [0, \bar{I}_{tl}])$ to encrypt the symmetric key *DEK*, based on the access policy $\mathcal{A}$ related to TL. The output ciphertext *CT* is

$$CT_i = \langle E_i = DEK \cdot \left(PK_-(TL_i, u)'' \right)^{R_i},$$
$$E_i' = P^{R_i},$$
$$E_i'' = \left(PK_-(TL_i, u)' \right)^{R_i} \rangle$$

where R_i is a random value and $R_i \in Z_p$. The ciphertext *CT* consists of the tuple $CT = \langle CT_1, \ldots, CT_m \rangle$, and it iterates over all $i = 1,\ldots,m$ $(m < \bar{I}_{tl})$, where m represents the number of selected TL in the access tree $\mathcal{A}$.

IssueTrustSK: The IssueTrustSK is executed after verifying user u' eligibility and if the TL of u' is equal or above the required TL level. It takes u' s pubic key as input and issues the CP-ABE private attribute key based on trust levels for the purpose of decrypting *DEK*:

$$SK_-(TL_i, u, u') = PK_-u'^{H_{SK_-u}(TL_i)} = g^{mk_{u'}H_{SK_-u}(TL_i)}(i = I)$$

Decrypt: The user decrypts the cipherkey to get the symmetric key DEK using its own secret key SK_-u' and $SK_-(TL_i, u, u')$ issued by the data owner. The output DEK is

$$DEK = E_i \cdot \frac{e\left(E_i', SK_-(TL_i, u, u')\right)}{e(E_i'', SK_-u')}$$

Then the user can decrypt the encrypted data using the *DEK*.

Figure 4 shows the detailed procedures of data access control based on trust assessment and CP-ABE encryption in mobile social networks.

Step 1: The data owner conducts trust evaluations based on activities, behaviors and user experiences in mobile social networking. After the trust evaluation, the data owner determines the requirements of trust levels specified in his/her access policy. The data owner then encrypts the data using a secret symmetric key *DEK*, and encrypts the *DEK* using its own CP-ABE public key *PK_TL* which is based on specified trust levels. The encrypted data and symmetric key are denoted as *E(DEK, data)* and *E(PK_TL, DEK)*. Then it uploads the encrypted data and *DEK* to the CSP and sends the access policy based on the required trust level TL.

Step 2: A user firstly sends an access request to both the data owner and CSP.

Step 3: The CSP verifies the user's ID in order to check if the user's ID is valid, or in a blacklist. If the user's ID is valid, the CSP forwards the access request to the data owner. Otherwise, the request is rejected.

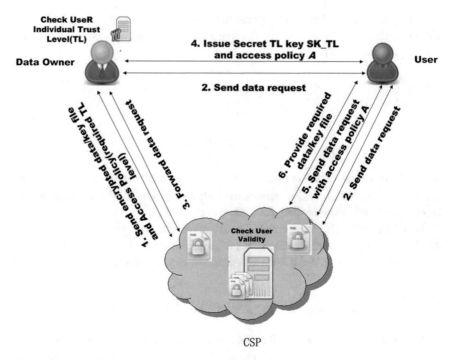

Fig. 4 Procedure of cloud data access control based on trust assessment

Step 4: The data owner evaluates the user's trust level TL based on previous
behaviors and activities. If the user is trustworthy and allowed to access the
data, the data owner issues a secret key *SK_TL* based on the user's trust
level TL, and also sends corresponding access policy $\mathcal{A}$. Otherwise, the
request is rejected.

Step 5: After receiving the secret key *SK_TL* and access policy $\mathcal{A}$ from the data
owner, the user again sends a data access request along with the access
policy $\mathcal{A}$ to the CSP.

Step 6: The CSP checks if the access policy from the user is the same as that
received from the data owner. If both of the policies match, the CSP sends
the required data to the user, so that the user can decrypt *DEK* with *SK_TL*
and get the plain data.

4.3 Scheme 3: Flexible Cloud Data Access Control Based on Trust and Reputation

This scheme proposed a flexible multi-dimensional control on cloud data access
[81]. It is a heterogeneous scheme that combines the techniques in Scheme 1 and
Scheme 2. There are four kinds of entities in the system model: *Data Owners,*

Cloud Service Providers (CSP), Reputation Center (RC), and *Users.* The *RCs* are fully trusted and employed for reputation management, and checking if the users meet the access policies. The *CSPs* are responsible for data storage, controlling data access including data re-encryption and issuing access right to eligible users. *Data Owners* are the ones who own the right of data access and alteration, and *Users* are the ones who request for data access.

4.3.1 The Proposed Scheme 3

This scheme realizes multi-dimensional control on cloud data access based on individual trust evaluated by the data owners, and/or public reputation evaluated by one or multiple RCs. More concretely, a data owner firstly encrypts its data with a symmetric key *DEK*, and then the data owner can divide the *DEK* into several segments K_0, K_1, K_2, ..., K_n. K_1, K_2, ..., K_n are encrypted with public keys from different RCs which are employed to evaluate reputations and control data access. K_0 can be encrypted with a public key *PK_TL* which is related to individual trust levels. After the data encryption, the data owner uploads the encrypted data and encrypted key segments to the CSP, and specifies the access policy to each of the RCs. In order to access data, a user needs to be authorized by all the RCs, and collect all encrypted key segments to recover *DEK* for decryption.

The size of the key segments *K0, K1, K2, ... Kn* can be flexibly set by data owners according to different application scenarios, data access policies and security requirements. If a data owner would like to control data access only by itself, the symmetric key *DEK* will not be divided, and is encrypted with a public key *PK_TL* that is related to individual trust levels. If a data owner would like the RCs to the control data access, all the key segments are encrypted with the RCs' public keys.

Notably, the data encryption key *DEK* can be divided into multiple parts in order to really support multi-dimensional data access control. For example, the data owner can set the access control strategy such as controlling its data access based on the reputations evaluated by multiple RCs in order to highly ensure data security and privacy, especially when it is off-line. In this case *DEK* is separated into multiple parts K_1, K_2, ..., K_n ($n \geq 2$). The data owner encrypts different parts of *DEK* with different RC's public keys pk_RC_n. Later on, the data access can be controlled by all RCs (e.g., with regard to different reputation properties or contexts) by issuing re-encryption keys to decrypt different parts of *DEK*. Obviously, our scheme can flexibly support controlling cloud data access by not only the data owner, but also a number of RCs.

The data owner manages the policy of data access control. It decides who has rights to control the data access. Our scheme is very flexible to handle many cases: data is controlled by the owner itself, or only one RC, or multiple RCs or all above at the same time. Their control relationship could be "or", not only "and", in order to ensure control availability. In case some control party is not always online, another party can delegate the duty. We achieve this by issuing the same partial key to multiple parties (e.g., RCs). Free control of access can also be supported without changing the system design by setting $K_n = K_0 = DEK = $ null. The proposed scheme doesn't request the data owner to be always online. In this case, one or multiple RCs can delegate the data owner to control data access based on reputation. Our scheme offers a comprehensive and flexible solution to support various scenarios and different data access policies or strategies.

User revocation is achieved by applying a blacklist that contains the ID of distrusted or ineligible users. The blacklist is managed by the CSP, and can be updated according to RC or data owners' notification and feedback.

Scheme Algorithms

The scheme consists of four main algorithms: Key Generation, Symmetric Key *DEK* Division and Combination, PRE, Modified CP-ABE that is proposed in [17].

Key Generation: three kinds of keys are generated: Symmetric key *DEK* for data encryption, public key pairs for PRE, and key pairs for CP-ABE. The key generation for CP-ABE consists of generating system public key *PK*, master key *MK*, user public key pairs and Individual Trust public key and secret key. The key generation can be conducted by users or by a trustworthy user agent.

Symmetric Key *DEK* Division and Combination: Key division is operated by the data owner based on its data access control policy. The symmetric encryption key *DEK* is divided into n + 1 parts, where n is the number of RCs which are employed by the data owner to control its data access based on the access policy. Key combination is operated by the user who receives all pieces of the partial symmetric keys of *DEK* and aggregates all of them together in order to get a complete *DEK* for decryption.

PRE: The data owner encrypts n pieces of partial symmetric key *DEK* using corresponding RC's PRE public key, and stores the encrypted data and key files in the CSPs. The RCs control data access right by evaluating the access policy and users' reputation, and conduct re-encryption key generation if a user is eligible for accessing the data. The CSPs conduct the re-encryption and send the re-encrypted keys to the user.

CP-ABE: CP-ABE is applied for the purpose of integrating the individual TL into the data access control mechanism, and controlling access right by the data owner itself. One piece of symmetric key *DEK* is encrypted using the data owner's Individual Trust public key, and is stored in the CSPs. After verifying the individual trust level of a user who requests the data, the data owner will then issue the user an Individual TL secret key, and inform the CSPs to send the encrypted data to the user.

Figure 5 illustrates the detailed procedures of two-dimensional data access control in the proposed scheme.

Step 1:   The data owner encrypts its data using a symmetric key *DEK*, and divides the *DEK* into two segments: K_0 and K_1. K_1 is encrypted with the RC's public key *pk_RC*, and K_0 is encrypted with a public key *PK_TL* which is related to individual trust levels. The encrypted data is denoted as E(*DEK*, data), and the key segments are denoted as E(*pk_RC*, K_1) and E(*PK_TL*, K_0). Then the data owner uploads the encrypted data to the CSP, and specifies an access policy to both the CSP and RC.

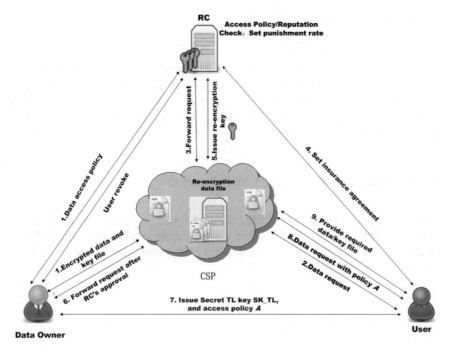

Fig. 5 An example procedure of cloud data access control based on heterogeneous scheme

Step 2: The user sends an access request to the CSP and waits for responses.

Step 3: The CSP verifies the user's ID and checks the blacklist in order to decide whether to forward the access request to the RC. If the user's ID is valid and it is not in the blacklist, the CSP will forward the request to the RC. Otherwise, the request is rejected.

Step 4: The RC evaluates the user's reputation, and decides if the user meets the access policy. If the user is eligible, the RC will set an insurance agreement with the user in case of illegal data disclosure. Otherwise, the request is rejected.

Step 5: The RC issues the re-encryption key $rk\_RC\text{-} > u = RG(sk\_RC, pk\_RC, pk\_u)$, in which RG is the re-encryption key generation function.

Step 6: After receiving the re-encryption key rk_RC from the RC, the CSP forwards the access request to the data owner.

Step 7: The data owner evaluates the user's trust level TL based on previous social experiences, behaviors and activities. If the user is trustworthy, the data owner issues a secret key SK_TL based on the user's trust level TL, and also sends corresponding access policy $\mathcal{A}$. Otherwise, the request is rejected.

Step 8:   After receiving the secret key *SK_TL* and access policy $\mathcal{A}$ from the data owner, the user again sends a data access request along with the access policy $\mathcal{A}$ to the CSP.

Step 9:   The CSP checks if the access policy from the user is the same as that received from the data owner. If both of the policies match, the CSP conducts the cipherkey re-encryption $R(rk\_RC\text{-} > u, E(pk\_RC, K_1)) = E(pk\_u, K_1)$, and sends $E(DEK, data)$, re-encrypted data $E(pk\_u, K_1)$ and $E(PK\_TL, K_0)$ to the user.

5   Performance Evaluation and Analysis

In this section, we conducted a number of tests in order to evaluate the performance of the three schemes. We provide performance analysis in terms of the computation complexity, data confidentiality, flexibility and key management and present the test results.

5.1   Performance Analysis

5.1.1   Computational Complexity

Computational complexity is a key indicator of scheme efficiency, and it is an essential factor for applying an access control scheme in practice. We evaluate the performance of three schemes by analyzing their computational complexity of each main operation. Table 3 presents their computational complexity and compares them with the HASBE scheme proposed in [64].

As shown in the table, our schemes are more efficient than HASBE in terms of System Setup and User Initiation, because we constrain the number of attributes and supplement the security level by providing reputation, Individual TL, and the punishment mechanism. Moreover, there are no pairing operations for Encryption in our schemes, thus it makes the computation faster. For Encryption and Decryption, Scheme 1 is the most computationally efficient since the algorithm are not affected by any variables. The computational complexity of Scheme 2, Scheme 3, and HASBE depend on the number of attributes specified in the access policy tree. Scheme 2 and HASBE have the same computational cost if they have the same number of attributes. Besides the number of attributes, the computational cost of Scheme 3 also depends on the number of symmetric key segments divided by the data owner. The computation of reputation evaluation and TL assessment is quite efficient, since there are neither complicated exponentiations nor pairing operations.

When considering the computational efficiency for each entity in the system, the devices of data owners tend to be most lightweight. Data owners expect sound user experiences and are limited by device capability, while service providers (e.g., CSP,

Table 3 Computation complexity

Operation	Scheme 1	Scheme 2	Scheme 3	HASBE [64]
System setup	$\mathcal{O}(1)$	$\mathcal{O}(1)$	$\mathcal{O}(1)$	$\mathcal{O}(1)$
User initiation	$\mathcal{O}(1)$	$\mathcal{O}(1)$	$\mathcal{O}(1)$	$\mathcal{O}(2M+s)$
Individual TL PK generation	N/A	$\mathcal{O}(2I)$	$\mathcal{O}(2I)$	N/A
Individual TL SK generation	N/A	$\mathcal{O}(1)$	$\mathcal{O}(1)$	N/A
Re-encryption key generation	$\mathcal{O}(n)$	N/A	$\mathcal{O}(n)$	N/A
Encryption	$\mathcal{O}(1)$	$\mathcal{O}(3w)$	$\mathcal{O}(3w+2j)$	$\mathcal{O}(2Y+X)$
Decryption	$\mathcal{O}(1)$	$\mathcal{O}(1)$	$\mathcal{O}(j+1)$	$\mathcal{O}(2Y+2X)$
Re-encryption	$\mathcal{O}(n)$	N/A	$\mathcal{O}(n)$	N/A
Reputation evaluation	$\mathcal{O}(m)$	N/A	$\mathcal{O}(m)$	N/A
Individual TL assessment	N/A	$\mathcal{O}(1)$	$\mathcal{O}(1)$	N/A

Notes
Scheme1 Reputation and PRE based data access control
Scheme2 Individual TL and CP-ABE based data access control
Scheme3 Heterogeneous data access control
n the number of authorized users for data access
m the number of votes
I the maximum number of Individual TLs
w the number of TLs specified in the access policy, $w \leq \bar{I}_{tl}$, $\bar{I}_{tl}$ is the maximum numbers of Trust levels
Y the number of leaf nodes in an access policy tree
S the attribute set
M the number of attributes in S
X translating nodes of access policy tree
j the number of divided key segments

RC, etc.) are believed to have adequate system capability for data storage and process. Table 3 shows that Scheme 1 is the most computationally efficient for data owners, while the others provide more fine-grained data access.

5.1.2　Data Confidentiality

The data confidentiality of the three schemes is evaluated through three factors: cryptographic security, collusion, and punishment mechanism. Cryptographic security is the basic security requirement, since all the schemes are based on cryptographic algorithms. Collusion and punishment mechanism are related to the scheme design in order to achieve good performance and improve system security level.

Cryptographic Security
The cryptographic security depends on the arithmetic security of the symmetric encryption algorithm, PRE and CP-ABE. For the symmetric encryption algorithm, we applied AES whose key size is beyond 128-bit. It is widely used in multiple cryptographic systems, and is believed to be secure for data encryption. The standard security and master key security of PRE are proved in [77] under the

assumption of extended Decisional Bilinear Diffie-Hellman (eDBDH). Additionally, PRE enables non-transitive property to prevent the re-delegation of the decryption rights from, for example, rk_RC-> A and rk_A-> B to produce rk_RC-> B. The arithmetic security of CP-ABE is proved in [80] under the assumption of Decisional BDH (DBDH).

Collusion

The problem of collusion is mainly concerned about the collusion of CSPs and all users in a system, because the RCs and other user authorities are fully trusted under the system model assumptions. CSPs are assumed to be semi-trusted, which do not disclose users' data. The plain data is hidden from the CSPs through symmetric key encryption. Although CSPs do not disclose stored data nor try to crack to obtain the plaintext, it is possible to collude with users to allow unauthorized access or extension of access right.

Scheme 1 depends on reputation and PRE to achieve data security. The collusion between CSP and users is controlled by hiding the plain data and any secret keys from the CSPs. The only encryption key a CSP has is the re-encryption key for an eligible user. Even though CSP colludes with the user who has been delegated the decryption rights, they can only recover the weak secret g^{a_1}, instead of the secret key of RC. However, since the AES key file is encrypted under the RC's public key and the re-encryption key will not change if the RC and the user's public key pairs remain unchanged, the collusion of CSP and user can break the property of fine-grained data access because the user with decryption right can access all the data from the same data owner. And the CSP can allow the extension of access right even if the data owner informs it to block a specific user.

The second and third schemes based on CP-ABE provide higher security by hiding either encryption/decryption operations or secret keys from the CSPs. The access authorization and secret key generation are both controlled by the data owner itself. Moreover, a data owner can re-encrypt the data by modifying the policy tree integrated in the data encryption, instead of updating all public key pairs to manage the collusions, thus reducing the complexity of key management. The third scheme further improves the data confidentiality by dividing the symmetric key into multiple segments, so that a data requester has to recombine all the key segments for decryption. However, the two schemes do not completely eradicate the problem of extending users' access right, since the user revocation is based on the blocking list controlled by CSPs that could not be fully trusted.

Punishment Mechanism

Punishment Mechanism is applied in the Scheme 1 and Scheme 3 to supplement the data confidentiality. It can reduce the possibility of security problems such as collusion, by monitoring performance and reputation, and carrying out several punishment actions. Driven by the business and profits, CSPs are encouraged to perform in an honest way and dedicate to provide secure data storage.

5.1.3 Flexibility

The scheme flexibility is evaluated in terms of User Flexibility, which relates to actions or online flexibility of users when dealing with access requests and delegations.

User Flexibility relates to complex operations and online requirements when dealing with data access requests, especially for data owners, because service users expect excellent user experiences and less operational complexity. The data owners in Scheme 1 depends on PRE and reputation are most flexible, since they do not need staying online to handle the re-encryption and access right authorization. On the contrary, the second scheme requires data owners to handle Individual TL assessment and TL secret key issuing when there is an access request. This mechanism increases the workload of the data owners, but provides higher security and looses the requirement of trust in CSPs. Although the third scheme relates to the most encryption keys, it provides flexibility for users to balance between operational complexity and security level. The data owners can choose any of the provided mechanisms, and decide whether they are willing to stay online and fully control the access right delegation. This heterogeneous property enables full flexibility to choose a proper access control mechanism according practical demands.

5.1.4 Scalability

The essential factors that affect system scalability are key management and user revocation. Table 4 lists the number of keys issued or managed by each entity in different schemes.

As shown in the table, Scheme 1 that is based on PRE and reputation has the least number of keys that need to be managed. All the entities in the system are responsible for generating and managing their own public key pairs, and the re-encryption algorithm only requires RC's secret key and user's public key. The data owner does not have to handle re-encryption no matter how many access rights it needs to delegate. The public keys in Scheme 2 include the system public key, user public key and TL public key. The secret keys in Scheme 2 include user's secret key and TL secret key that is issued to an authorized data requester. The reason why Scheme 3 has the maximum number of keys is that it enables both PRE and CP-ABE based encryption algorithms, in order to provide flexible access control and achieve high security. HASBE [64] grants all the key issuing and management to domain authorities that makes the domain authorities the bottleneck of system performance in terms of scalability.

One improvement of the three schemes for reducing key management load is applying reputation and Individual TL evaluation, as well as blacklist for user revocation. This improvement supplements the security level of data confidentiality while decreases the computational cost compared to other cryptographic methods [17].

Table 4 Number of keys owned by different entities

	Scheme 1	Scheme 2	Scheme 3	HASBE [64]
Data owner	1 PK; 1 SK; 1 Symmetric key	3 PK; 2 SK; 1 Master key; 1 Symmetric key	4 PK; 3 SK; 1 Master key; 1 Symmetric key	1 PK; 1 SK; 1 Master key; 1 Symmetric key
User	1 PK; 1 SK; 1 Symmetric key	2 PK; 2 SK; 1 Master key; 1 Symmetric key	3 PK; 3 SK; 1 Master key; 1 Symmetric key	1 SK; 1 Symmetric key
RC	1 PK; 1 SK; N * Re-encryption key	N/A	1PK; 1SK; N * Re-encryption key	N/A
CSP	N/A	N/A	N/A	N/A
Trust authority	N/A	N/A	N/A	1 PK; n * Master key; m * SK

Notes
Scheme1 Reputation and PRE based data access control
Scheme2 Individual TL and CP-ABE based data access control
Scheme3 Heterogeneous data access control
N number of authorized user
n number of sub-domain authorities
m number of users in the domain

In our proposed three schemes, all system entities' performance is monitored for trust evaluation and reputation generation, which is a effective way to encourage good behaviors and performance.

5.2   Performance Test

5.2.1   Performance Test of Scheme 1

The performance test for Scheme 1 was based on the function blocks of PRE and Reputation Model described in Sect. 4. We evaluated the scheme by analyzing the operations conducted by different entities: *Data Owner, User, CSP, RC*, and tested their performance for each operation. Table 5 lists the main operations and computation efforts managed by different entities in the system.

The *Data Owner* is only responsible for its public key pair generation, data encryption with AES and *DEK* encryption with PRE. The computational cost for AES depends on the size of the underlying data, and it is inevitable in any cryptographic method. The PRE encryption requires 2 exponentiations, and the key generation requires 1 pairing and 1 exponentiation. The computation for a data owner is quite lightweight since it does not need to handle re-encryption or key management no matter how many authorizations of data access right need to be

Table 5 Operation and computations in Scheme 1

Role	Operations	Computations
Data owner	Public key pair generation	1 Pairing + 1 Exp
	Encryption	2 Exp
User	Public key pair generation	1 Pairing + 1 Exp
	Decryption	1 Exp
CSP	Re-encryption	n Pairing
RC	Public key pair generation	1 Pairing + 1 Exp
	Re-encryption key generation	N Exp
	Reputation generation	(m Mult + m Exp)K

Notes
Exp Exponentiation
Pairing Bilinear Pairing
Mult Multiplication
n Number of authorized users for data access
m Number of votes
K the total number of users who request data access

issued. The *User* that requests to access data in the *CSP* only needs to decrypt key file using its own secret key and gain plain data, no matter which owner the encryption key comes from.

The *CSP* in this scheme is responsible for data re-encryption and user revocation. For data re-encryption, it takes 1 pairing, and the computational effort is linear to the number of users who are authorized to access data. User revocation does not require any computational effort. Instead, it does the database lookup and blocks the user's access right according to the data owner's demand.

The *RC* is a TTP responsible for checking data's access policy and a user's reputation level, in order to determine if the user has the right to access the required data. If it is the case, the *RC* then issues the re-encryption key to the *CSP* using the authorized user's public key. The computational effort consists of two parts: re-encryption key generation and reputation evaluation. Re-encryption key generation requires 1 exponentiation for every request, so that the computation is linear to the number of users who request data access at any time point. Reputation evaluation depends on the number of users who request for data access, and the number of the other entities that provide votes.

Figure 6 presents the execution time of each operation conducted by different entity, in the cases of 128-bit, 192-bit and 256-bit sized AES keys. The reputation evaluation contains 10,000 user votes. As shown in Fig. 5, the algorithm of PRE encryption, decryption and Delegation (re-encrypt key generation) consumes less computational power, because they do not contain pairing operations. We can observe that the Re-encryption handled by CSP is the most time consuming process because it contains pairing operations, which are the most expensive computations in PRE. Additionally, the execution time of PRE over AES symmetric key does not differ much with the tested three-sized AES keys. This fact could greatly benefits data owners to choose a long symmetric key to ensure a high level of data security.

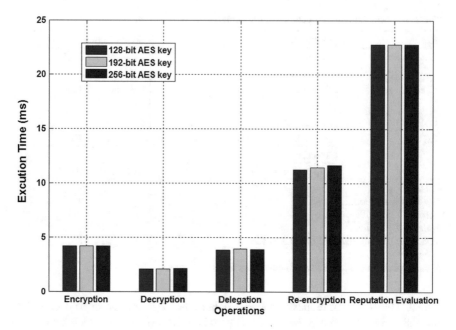

Fig. 6 Execution time of Scheme 1 operations

The execution time of Delegation, which herein refers to the operation of issuing re-encryption key *rk(RC- > u)* to the authorized user, also remains the same with different sized AES keys, since it only operates on the private key of RC and the public key of the user. The reputation evaluation time is not influenced by AES key size.

One improvement we have in our scheme for PRE is to skip the re-encryption key generation process at RC, if the key of the user has been already generated before and the user's information remains unchanged. Figure 7 shows the request processing time in two cases at RC, including data policy checking, reputation evaluation and re-encryption key generation. In the first case, the RC generates a re-encryption key in every request if the requester satisfies the policy and reputation level. In the second case, RC directly fetches the already generated re-encryption key from the granted user's record. As shown in Fig. 7, the higher the number of requests from the users who have the records at RC, the more efficiency improvement our scheme can achieve. In practice, the user could access cloud services multiple times. Thus, utilizing the existing re-encryption keys greatly helps the RC to improve its efficiency and capacity.

We compare our scheme with the personal health records sharing scheme over cloud proposed in [82] based on ABE (combination of Multi-Authority ABE and Key-Policy ABE). Table 6 presents the time comparison between our scheme and [82]. The encryption and decryption time (containing 50 attributes) in [82] is 264 ms and 25 ms respectively, which is much longer than that in our scheme.

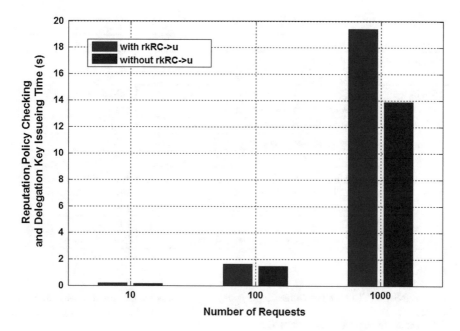

Fig. 7 Operation time of $rk\_RC\text{-} > u$ generation and without $rk\_RC\text{-} > u$ key generation for old users

Table 6 Operation time comparison

Time	Scheme 1 (tested on a workstation with 3.2 GHz processor, 256-bit AES key and 512-bit prime field)	Records sharing scheme in [82] (tested on a workstation with 3.4 GHz processor, 256-bit AES key and 512-bit prime field)
Key encryption	4.2 ms	264 ms
Key decryption	2.1 ms	25 ms
User revocation	Included in reputation and data policy checking	32 ms
Re-encryption	11.7 ms	N.A
Reputation evaluation	22.7 ms (for 10,000 votes)	N.A
Total sum of operations	40.7 ms	321 ms

Although the experiment workstations are different, the apparent time differences can still indicate that our scheme is more efficient. In addition, sophisticated key management and user revocation are replaced by various access policies regarding user reputation, which has much less time consuming computations.

Table 7 Operation and computations

Role	Operations	Computations
User agent	Master key generation	1 Exp
	System public key generation	1 Pairing
Data owner	CSP reputation evaluation	(mMult + mExp)K
	Public key pair generation	2 Exp
	Individual TL public key generation	I * 2 Exp
	Issue TL secret key	1 Exp
	Encryption	w * 3 Exp
	Trust assessment	K * 4 Exp
User	CSP reputation evaluation	(mMult + mExp)K
	Public key pair generation	2 Exp
	Decryption	2 Pairing
CSP	Checking user ID validity and blacklist	User data search

Notes

Exp Exponentiation

Pair Bilinear Pairing

w the number of TLs specified in the access policy, $w \leq \bar{I}_{tl}$, $\bar{I}_{tl}$ is the maximum number of Trust levels

I the maximum number of Individual TLs

m Number of votes

K the total number of users who request data access

5.2.2 Performance Test of Scheme 2

We evaluated Scheme 2 by analyzing the operations conducted by different system entities: Data Owner, User, and CSP, and tested their performance for each operation. Table 7 lists the main operations and computation efforts managed by different entities in the system.

The system setup, including Master Key (MK) and system Public Key (PK) generation, can be conducted in User Agent, or other trusted authorities in the system. The MK and PK are global parameters that are shared between users to initiate their own public key pairs. And they do not change with either the number of users in the system, or the specified maximum Individual TLs for access policy.

In this scheme, either Data Owner or User who requests for data service can firstly evaluate the CSP's reputation through the Reputation Evaluation Model, in order to ensure if the CSP is trustworthy enough to provide data storage service. Data Owner is responsible for encryption, verifying requester's eligibility through individual trust assessment, and issuing a TL based secret key. Encryption consists of data encryption with AES and CP-ABE encryption on *DEK*. The computational cost for AES encryption depends on the size of the underlying data, and it is inevitable in any cryptographic method. For CP-ABE encryption, the data owner firstly generates the TL public key *PK_TL*, which covers each of the TLs specified in the system. The computational cost for generating *PK_TL* is I * 2Exp, where I is the maximum levels of TL. Next, the data owner encrypts the AES key using

PK_TL, according to the required TL specified in the access policy. The computational cost is w * 3Exp, where w is the number of TLs enabled in the access policy. For verifying a requester's access right, the data owner evaluates the requester's individual TL through the Trust Assessment Model, which contains 4 exponentiations for each evaluation. The TL secret key *SK_TL* is issued by the data owner, and implies the requester's individual TL. The user who requests data can use the issued *SK_TL* to decrypt the AES key file. The CP-ABE Decryption algorithm contains two pairing operations no matter how complex the access policy is, because the algorithm conducts the decryption algorithm only if the TL of the data requester meets one of the TL attributes in the access policy.

CSP is computationally lightweight since it does not need to carry any data encryption or key issuing. Besides of the data storage, the CSP is responsible for checking all users' identities and user revocation by searching in the blacklist.

In the performance test, we specified the Individual TLs as attributes and divided them into 5 levels ($\bar{I}_{tl} \leq 5$). Figure 8 presents the execution time of Setup, User Initiation including user's public key pair generation, Individual TL public key and secret key generation, Decryption, Trust Assessment and CSP's reputation evaluation. As introduced in Table 7, the required TLs in access policy have no impact on the performance of the above operations presented in Fig. 8. In our test, the TL public key *PK_TL* generation takes about 17 ms, and TL secret key *SK_TL* issuing takes less than 5 ms. In various applications, *PK_TL* generation process can be different depending on the number of specified levels, as shown in Fig. 9. The execution time of *SK_TL* generation stays constant and it is about 3 ms, which implies that the *SK_TL* issuing process should be very efficient. The decryption time is about 7 ms. The data owner conducts trust assessment, which costs around 3 ms to evaluate the TLs with 10,000 trust assessment requests. The reputation evaluation on CSP contains 100,000 votes and took about 10 ms, which is reasonable in terms of system performance.

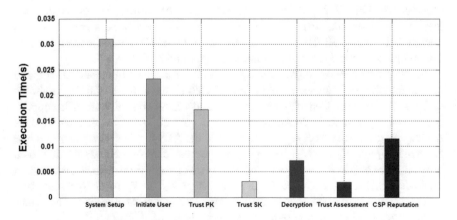

Fig. 8 Execution time of operations in Scheme 2

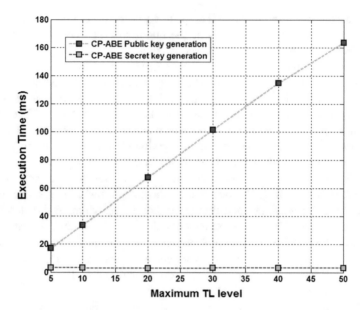

Fig. 9 Execution time of *PK_TL* and *SK_TL*

Figure 10 shows the performance of Encryption algorithm with different required TLs. It illustrates that the higher the required trust level, the less time the Encryption spends, because there are less authorized access conditions to be enabled.

5.2.3 Performance Test of Scheme 3

Scheme 3 provides a flexible way for the data owner to choose either PRE/Reputation based or CP-ABE/TL based data access control. In the performance evaluation, we implemented four scenarios for data access control and tested the performance of each operation in both PRE/Reputation-based and CP-ABE/TL-based access control. Table 8 presents the four scenarios that are supported by Scheme 3.

In our performance test, we employed one RC for reputation management, and proved that the performance is negligibly affected by the number of RCs. As shown in Table 8, Scenario 1 enables reputation based access control with PRE support. Scenario 2 supports Individual TL based access control. Therefore, data owner does not need to divide the AES key. In Scenario 3, the AES key is divided into multiple segments K_0 and K_n in order to enable both access control methods for dual data protection. Scenario 4 provides an option to use either of the two access control methods, so that a data requester can get either K_0 or K_n to decrypt the data. Figure 11 shows the packet format of encrypted key file. The sequence No. indicates the position of the key segment, in order to correctly recombine K_n.

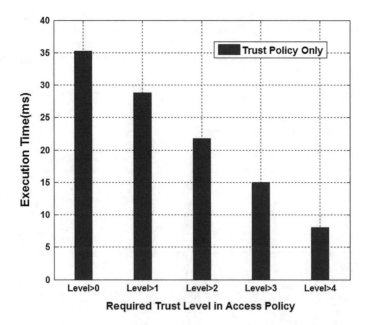

Fig. 10 Execution time of encryption with different maximum TL levels with different required TL

Table 8 Scenarios of heterogeneous data access control

	Access control method	Key division
Scenario 1	Only PRE and reputation evaluation based access control	$K_0 = null$; $K_1 + K_2 + \cdots + K_n = DEK$; ($n$ is the number of RCs)
Scenario 2	Only CP-ABE and individual TL assessment based access control	$K_0 = DEK$; $K_n = null$
Scenario 3	Both PRE-reputation and CP-ABE-TL based access control	$K_0 + K_1 + \cdots + K_n = DEK$; $K_n \neq null$; (n is the number of RCs)
Scenario 4	Either PRE-reputation or CP-ABE-TL based access control	$K_0 = K_1 + K_2 \cdots + K_n = DEK$; ($n$ is the number of RCs)

Notes
K_n the nth key segment for PRE and Reputation based access control
K_0 key segment for CP-ABE and TL based access control

Figure 12 shows the execution time of CP-ABE setup, CP-ABE key pair generation, PRE key pair setup and generation, TL public/secret key generation, and Re-encryption key generation. The performance of the operations is not affected by either the user's preference of key division or the required individual Trust Level. In the performance test, we assumed 5 trust levels, and the TL public key generation process can vary with different number of maximum trust levels, as shown in Fig. 9.

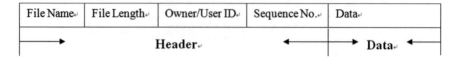

File Name	File Length	Owner/User ID	Sequence No.	Data

Fig. 11 Encrypted key packet format

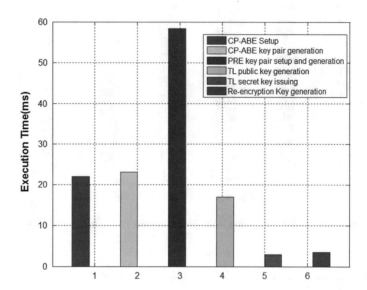

Fig. 12 Execution time of operations in Scheme 3

Figure 13 shows the execution time of reputation evaluation and policy checking conducted in RC and trust assessment processed by the data owner. The performance of reputation evaluation depends on the number of votes received by RC, and the execution time of Individual TLs depends on the number of access requests received by the data owner. The computational cost for trust assessment is quite low, which implies that the computation load of data owner is very low.

We further tested three different sized AES keys: 128-bit, 192-bit and 256-bit. Figure 14 illustrates that the AES key size has little effect on the performance of CP-ABE encryption, as well as the decryption, which takes around 6.50 ms in the test, shown in Fig. 15. For different required Individual TLs, the encryption time varies because different numbers of authorized levels are enabled in the access policy. The higher the required trust level is, the less time the encryption process spends.

Figure 16 shows the performance of the PRE process. We observe that the PRE operations are not clearly affected by the size of AES key. This fact benefits the data owners to choose a long symmetric key if they need high level of data security and

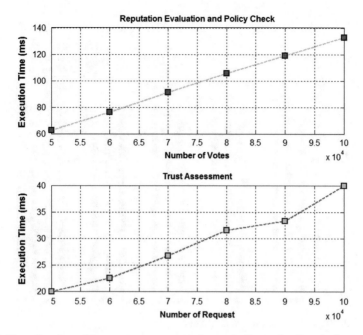

Fig. 13 Execution time of trust assessment, reputation evaluation and policy check

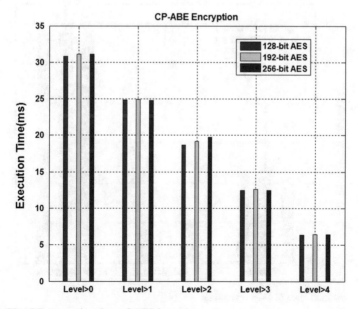

Fig. 14 CP-ABE encryption time of AES keys

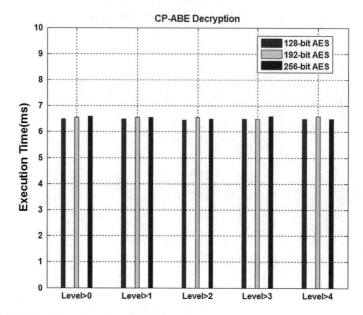

Fig. 15 CP-ABE decryption time of AES keys

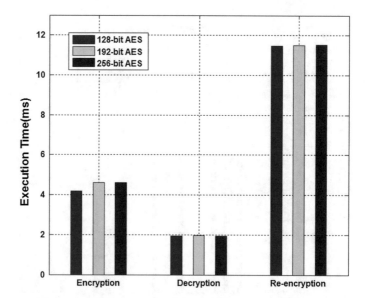

Fig. 16 Execution time of PRE operations

the selection of access control mechanisms can depend on user's requirements in terms of security and efficiency, instead of the size of key segment. We can also observe that the computational cost of CP-ABE is higher than that of PRE algorithm, because CP-ABE carries more exponentiation and pairing operations.

6 Discussions and Future Research Trends

6.1 Usage Scenarios and Limitations for Practical Deployment

Upon the introduction of the data access control schemes and the performance evaluation, we now provide some discussions in terms of usage scenarios and deployment limitations in practice.

Through the performance evaluation, Scheme 1 is the most computational efficient scheme, and it frees data owners from heavy computations by distributing most of the computational operations on the RC and CSP. Simplifying operations on client's (data owners and requesters) side is an essential factor to determine if mobile users are willing to use a service using their resource-constrained mobile devices. Moreover, simpler operations and lightweight computations can offer good user experiences. However, simple scheme sometimes implies less security. In Scheme 1, the RC should be fully trusted in order to ensure that the access rights can be delegated only to eligible users. In practice, it is difficult and risky to fully trust one entity in a public cloud. Therefore, Scheme 1 is more suitable for a private cloud environment, or a cloud data storage system in side an enterprise, because the infrastructure and authority is trusted in these situations. Scheme 2 based on CP-ABE and trust assessment can be applied in public cloud, because the access rights delegation are handled by the data owners themselves and it enables a higher security level than Scheme 1. For example, in a professional social network, a user can decide which company is eligible to access his/her resume by evaluating the company's individual trust level. Scheme 3 which combines Scheme 1 and 2 can also be deployed in public cloud, since it provides high security and flexibility. For example, it can be applied in a public health-care system, in which a patient can delegate the access right of the personal information to unknown doctors in urgent situations by choosing the PRE and reputation based access control mechanism. In addition, the patient can choose CP-ABE and Individual TL based mechanism to issue the access right to other research organizations or third parties. However, Scheme 2 and Scheme 3 consume much computational costs, especially in Scheme 3 since it applies two encryption algorithms. In MCC, it is important to balance between energy efficiency and data security in different application scenarios with different security demands.

6.2 *Future Research Trends*

Based on the review of current literature, we can see that there are a number of unsolved issues towards data security in MCC. Existing research and industrial products provide valuable solutions to achieve connection and communication security, secure data access control and privacy assurance, mainly for cloud computing, but not targeting well for MCC. Due to the distinct characteristics of MCC, additional research problems should be solved and many challenges need to be overcome. Herein, we list some important issues in order to propose future research trends.

First, usability is an important issue in MCC due to the use of mobile devices to access cloud services. MCC client solution should concern usage experience during mobile cloud service consumption. Lightweight data security solutions should be applied in order to reduce power consumption at the same time to enhance user privacy and ensure user data security. This issue hasn't been deeply studied in the past.

Second, it lacks a flexible and comprehensive solution to ensure data security and privacy in all MCC scenarios. So far, there is no serious study on how to ensure data security and privacy if the data need to be transferred in hybrid clouds, especially among mobile and pervasive clouds. It is a challenge to ensure data security and privacy if heterogeneous networks that contain multiple trusted domains carry user access.

Third, deep investigation hasn't yet been done for the purpose of providing MCC client intelligence. MCC aims to fully use all kinds of computing resources in order to achieve the best quality of services. But this requires more on a mobile device to be capable of detecting its environment and adapt cloud computing decision accordingly. How to integrate MCC with other technologies, such as internet of things, could be an interesting research topic.

Fourth, how to enable MCC for big data will become a hot topic. Mobile devices, especially the devices with sensing and processing capability can play as a pre-processor for big data analysis and process and it could play a crucial role for personalized services based on big data. On the other hand, big data security and trust in MCC bring tremendous challenges due to its 4 V characteristics: Volume, Velocity, Variety, and Veracity.

Finally, data trust auditing and verifiable computing will become a significant research area. How to verify and audit the trust of data collection, transmission, process, computation and analytics in MCC will motivate us to make efforts to investigate deeply.

7 Conclusions

In order to optimize the MCC systems and to promote various MCC applications, it is important to provide privacy preservation and data security for MCC users. In this book chapter, we firstly introduced a MCC system model, and the importance of data access control to provide data confidentiality and protect from privacy violation. We then reviewed cryptographic algorithms, reputation and trust management models, and other techniques related to controlling data access in cloud computing. Three data access control schemes based on trust and reputation for MCC were further introduced attempting to overcome the open research issues for MCC data access control. We presented the performance of three schemes in terms of computational efficiency, security, and flexibility, and discussed their pros and cons for usage in different application scenarios. Finally we propose future research trends and directions in MCC data security.

Acknowledgements This work was supported in part by the National Key Foundational Research and Development on Network and Space Security, China, under Grant 2016YFB0800704, in part by NSFC under Grant 61672410 and U1536202, in part by the 111 Project under Grant B08038, in part by the Ph.D. Grant of Chinese Educational Ministry under Grant JY0300130104, in part by the Natural Science Basic Research Plan in Shaanxi Province of China under Program 2016ZDJC-06, and in part by Aalto University.

References

1. The NIST Definition of Cloud Computing, http://www.nist.gov/itl/cloud/
2. Cisco Global Cloud Index: Forecast and Methodology 2013–2018 White Paper
3. Cisco Visual Networking Index: Global Mobile Data Traffic Forecast Update, 2014–2019
4. Doukas, C., Pliakas, T. Magloginannis, I.: Mobile healthcare information management unitizing Cloud Computing and Android OS. In: 32nd Annual International Conference of the IEEE EMBS. IEEE, Argentina, pp. 1037–1040 (2010) doi:10.1109/IEMBS.2010.5628061
5. https://www.onehourtranslation.com/
6. Popa, L., Yu, M., Steven, Y.K., Ratnasamy, S., Stoica, I.: CloudPolice: taking access control out of the network. In: Proceedings of the 9th ACM SIGCOMM Workshop on Hot Topics in Networks, Article No.7, New York (2010). ISBN:978-1-4503-0409-2
7. Ardagna, C., Conti, M., Leone, M., Stefa, J.: An anonymous end-to-end communication protocol for mobile cloud environments. IEEE Trans. Serv. Comput. **7**(3), 373–386 (2014)
8. Kallahalla, M., et al.: Plutus: Scalable secure file sharing on untrusted storage. In: Proceedings of the USENIX Conference on File and Storage Technologies (FAST), pp. 29–42 (2003)
9. Goh, E., Shacham, H., Modadugu, N., Boneh, D.: Sirius: Securing remote untrusted storage. In: Proceedings of NDSS, pp. 131–145 (2003)
10. Zhou, L., Varadharajan, V., Hitchens, M.: Achieving secure role-based access control on encrypted data in cloud storage. IEEE Trans. Inf. Forensics Secur. **8**(12), 1947–1960 (2013)
11. Wang, G., Liu, Q., Wu, J., Guo, M.: Hierarchical attribute-based encryption and scalable user revocation for sharing data in cloud servers. Comput. Secur. **30**(5), 320–331 (2011)
12. Yu, S., Wang, C., Ren, K., Lou, W.: Attribute based data sharing with attribute revocation. In: Proceedings of the ACM ASIACCS, pp. 261–270 (2010)
13. Wang, G., Liu, Q., Wu, J.: Hierarchical attribute-based encryption for fine-grained access control in cloud storage services. In: Proceedings of the 17th ACM CCS, pp. 735–737 (2010)

14. Zhou, M., Mu, Y., Susilo, W., Yan, J.: Piracy-preserved access control for cloud computing. In: Proceedings of IEEE TrustCom11, pp. 83–90 (2011)

15. Yu, S., Wang, C., Ren, K., Lou, W.J.: Achieving secure, scalable, and fine-grained data access control in cloud computing. In: Proceedings of the 29th conference on Information communications, p. 1. IEEE, San Diego, Mar 2010. ISBN 978-1-4244-5836-3

16. Yan, Z.: Chapter 4: trust management in mobile environments, trust management in mobile environments—usable and autonomic models. IGI Global, Hershey, Pennsylvania, USA (2013). doi:10.4018/978-1-4666-4765-7, ISBN 13: 9781466647657, ISBN 10: 1466647655, EISBN13: 9781466647664

17. Yan, Z., Li, X.Y., Kantola, R.: Personal data access based on trust assessment in mobile social networking. In: 2014 IEEE 13th International Conference on Trust, Security and Privacy in Computing and Communications, p. 989. IEEE, Beijing, Sept 2014. doi:10.1109/TrustCom. 2014.131

18. Yan, Z., Li, X.Y., Kantola, R.: Controlling cloud data access based on reputation. Mob. Netw. Appl. March 2015. Springer, ISSN 1572-8153

19. www.mobilecloudcomputingforum.com

20. Sanaei, Z., Abolfazli, S., Gani, A., Buyya, R.: Heterogeneity in mobile cloud computing: taxonomy and open challenges. IEEE Commun. Surv. Tutor. **16**(1), 369 (2014)

21. Khan, A.R., Othman, M., Madani, S.A., Khan, S.U.: A survey of mobile cloud computing application models. IEEE Commun. Surv. Tutor. **16**(1), 393 (2014)

22. Xiong, L.Z., Xu, Z.Q.: Re-encryption security model over outsourced cloud data. In: 2013 International Conference on Information and Network Security, Beijing. Nov. 2013, IET P.1, ISBN 978-1-84919-729-8

23. Fugkeaw S.: Achieving privacy and security in multi-owner data outsourcing. In: 2012 Seventh International Conference on Digital Information Management, p. 239. IEEE, Macau, Aug. 2012. ISBN 978-1-4673-2428-1

24. Durga Priya, G., Prathibha, S.: Assuring correctness for securing outsourced data repository in cloud environment. In: 2014 International Conference on Advanced Communication Control and Computing Technologies, Ramanathapuram, p. 1745. IEEE, May 2014. ISBN 978-1-4799-3913-8

25. Wang, C., Cao, N., Ren, K., Lou, W.J.: Enabling secure and efficient ranked keyword search over outsourced cloud data. IEEE Trans. Parallel Distrib. Syst. **23**(8), 1467. IEEE, Dec. 2011. ISSN 1045-9219

26. Habib, S.M., Ries, S., Muhlhauser, M.: Cloud computing landscape and research challenges regarding trust and reputation. In: 2010 7th International Conference on Computing (UIC/ATC), Xi'an, p. 410. IEEE. Oct. 2010. ISBN 978-0-7695-4272-0

27. Wu, Q.T., Zhang, X.L., Zhang, M.C., Lou, Y., Zhang, R.J., Wei, W.Y.: Reputation revision method for selecting cloud services based on prior knowledge and a market mechanism. Sci. World J. **2014**(617087). ISSN 1537-744X

28. Wang, M., Wang, G.L., Tian, J., Zhang, H.W., Zhang, Y.J.: An accurate and multi-faceted reputation scheme for cloud computing. In: The 11th International Conference on Mobile Systems and Pervasive Computing (MobiSPC'14), Procedia Computer Science 2014, vol. 34, p.466. ISSN 1877-0509

29. Raghebi, Z., Hashemi, M.R.: A new trust evaluation method based on reliability of customer feedback for cloud computing. In: 2013 10th International ISC Conference on Information Security and Cryptology, Yazd, p. 1. IEEE, Aug. 2013, doi:10.1109/ISCISC.2013.6767353

30. Ristenpart, T., Tromer, E., Shacham, H., Savage, S.: Hey, you, get off of my cloud: exploring information leakage in third-party compute clouds. In: Proceedings of the 16th ACM Conference on Computer and Communications security, p. 199. ACM, New York (2009). ISBN 978-1-60558-894-0

31. Factor, M., Hadas, D., Hamama, A., Har'el, N., Kolodner, E.K., Kurmus, A., Shulman-Peleg, A., Sorniotti, A.: Secure logical isolation for multi-tenancy in cloud storage. In: 2013 IEEE Symposium on Mass Storage Systems and Technologies, p. 1. IEEE. Long Beach, May 2013. ISBN 978-1-4799-0217-0

32. Yang, T.C., Guo, M.H.: An A-RBAC mechanism for a multi-tenancy cloud environment. In: 2014 4th International Conference on Wireless Communications, Vehicular Technology, Information Theory and Aerospace & Electronics Systems, p. 1. Aalborg, May2014. ISBN 978-1-4799-4626-6

33. Li, X.Y., Shi Y., Guo, Y., Ma, W.: Multi-tenancy based access control in cloud. In: 2010 International Conference on Computational Intelligence and Software Engineering, p. 1. IEEE, Wuhan, Dec. 2010. ISBN 978-1-4244-5392-4

34. Tang, B., Li, Q., Sandhu, R.: A multi-tenant RBAC model for collaborative cloud services. In: 2013 Eleventh Annual International Conference on Privacy, Security and Trust, p. 229. IEEE, Tarragona, July 2013. doi:10.1109/PST.2013.6596058

35. Wang, J., Zhao, Y., Jiang, S., Le, J.J.: Providing privacy preserving in Cloud computing. In: 2010 3rd Conference on Human System Interactions, p. 472. IEEE, Rzeszow, May 2010. ISBN 978-1-4244-7560-5

36. Yang, P., Gui, X.L., An, J., Yao, J., Lin, J.C., Tian, F.: A retrievable data perturbation method used in privacy-preserving in cloud computing. Communications, 11(8), 1637. IEEE, Aug. 2014. ISSN 1673-5447

37. Haas, S., Wohlgemuth, S., Echizen, I., Sonehara, N., Muller, G.: Aspects of privacy for electronic health records. Int. J. Med. Inf. 80(2), 26. Epub. Feb. 2011. doi:10.1016/j.ijmedinf. 2010.10.001

38. Wang, C., Wang, Q., Ren, K., Lou, W.J.: Privacy-preserving public auditing for data storage security in cloud computing. In: 2010 IEEE Proceedings of INFOCOM, p. 1. IEEE, San Diego, March 2010. ISBN 978-1-4244-5836-3

39. Wang, C., Cao, N., Ren, K., Lou, W.J.: Enabling secure and efficient ranked keyword search over outsourced cloud data. IEEE Trans. Parallel Distrib. Syst. 23(8), 1467. IEEE, Dec. 2011. ISSN 1045-9219

40. Leng, C.X., Yu, H.Q., Wang, J.M., Huang, J.H.: Securing personal health records in the cloud by enforcing sticky policies. TELKOMNIKA Indones. J. Electr. Eng. 11(4), 2200 (2013). doi:10.11591/telkomnika.v11i4.2406

41. Narayan, S., Gagne, M., Safavi-Naini, R.: Privacy preserving EHR system using attribute-based infrastructure. In: Proceeding of 2010 ACM workshop on Cloud Computing Security, p. 47. ACM, New York (2010). ISBN 978-1-4503-0089-6

42. Pearson, S.: Taking account of privacy when designing cloud computing services. In: Proceedings of the 2009 ICSE Workshop on Software Engineering Challenges of Cloud Computing, p. 44. IEEE, Washington (2009). ISBN 978-1-4244-3713-9

43. Ferraiolo, D.F., Kuhn, D.R.: Role-based access control. In: 15th National Computer Security Conference, p. 554, Baltimore, Oct 1992

44. Bertino, E., Bonatti, P.A., Ferrari, E.: TRBAC: a temporal role-based access control model. J. ACM Trans. Inf. Syst. Secur. 4(3), p.191, Aug 2001

45. Joshi, J.B.D., Bertino, E., Latif, U., Ghafoor, A.: A generalized temporal role-based access control model. IEEE Trans. Knowl. Data Eng. 17(1), 4 (2005). ISSN 1041-4347

46. Yu, D.G.: Role and task-based access control model for web service integration. J. Comput. Inf. Syst. 8(7), 2681 (2012)

47. Barati, M., Mohammad, S.K., Lotfi, S., Rahmati, A.: A new semantic role-based access control model for cloud computing. In: 9th International Conference on Internet and Web Applications and Services, Paris, July 2014. ISBN 978-1-61208-361-2

48. Tang, Z., Wei, J., Sallam, A., Li, K., Li, R.X.: A new RBAC based access control model for cloud computing. In: Proceeding of the 12th International Conference on Advances in Grid and Pervasive Computing, p. 279. Springer, Berlin (2012). ISBN 978-3-642-30766-9

49. Na, S.Y., Cheon, S.H.: Role delegation in role-based access control.In: Proceeding of the 5th ACM workshop on Role-based access control, p. 39. ACM, NewYork (2000). ISBN 1-58113-259-X

50. Lin, G.Y., Bie, Y.Y., Lei, M.: Trust based access control policy in multi-domain of cloud computing. J. Comput. 8(5) (2013. ISSN 1796-203X

51. Gitanjali, S., Sehra, S., Singh, J.: Policy specification in role based access control on clouds. Int. J. Comput. Appl. **75**(1) (2013). ISSN 0975-8887
52. Meghanathan, N.: Review of access control models for cloud computing. In: Proceedings of the 3th International Conference on Computer Science, Engineering & Applications, ICCSEA, India, p. 77, May 2013. doi:10.5121/csit.2013.3508
53. Sahai, A., Waters, B.: Fuzzy identity based encryption. In: Proceedings of 24th Annual International Conference on the Theory and Applications of Cryptographic Techniques, Denmark, p. 457. Springer, May 2005. ISBN 978-3-540-32055-5
54. Goyal, V., Pandey, O., Sahai, A., Waters, B.: Attribute-based encryption for fine-grained access control of encrypted data. In: Proceedings of the 13th ACM conference on Computer and communications security, p. 89. ACM, New York (2006). ISBN 1-59593-518-5
55. Beimel, A.: Secure schemes for secret sharing and key distribution, 115p. Ph.D. thesis, Israel Institute of Technology, Technion, Haifa, Israel (1996)
56. Chase, M.: Multi-authority attribute based encryption. In: Proceeding of the 4th Conference on Theory of Cryptography, p. 515. Springer, Berlin (2007). ISBN 978-3-540-70935-0
57. Wang, C.J., Luo, J.F.: A key-policy attribute-based encryption scheme with constant size ciphertext. In: the 8th International Conference on Computational Intelligence and Security, Guangzhou, p. 447. IEEE (2012). ISBN 978-1-4673-4725-9
58. Yan, H.Y., Li, X., Li, J.: Secure personal health record system with attribute-based encryption in cloud computing. In: 9th International Conference on P2P, Parallel, Grid, Cloud and Internet Computing, p. 329. IEEE, Guangdong, Nov 2014. doi:10.1109/3PGCIC.2014.138
59. Lv, Z.Q., Chi, J.L., Zhang, M., Feng, D.G.: Efficiently attribute-based access control for mobile cloud storage system. In: IEEE 13th International Conference on Trust, Security and Privacy in Computing and Communications (TrustCom), p. 292. IEEE, Beijing, Sept 2014. doi:10.1109/TrustCom.2014.40
60. Bethencourt, J., Sahai, A., Waters, B.: Ciphertext-policy attribute based encryption. In: Proceedings of SP'07. IEEE Symposium on Security and Privacy, p. 321. IEEE, Washington, May 2007. ISBN 0-7695-2848-1
61. Lewko, A., Waters, B.: Decentralizing Attribute-Based Encryption. The 30th Annual International Conference on the Theory and Applications of Cryptographic Techniques, p. 568. Springer, Tallinn May 2011. ISBN 978-3-642-20465-4
62. Horvath, M.: Attribute-based encryption optimized for cloud computing. The 41st International Conference on Current Trends in Theory and Practice of Computer Science, p. 566. Springer, Czech. ISBN 978-3-662-46078-8
63. Xu, D.Y., Luo, F.Y., Gao, L., Tang, Z.: Fine-grained document sharing using attribute-based encryption in cloud servers. The 3rd International Conference on Innovative Computing Technology, p. 65. IEEE, London, Aug 2013. ISBN 978-1-4799-0047-3
64. Wan, Z., Liu, J., Deng, R.H.: HASBE: a hierarchical attribute-based solution for flexible and scalable access control in cloud computing. IEEE Trans. Inf. Forensics Secur. **7**(2), 743. IEEE, Oct 2011. ISSN 1556-6013
65. Hussain, F.K., Chang, E.: An overview of the interpretations of trust and reputation. In: The Third Advanced International Conference on Telecommunications, p. 30. IEEE, Morne AICT (2007). ISBN 0-7695-2843-0
66. Dasgupta, A., Prat, A.: Reputation and asset prices: a theory of information cascades and systematic mispricing. Manuscript, London School of Economies (2005)
67. Abdul-Rahman, A., Hailes, S.: Supporting trust in virtual communities. In: Proceedings of the 33rd Annual Hawaii International Conference on System Sciences (2000)
68. Sato, H., Kanai, A., Tanimoto, S.: A cloud trust model in a security aware cloud. In: 2010 10th IEEE/IPSJ International Symposium on Applications and the Internet, p. 121. IEEE, Seoul, July 2010. ISBN 978-1-4244-7526-1
69. Yao, J.H., Chen, S.P., Chen, W., Levy, D., Zic, J.: Accountability as a service for the cloud. In: 2010 IEEE International Conference on Services Computing (SCC), p. 81. IEEE, Miami, July 2010. ISBN 978-1-4244-8147-7

70. Pawar, P.S., Rajarajan, M., Nair, S.K., Zisman, A.: Trust model for optimized cloud services. In: Proceedings of 8th IFIP WG11.11 International Conference, IFIPTM 2014, p. 97. Springer, Singapore, July 2014

71. Prajapati, S.K., Changder, S., Sarkar, A.: Trust management model for cloud computing environment. In: Proceedings of the International Conference on Computing, Communication and Advanced Network (ICCCAN 2013), p. 1–5. India, March 2013

72. Habib, S.M., Ries, S., Muhlhauser, M.: Towards a trust management system for cloud computing. In: 2011 IEEE 10th International Conference on Trust, Security and Privacy in Computing and Communications (TrustCom), p. 933. IEEE, Changsha, Nov 2011. ISBN 978-1-4577-2135-9

73. Bradai, A., Ben-Ameur, W., Afifi, H.: Byzantine resistant reputation-based trust management. In: 2013 9th International Conference on Collaborative Computing: Networking, Applications and Worksharing, p. 269. IEEE, Austin, Oct 2013

74. Koneru, A., Venugopal Rao, K.: Reputation management in distributed community clouds. Int. J. Adv. Res. Comput. Commun. Eng. 1(10), Dec 2012. ISSN 2278-1021

75. Muralidharan, S.P., Kumar, V.V.: A novel reputation management system for volunteer clouds. In: 2012 International Conference on Computer Communication and Informatics, p. 1. IEEE, Coimbatore, Jan 2012. ISBN 978-1-4577-1580-8

76. Zhu, C.S., Nicanfar, H., Leung, V.C.M., Yang, L.T.: An authenticated trust and reputation calculation and management system for cloud and sensor networks integration. IEEE Trans. Inf. Forensics Secur. 10(1), 118. IEEE, Oct 2014. ISSN 1556-6013

77. Ateniese, G., Fu, K., Green, M., Hohenberger, S.: Improved proxy re-encryption schemes with applications to secure distributed storage. ACM Trans. Inf. Syst. Secur. 9(1), 1. ACM, Feb 2006. doi:10.1145/1127345.1127346

78. Yan, Z., Zhang, P., Deng, R.H.: TruBeRepec: a trust-behavior-based reputation and recommender system for mobile applications. J. Pers. Ubiquitous Comput. 16(5), 485–506 (2012)

79. Yan, Z., Chen, Y., Shen, Y.: PerContRep: a practical reputation system for pervasive content services. Supercomputing, 70(3), 1051–1074 (2014)

80. Bethencourt, J., Sahai, A., Waters, B.: Ciphertext-policy attribute based encryption. In: 2007 SP'07. IEEE symposium on Security and Privacy, p. 321. IEEE, May 2007. ISBN 0-7695-2848-1

81. Yan, Z., Li, X.Y., Wang, M.J., Vasilakos, A.V.: Flexible data access control based on trust and reputation in cloud computing. IEEE Trans. Cloud Comput. Submitted

82. Li, M., Yu, S.C., Zheng, Y., Ren, K., Lou, W.J.: Scalable and secure sharing of personal health records in cloud computing using attribute-based encryption. IEEE Trans. Parallel Distrib. Syst. 24(1), 131. IEEE, Jan 2013, ISSN 1045-9219

83. Dinh, H.T., Lee, C., Niyato, D., Wang, P.: A survey of mobile cloud computing: architecture, applications, and approaches. Wirel. Commun. Mob. Comput. 13(18), 1587. 25th Dec. 2013. doi:10.1002/wcm.1203

A Survey on IoT: Architectures, Elements, Applications, QoS, Platforms and Security Concepts

Gonçalo Marques, Nuno Garcia and Nuno Pombo

Abstract The recent development of communication devices and wireless network technologies continues to advance the new era of the Internet and the telecommunications. The vision for the Internet of Things (IoT) states that various "things", which include not only communication devices but also every other physical object on the planet, are going to be connected and will be controlled across the Internet. The concept of the IoT has attracted significantly attention from many investigators in recent years. The incessant scientific improvements make possible to construct smart devices with huge potentials for sensing and connecting, allowing several enhancements based on the IoT paradigm. This chapter presents a review on research on IoT and analyses several IoT projects focused on IoT architectures, elements, Quality of Service (QoS) and currently open issues. The main objective of this chapter is to allow the reader to have an overview on the most important concepts and fundamental knowledge in IoT.

Keywords AAL · Smart homes · Sensors · IoT · Mobile computing · Design challenges · Social and ethical challenges · System architectures · Security · QoS

1 Introduction

The basic idea of the Internet of Things (IoT) is the pervasive presence of a variety of objects with interaction and cooperation capabilities among them to reach a common objective [1].

G. Marques · N. Garcia (✉) · N. Pombo
Faculty of Engineering, Computer Science Department,
Universidade da Beira Interior, Covilhã, Portugal
e-mail: ngarcia@di.ubi.pt

G. Marques · N. Garcia · N. Pombo
Assisted Living Computing and Telecommunications Laboratory (ALLab),
Instituto de Telecomunicações, Covilhã, Portugal

N. Garcia
Universidade Lusófona de Humanidades e Tecnologias, Lisbon, Portugal

© Springer International Publishing Switzerland 2017
C.X. Mavromoustakis et al. (eds.), *Advances in Mobile Cloud Computing and Big Data in the 5G Era*, Studies in Big Data 22,
DOI 10.1007/978-3-319-45145-9_5

Is expected that the IoT will have a great impact on several aspects of everyday-life and this concept will be used in many applications such as domotics, assisted living, e-health and is also an ideal emerging technology to provide new evolving data and computational resources for create revolutionary software applications (also known as "apps") [2]. Systems based on the IoT interact via wireless technologies such as RFID (Radio-Frequency Identification), NFC (Near Field Communication), ZigBee, WSN (Wireless sensor network), DSL (Digital Subscriber Line), WLAN (wireless local area network), WiMax (Worldwide Interoperability for Microwave Access), UMTS (Universal Mobile Telecommunications System), GPRS (General Packet Radio Service), or LTE (Long-Term Evolution).

Moreover, IoT presents several challenges to be solved such as security and privacy, participatory sensing, big data, and architectural issues apart of the known WSN challenges, including energy efficiency, protocols, and Quality of Service (QoS) [2]. In the one hand, technological standardization of the IoT is now starting to be missed, so collaboration among IEEE (Institute of Electrical and Electronics Engineers), ISO (International Organization for Standardization), ETSI (European Telecommunications Standards Institute), IETF (Internet Engineering Task Force), ITU (International Telecommunication Union) and other related organizations is very important and urgent [3]. In the order hand, some authors define that premature standardization could risk stifling innovation [4]. Industry applications, monitoring and water control, smart homes' architecture, estimation of natural disaster, medical applications, agriculture application, intelligent transport system design, design of smart cities, smart metering and monitoring, smart security are examples of interesting applications of IoT [5].

This document is organized as follows. This paragraph concludes the introduction in Sect. 1. In Sect. 2, IoT visions and architecture are introduced. The essential elements of IoT are the subject of discussion in Sect. 3 and IoT applications, smart homes and health projects are addressed in Sect. 4. Section 5 gives an overview about IoT platforms and their open issues. In Sect. 6 important aspects about quality of service issues in IoT are introduced and Sect. 7 focuses on security and privacy concepts. Conclusions and future research topics are presented in Sect. 8

2 IoT Visions and Architecture

The paradigm of the IoT is referred to as the result of the merging of different views: things oriented vision, Internet oriented vision, and semantic oriented vision [6]. Under the same article, the IoT semantic oriented vision means a global network of interconnected objects that have a unique address based on standard communication protocols. The things oriented vision focuses on intelligent autonomous devices that use technologies such as NFC and RFID objects, applied to our daily lives. The Internet oriented vision focuses on the idea of keeping the devices connected to the network, having a single address and using standard

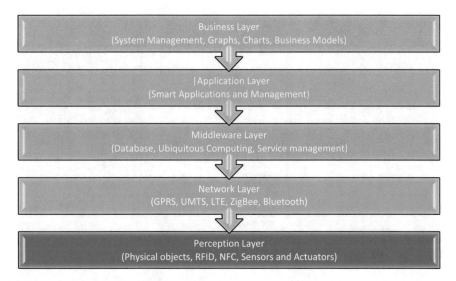

Fig. 1 IoT architecture layers adapted from [7]

protocols. Semantic oriented vision focuses on storage, searching and organizing information generated by IoT, seeking to create data and environment architectures modelling solutions to handle efficiently with the produced information.

The architecture of the IoT can be divided into five layers such as: Objects Layer or Perception Layer, Object Abstract layer or Network Layer, Service Management Layer or Middleware Layer, and Application Layer and Business layer (Fig. 1). On the one hand, the Perception Layer refers to physical sensors and actuators that IoT systems absorb [7]. On the other hand, the Network Layer transfers data produced by the Perception Layer to the Middleware Layer through secure channels using technologies such as RFID, ZigBee, WPAN, WSN, DSL, UMTS, GPRS, WiFi, WiMax, LAN, WAN, 3G and LTE. Furthermore, the Middleware Layer pairs a service with its requester based on addresses and names in order to maintain independence from the hardware. On the contrary, the Application Layer provides the services requested by customers providing the system output information to the user that requests that information.

Finally, the Business Layer manages the overall IoT system activities and services to build a business model, graphs, flowcharts, etc. based on the received data from the Application layer.

3 Elements of the Internet of Things

This section will revisit the elements of IoT such as identification, sensing, communication, computation, services and semantic (Fig. 2).

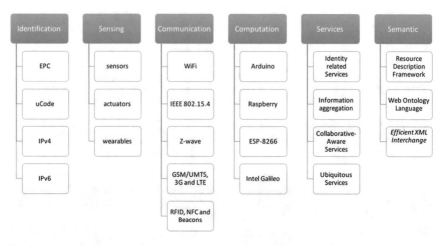

Fig. 2 IoT elements

The identification is essential for the development of the IoT and is important to ensure correct identification of objects in order to match services with their demand. Actually many identification methods exist such as electronic product codes (EPC) and ubiquitous code (uCode). Object identification refers to its name or designation and addressing refers to its IP address for communication on the network. Addressing methods include today's IPv4, IPv6 and 6LoWPAN that provides compression on IPv6 headers [8]. With the large address space provided by IPv6, all the addressing needs of the IoT are thought to be taken care of.

In IoT, sensing refers to acquire data from the environment and send it to a database, remote, local or in a cloud, and as example of IoT sensors one can find smart sensors, actuators or wearable sensors.

Moreover, the communication is an integral part of all IoT devices. Limited by the nature of the "things" themselves, as for example, battery life or limited range of data transmission, protocols such as WiFi, ZigBee, GSM, GPRS, UMTS, 3G, LTE and 5G [9], among others, are common. The IEEE 802.15.4e standard was released by IEEE in 2012 to enhance and add functionality to the previous 802.15.4 standard, as to address the emerging needs of embedded industrial applications [10]. There are also other communications technologies used for proximity communications like RFID [11], NFC [12] and Beacons (Bluetooth Low Energy) [13]. An memoryless-based collision window tree plus (CwT+) protocol for simplified computation in anti-collision radio frequency identification (RFID) is proposed by [14], the authors concluded the outperformance of the CwT+ compared with earlier protocols. A low-cost flexible NFC tags to allow everyday objects to communicate to smartphones and computers and thus participate in the IoT is proposed by [15], according to the authors the most important NFC regulatory standards are met, even with relaxed 5 micron design rules, using optimized design topologies. Bluetooth 4.2 offers features that makes Bluetooth Low Energy an appropriate protocol for low-power communication technologies in the IoT and are applied in 6LoWPAN networks [16].

In addition, IoT are deployed using several hardware platforms applications such as Arduino [17], Intel Galileo [18], Raspberry Pi [19] or ESP8266 [20]. Cloud computing platforms are also an important computational part of the IoT paradigm because they provide facilities for storing and/or processing data in real time.

The different IoT services can be categorized as Identity related services, Information aggregation services, Collaborative-Aware services, and Ubiquitous services [21]. Identity-related services are focused on the identification of objects, whereas Aggregation services collet and summarize sensory data, and send them to the backend application. Furthermore, Collaborative-Aware services are used to turn the obtained data into a decision in order to react accordingly, while Ubiquitous services aim to provide Collaborative-Aware services anytime to anyone anywhere.

Semantic in the IoT is considered as the ability to extract knowledge from machines to provide the required services by discovering and using resources and modelling information. Thus, Semantic Web technologies examples are the Resource Description Framework (RDF) [22] and the Web Ontology Language (OWL) [23].

4 Applications

IoT turn into reality several important applications like smart homes, ambient assisted living and health domains, which are detailed in the next sections.

4.1 Smart Homes

Smart homes have been researched for decades, being the first Smart Rooms project implemented by the MIT Media Lab [24]. Nowadays, smart homes may be classified into three different categories: The first category aims to detect and to recognize the actions of its residents in order to determine their health condition. The second category, aims at storing and retrieving of multi-media captured within the smart home, in different levels from photos to experiences. The third category is focused on the surveillance, where the data captured in the environment are processed to obtain information that may help to raise alarms, in order to protect the home and the residents. There is also a type of smart homes that have the objective to reduce the energy consumption by monitoring and controlling electric devices [25].

Recent advances in information technology allowed lower prices of smart homes but provide them intelligence environments to make complex decisions remains a challenge. Thus, is expected an increasingly amount of interconnections among sensors for collect data in real time.

Three broad views about smart homes are introduced by [26]: a functional view; an instrumental view; and a socio-technical view. The functional view sees smart homes as a way of better managing the demands of daily living through technology. The instrumental view emphasises smart homes' potential for managing and reducing energy demand in households as part of a wider transition to a low-carbon future. The socio-technical view sees the smart home as the next wave of development in the ongoing electrification and digitalisation of everyday life.

A really interesting smart home application is introduced by [27], called Vital-Radio, which consists in a wireless sensing technology that monitors breathing and heart rate without body contact. This method demonstrates through a user study that it can track users' breathing and heart rates with a median accuracy of 99 %. In Europe, smart home projects include iDorm [28], Gloucester Smart Home [29], and CareLab [30], which is distinguished by its importance for the development of state of the art and due to innovative features.

Several challenges are related to IoT and AAL such as security, privacy, and legal. IoT devices are typically wireless and exposed to public range, the ownership of data collected from IoT devices must be clearly established. In fact, IoT devices should use encryption methods and be equipped with privacy policies.

In order to try to resolve privacy issues, the ambient sensing system AmbLEDs project, proposes the use of LEDs instead of other types of more invasive sensors such as cameras and microphones. In fact, this type of applications reveal the importance of using simple and intuitive interfaces for the interaction with people in the ambient assisted living [31].

Humans will often be the integral parts of the IoT system and therefore IoT will affect every aspect of human lives. In addition, due to the large scale of devices arise continuing problems of privacy and security so as consequence cooperation between the research communities is essential [32].

An really interesting example of IoT combined with ALL is proposed by [33] where an integrated platform for monitoring and controlling of a household that uses ZigBee Wireless network is reported which is distinguished by the use of open-source technologies.

The SPHERE Project [34] aims to build a generic platform that fuses complementary sensor data to generate rich datasets that support the detection and management of various health conditions. This project uses three sensing technologies: environment, video, and wearable sensing.

Furthermore, a cloud-based IoT platform for AAL proposed by [35], aims to manage the integration and behaviour-aware orchestration of devices as services stored and accessed on the cloud.

Well-known technologies like RFID, is still used to match IoT and ALL, allowing the creation of intelligent systems that can detect user-object interactions using for example supervised machine learning algorithms. In line with this, the project described in [36] where the authours present weighted Information Gain (wIG), an empirical method for reliably detecting unassisted, deviceless, and real time—user-object interaction using RFID.

In addition. the Home Health Hub Internet of Things (H3IoT) consists in an architectural framework for monitoring health of elderly people at his home. This framework presents several advantages such as: mobility, affordable price, usability, simple layered design, and delay tolerant [37].

4.2 Health Projects Based on IoT

IoT is a suitable approach to build health care systems, based on the technology advancements that allows to define new advanced methods for the treatment of many diseases e.g. by monitoring of chronic diseases to help doctors to determine the best treatments as proposed by [38].

A solution for diabetes therapy based on the IoT is proposed by [39]. This solution supports a patient's profile management architecture based on the personal RFID cards. A global connectivity among the patient's personal device, based on 6LoWPAN, the nurses/physicians' desktop application to manage personal health records, the glycaemic index information system, and the patient's web portal is provided.

On the last few years, the IoT has been proposed on several projects for remote health care aiming at to improve acquisition and processing of data [40]. Despite the potential of the IoT paradigm and technologies for health systems, there are still room for improvement on different topics. The direction and impact of the IoT economy is not yet clear, there are barriers to the immediate and ubiquitous adoption of IoT products and services, and these solutions may sound feasible for implementation, the timing may be too early [41].

In addition, IoT technologies provide many benefits to the healthcare domain in activities such as tracking of objects, patients and staff, identification and authentication of people, automatic data collection and sensing [6]. Thus, IoT gives a considerable solution as a platform to ubiquitous healthcare using wearable sensors to upload the data to servers and smartphones for communication along with Bluetooth for interfacing sensors measuring physiological parameters [2].

Health-care applications should incorporate several mechanisms that should be used to provide privacy of personal and/or sensitive information has harnessed the adoption of IoT Technologies. The interconnection of many IoT systems and sensors could trigger an intervention by the medical staff upon detection of conditions that otherwise unattended could lead to health and wellbeing deterioration, thus realizing preventive care [42].

In fact, security vulnerabilities exist in a Machine-to-machine (M2M)/IoT communication, aiming at to ensure the proper access to the right entities at the right time, and supported by a secure architecture. Other big challenge is that M2M/IoT devices may not have enough capabilities to execute encryption methods on the device [43]. This challenge must be solved if a secure health systems based on IoT is provided.

The Health-IoT (in-home health care service based on the IoT technologies) has promising prospects. A business-technology co- design methodology is proposed for cross-boundary integration of in-home health care devices and services based in IoT [44].

An IoT based sensing architecture facilitates improving energy efficiency by permit dynamic utilization of sensors and the use of IPv6 over Low Power Wireless Personal Area Networks (6LoWPAN) has been proposed to connect energy constrained WPAN devices to the Internet [45, 46].

The IoT paradigm specifies a way to monitor, store and utilize health and wellbeing related data on a 24/7 basis [47] and also provide services to be ubiquitous and customized for personal needs [42, 48].

An proof of concept implementation of an IoT-based remote health monitoring system includes an demo of a Smart e-Health Gateway called UT-GATE is proposed by [49]. UT-GATE provides local services for health monitoring applications such as local repository, compression, signal processing, data standardization, WebSocket server, protocol translation and tunnelling, firewall, and data mining and notification.

5 IoT Platforms

A IoT platform can be defined as the middleware infrastructure that supports the interactions between devices and users. These platforms can be divided in general into Cloud-based platforms (Fig. 3) and Local platforms (Fig. 4).

The most important characteristics of an IoT platform are the support to heterogeneous devices, the data privacy and security, data fusion and sharing, the support to APIs and to the IoT ecosystem. In line with this, this section will briefly summarize a number of platforms for the Internet of Things. It is intended to

Fig. 3 Cloud platform

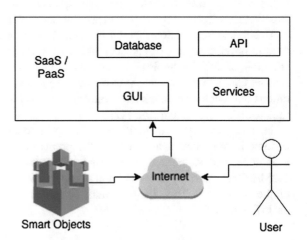

Fig. 4 Local platform

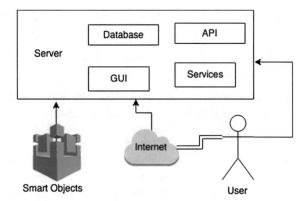

highlight the most important platforms, so those who have more advantages and which are distinguished by its importance for the development of state of the art and due to innovative features.

1. **OpenIoT**: supports heterogeneous devices have a decentralized architecture, is open-source (LGPLv3), have user-based privileges and service discovery [50].
2. **The thing system**: have a centralized architecture, provides service discovery, is open-source (M.I.T. License), does not provide storage functionalities and are designed to only provide access remotely to home's smart devices [51].
3. **Xively**: is a proprietary cloud-based platform that provides open-source API's and supports RESTful API [52].
4. **ARM mbed**: is a proprietary platform with centralized/Cloud-based architecture designed for embedded devices that supports security functionalities such as Transport Layer Security (TLS), it supports CoAP and RESTful API for create M2M networks [53].
5. **H.A.T.—Hub of All Things**: is an open-source platform that have a decentralized architecture that aim to create economic and business opportunities using generated data by IoT home's systems and supports RESTful API [54].
6. **Ericsson IoT Framework**: is a PaaS that includes a REST API, data storage functionalities and OpenId access control. It is a open-source (Apache license 2.0) platform with a centralized architecture [55].
7. **Calvin-Base**: is an open-source platform with a centralized architecture that supports REST API and is main goal is to be extremely extendable and for that it have a large amount of plugins applications to ensure interoperability [56].
8. **OpenRemote**: is an open-source platform with a centralized architecture that supports REST API and have local store system. This platform supports home and domotics environments [57].
9. **ThingWorx**: is a proprietary cloud-based architecture M2M platform (PaaS) that supports REST API and service discovery [58].

10. **Sense Tecnic WoTkit**: is a proprietary cloud-based architecture platform supports REST API and service discovery and have secured access [59].

As referred by [60] the success of the platforms and frameworks is based on different topics, as follows:

1. Enable devices, applications and systems to securely expose API's for 3rd party systems and to facilitate API management;
2. Enable systems to have protocol interoperability with other 3rd party API's and ensure they are extendable for new protocols;
3. Enabling constrained devices to participate into application networks. That is size, bandwidth, power supply(battery) and processing power constraints;
4. Governance—Enabling management and governance of heterogeneous networks of devices and applications.

There are challenges and problems that cross all these IoT platforms including the following: standardization, power and energy efficiency, Big Data, security and privacy, intelligence, integration methodology, pricing, network communications, storage and scalability and flexibility as showed in Fig. 5.

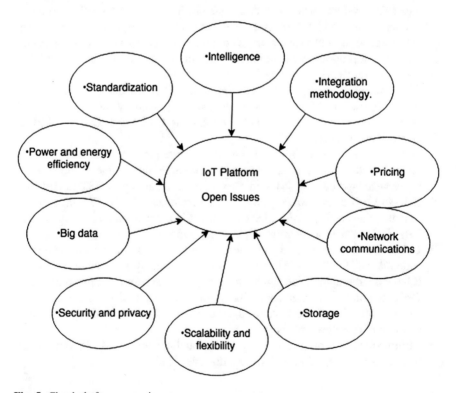

Fig. 5 Cloud platforms open issues

6 QoS in IoT

IoT presents several QoS (Quality of Service) issues such as availability, reliability, mobility, performance, scalability and interoperability, most of this issues are identified by [5–7, 42].

On the one hand, the availability of the IoT systems refers to provide anywhere and anytime services to customers. The IoT must try to be compatible with all devices and follow the protocols such as IPv6, and 6LoWPAN. On the other hand, the reliability refers to provide a high success rate for the IoT service delivery, and implemented in software and hardware throughout all the IoT layers. In addition, the mobility refers to connecting users with their desired services continuously while on the move, interruption for mobile devices can occur when these devices transfer from one gateway to another. An architecture based on IoT to support mobility and security in medical environments is proposed by [61].

Moreover, the performance of IoT services is a big challenge because it depends on the performance of many components of the IoT systems that needs to continuously develop and improve the performance of its services to meet the requirements.

Furthermore, the scalability of the IoT refers to the ability to add new devices, services and functions without affecting the quality of service. In the presence of diverse hardware platforms and communications protocols adding new operations and supporting new devices are challenging tasks. In addition, is not easy to assure the interoperability on these large-scale applications composed by a large number of heterogeneous things that belong to different platforms [62].

7 Security and Privacy

Security is the most significant challenge for IoT applications, and its architecture should fit the lifecycle of IoT, and its potentials. Thus, IoT systems must take into account the effect of packet fragmentation on security, with particular focus on possible DoS attacks [63].

Manufacturing IoT must not only address technical problems, but also consider planning, infrastructure, management, and security problems [64]. At the network layer, the IoT systems must use encryption and enhance the capacity against DoS attack.

In 2008 the ISO/IEC 29192 standards were created in order to provide light-weight cryptography for constrained devices, including block and stream ciphers and asymmetric mechanisms. Lightweight cryptography contributes to the security of smart objects networks because of its efficiency and smaller footprint [65].

The attacks on IoT systems can be categorized as physical attacks, side channel attacks, cryptanalysis attacks, software attacks and network attacks [66]. Physical attacks refer to attack the physical hardware and are harder to perform. Side channel attacks makes use of information to recover the key the device is using. Crypt-analysis attacks are focused on the cipher text with the objective to break the encryption. Software attacks exploit vulnerabilities in the systems through its own communication interface. Networks communications are vulnerable to networks security attacks due to the broadcast nature of the transmission medium.

In fact, IoT systems that use wireless technologies may experience several security issues like attacks on secrecy and authentication, silent attacks on service integrity and attacks at network availability. Network availability attacks can be catalogued as DoS (Denial of Service) attacks an occur at physical, link, network, transport and application layers (Fig. 6) [67].

Although a lot of research has been done in order to increase the security in the IoT, open problems remain in a number of areas, such as cryptographic mecha-nisms, network protocols, data and identity management, user privacy, self-management, and trusted architectures [68]. Several proposals for security arrangements for IoT can be found in [69–74]. An overview of low-complexity physical-layer security schemes that are suitable for the IoT is proposed by [69]. A novel system architecture, called the Unit and Ubiquitous IoT (U2IoT) is pro-posed by [70] to face security issues. On the contrary, a two way authentication security scheme for the IoT based on existing Internet standards, especially the Datagram Transport Layer Security (DTLS) protocol is proposed by [71]. The possibility of reducing the over- head of DTLS by means of 6LoWPAN header compression is proposed by [72]. A new security solution for integrating WSNs into the Internet as part of the IoT is presented by [73]. Authors of [74] demon-strates the feasibility of using the Generic Bootstrapping Architecture (GBA) de-fined by the 3rd Generation Partnership Project (3GPP) to perform secure authentication of resource-constrained for IoT devices.

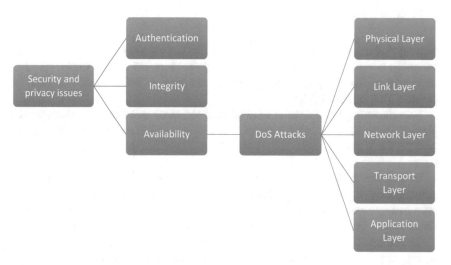

Fig. 6 Hierarchical diagram of security issues in IoT systems that incorporate wireless technologies adapted from [68]

8 Conclusions

The IoT is a paradigm that aims to improve the quality of human live. There are several visions of the IoT such as things oriented vision, Internet oriented and semantic oriented vision. The IoT can be divided into 5 layers such as Objects Layer or Perception Layer, Object Abstract layer or Network Layer, Service Management Layer or Middleware Layer, Application Layer and Business layer.

The main elements of the IoT are identification, sensing, communication, computation, services and semantic and exist several examples of fascinating applications like prediction of natural disaster, industry applications, design of smart homes, health applications, agriculture applications, intelligent transport system design, design of smart cities, smart security, smart metering and monitoring.

The IoT systems and ALL will continue side-by-side mutually contributing scientific advances in assisted living allowing also lower the cost of assisted living systems.

The IoT platforms must support heterogeneous devices, data fusion and sharing, data privacy and security, API's for interoperability and standardization. There are several open issues related with the IoT, such as standardization, security and privacy, power and energy efficiency, intelligence, integration methodology, big data, pricing, storage, network communications, scalability and flexibility. The IoT continues to present several QoS issues such as availability, reliability, mobility, performance, scalability and interoperability.

This paper presented an overview of the IoT concepts such as architecture, vision, elements, main applications focus on smart homes and heath systems, platforms, QoS and security issues, which should provide a good transversal sense of IoT technologies.

Despite the numerous technologic enhancements, some issues in the design of IoT systems continue to exist, namely corresponding to privacy, confidentiality, security, and interoperability of such systems.

Acknowledgements The authors would like to acknowledge the contribution of the COST Action IC1303—Architectures, Algorithms and Platforms for Enhanced Living Environments (AAPELE). Contributing to this research, the Authors affiliated with the *Instituto de Telecomunicações* also acknowledge the funding for the research by means of the program FCT project UID/EEA/50008/2013. (*Este trabalho foi suportado pelo projecto FCT UID/EEA/50008/2013*).

References

1. Giusto, D. (ed.): The Internet of Things: 20th Tyrrhenian Workshop on Digital Communications. Springer, New York (2010)
2. Gubbi, J., Buyya, R., Marusic, S., Palaniswami, M.: Internet of Things (IoT): a vision, architectural elements, and future directions. Future Gener. Comput. Syst. **29**(7), 1645–1660 (2013)
3. Tan, L., Wang, N.: Future internet: the Internet of Things, pp. V5–376–V5–380 (2010)

4. Blackstock, M., Lea, R.: IoT interoperability: a hub-based approach, pp. 79–84 (2014)
5. Khan, R., Khan, S.U., Zaheer, R., Khan, S.: Future internet: the Internet of Things architecture, possible applications and key challenges, pp. 257–260 (2012)
6. Atzori, L., Iera, A., Morabito, G.: The Internet of Things: a survey. Comput. Netw. **54**(15), 2787–2805 (2010)
7. Al-Fuqaha, A., Guizani, M., Mohammadi, M., Aledhari, M., Ayyash, M.: Internet of Things: a survey on enabling technologies, protocols, and applications. IEEE Commun. Surv. Tutorial **17**(4), 2347–2376 (2015)
8. Jara, A.J., Zamora, M.A., Skarmeta, A.: Glowbal IP: an adaptive and transparent IPv6 integration in the Internet of Things. Mob. Inf. Syst. **8**(3) (2012)
9. Jermyn, J., Jover, R.P., Murynets, I., Istomin, M., Stolfo, S.: Scalability of machine to machine systems and the Internet of Things on LTE mobile networks. In: 2015 IEEE 16th International Symposium on a World of Wireless, Mobile and Multimedia Networks (WoWMoM), pp. 1–9 (2015)
10. De Guglielmo, D., Anastasi, G., Seghetti, A.: From IEEE 802.15.4 to IEEE 802.15.4e: a step towards the Internet of Things. In: Gaglio, S., Lo Re, G. (eds.) Advances onto the Internet of Things, vol. 260, pp. 135–152. Springer (2014)
11. Falk, R., Kohlmayer, F., Köpf, A.: Device and method for providing RFID identification data for an authentication server. Google Patents (2015)
12. Curran, K., Millar, A., Mc Garvey, C.: Near field communication. Int. J. Electr. Comput. Eng. **2**(3), 371 (2012)
13. Gomez, C., Oller, J., Paradells, J.: Overview and evaluation of bluetooth low energy: an emerging low-power wireless technology. Sensors **12**(12), 11734–11753 (2012)
14. Garcia Zuazola, I.J., Bengtsson, L., Perallos, A., Landaluce, H.: Simplified computation in memoryless anti-collision RFID identification protocols. Electron. Lett. **50**(17), 1250–1252 (2014)
15. Myny, K., Cobb, B., van der Steen, J.-L., Tripathi, A.K., Genoe, J., Gelinck, G., Heremans, P.: 16.3 Flexible thin-film NFC tags powered by commercial USB reader device at 13.56 MHz, 1–3 (2015)
16. Raza, S., Misra, P., He, Z., Voigt, T.: Bluetooth smart: an enabling technology for the Internet of Things, pp. 155–162 (2015)
17. Doukas, C.: Building Internet of Things with the Arduino. CreateSpace Independent Publishing Platform (2012)
18. Ramon, M.C.: Intel Galileo and Intel Galileo Gen 2. Springer (2014)
19. Upton, E., Halfacree, G.: Raspberry Pi User Guide. Wiley (2014)
20. Raspaile, P., Keswani, V.: Integrating wireless sensor network with open source cloud for application of smart home
21. Gigli, M., Koo, S.: Internet of Things: services and applications categorization. Adv. Internet Things **01**(02), 27–31 (2011)
22. Vertan, C., Merkmale, R.-F.: Resource Description Framework (rdf) (2004)
23. McGuinness, D.L., Van Harmelen, F.: OWL web ontology language overview. W3C Recomm. **10**(10), 2004 (2004)
24. Moukas, A., Zacharia, G., Guttman, R., Maes, P.: Agent-mediated electronic commerce: an mit media laboratory perspective. Int. J. Electron. Commer. **4**(3), 5–21 (2000)
25. De Silva, L.C., Morikawa, C., Petra, I.M.: State of the art of smart homes. Adv. Issues Artif. Intell. Pattern Recognit. Intell. Surveill. Syst. Smart Home Environ. **25**(7), 1313–1321 (2012)
26. Wilson, C., Hargreaves, T., Hauxwell-Baldwin, R.: Smart homes and their users: a systematic analysis and key challenges. Pers. Ubiquit. Comput. **19**(2), 463–476 (2015)
27. Adib, F., Mao, H., Kabelac, Z., Katabi, D., Miller, R.C.: Smart homes that monitor breathing and heart rate. In: Proceedings of the 33rd Annual ACM Conference on Human Factors in Computing Systems, Seoul, Republic of Korea, pp. 837–846 (2015)
28. Pounds-Cornish, A., Holmes, A.: The iDorm—a practical deployment of grid technology. In: 2002. 2nd IEEE/ACM International Symposium on Cluster Computing and the Grid, pp. 470–470 (2002)

29. Orpwood, R., Gibbs, C., Adlam, T., Faulkner, R., Meegahawatte, D.: The gloucester smart house for people with dementia—user-interface aspects. In: Keates, S., Clarkson, J., Langdon, P., Robinson, P. (eds.) Designing a More Inclusive World, pp. 237–245. Springer, London (2004)

30. Henkemans, O.B., Caine, K.E., Rogers, W.A., Fisk, A.D.: Medical monitoring for independent living: user-centered design of smart home technologies for older adults, in Proc, pp. 18–20. Telemedicine and Health Information and Communication Technologies, Med-e-Tel Conf. eHealth (2007)

31. Cunha, M., Fuks, H.: AmbLEDs para ambientes de moradia assistidos em cidades inteligentes. In: Proceedings of the 13th Brazilian Symposium on Human Factors in Computing Systems, Foz do Igua and ccedil; u, Brazil, 2014, pp. 409–412

32. Stankovic, J.A.: research directions for the Internet of Things. Internet Things J. IEEE 1(1), 3–9 (2014)

33. Suryadevara, N.K., Kelly, S., Mukhopadhyay, S.C.: Ambient assisted living environment towards internet of things using multifarious sensors integrated with XBee platform. In: Mukhopadhyay, S.C. (ed.) Internet of Things, vol. 9, pp. 217–231. Springer (2014)

34. Zhu, N., Diethe, T., Camplani, M., Tao, L., Burrows, A., Twomey, N., Kaleshi, D., Mirmehdi, M., Flach, P., Craddock, I.: Bridging e-health and the Internet of Things: the SPHERE project. Intell. Syst. IEEE 30(4), 39–46 (2015)

35. Cubo, J., Nieto, A., Pimentel, E.: A cloud-based Internet of Things platform for ambient assisted living. Sensors 14(8), 14070–14105 (2014)

36. Parada, R., Melia-Segui, J., Morenza-Cinos, M., Carreras, A., Pous, R.: Using RFID to detect interactions in ambient assisted living environments. Intell. Syst. IEEE 30(4), 16–22 (2015)

37. Ray, P.P.: Home health hub Internet of Things (H3IoT): an architectural framework for monitoring health of elderly people. In: 2014 International Conference on Science Engineering and Management Research (ICSEMR), pp. 1–3 (2014)

38. Chui, M., Löffler, M., Roberts, R.: The Internet of Things. McKinsey Q. 2(2010), 1–9 (2010)

39. Jara, A.J., Zamora, M.A., Skarmeta, A.F.: An Internet of Things—based personal device for diabetes therapy management in ambient assisted living (AAL). Pers. Ubiquit. Comput. 15(4), 431–440 (2011)

40. Luo, J., Chen, Y., Tang, K., Luo, J.: Remote monitoring information system and its applications based on the Internet of Things. In: FBIE 2009. International Conference on Future BioMedical Information Engineering, 2009, pp. 482–485 (2009)

41. Swan, M.: Sensor mania! The Internet of Things, wearable computing, objective metrics, and the quantified self 2.0. J. Sens. Actuator Netw. 1(3), 217–253 (2012)

42. Miorandi, D., Sicari, S., De Pellegrini, F., Chlamtac, I.: Internet of Things: vision, applications and research challenges. Ad Hoc Netw. 10(7), 1497–1516 (2012)

43. Lake, D., Milito, R., Morrow, M., Vangheese, R.: Internet of Things: architectural framework for eHealth security. J. ICT 3, 301–330 (2014)

44. Pang, Z., Zheng, L., Tian, J., Kao-Walter, S., Dubrova, E., Chen, Q.: Design of a terminal solution for integration of in-home health care devices and services towards the Internet-of-Things. Enterp. Inf. Syst. 9(1), 86–116 (2015)

45. Hassanalieragh, M., Page, A., Soyata, T., Sharma, G., Aktas, M., Mateos, G., Kantarci, B., Andreescu, S.: Health monitoring and management using Internet-of-Things (IoT) sensing with cloud-based processing: opportunities and challenges, pp. 285–292 (2015)

46. Talukder, A.K., Garcia, N.M., Jayateertha, G.M.: Convergence Through All-IP Networks. CRC Press (2013)

47. Dohr, A., Modre-Opsrian, R., Drobics, M., Hayn, D., Schreier, G.: The Internet of Things for ambient assisted living, pp. 804–809 (2010)

48. Domingo, M.C.: Review: an overview of the Internet of Things for people with disabilities. J. Netw. Comput. Appl. 35(2), 584–596 (2012)

49. Rahmani, A.-M., Thanigaivelan, N.K., Gia, T.N., Granados, J., Negash, B., Liljeberg, P., Tenhunen, H.: Smart e-Health gateway: bringing intelligence to Internet-of-Things based ubiquitous healthcare systems, pp. 826–834 (2015)

50. Soldatos, J., Kefalakis, N., Hauswirth, M., Serrano, M., Calbimonte, J.-P., Riahi, M., Aberer, K., Jayaraman, P.P., Zaslavsky, A., Žarko, I.P., Skorin-Kapov, L., Herzog, R.: OpenIoT: open source Internet-of-Things in the cloud. In: Podnar Žarko, I., Pripužić, K., Serrano, M. (eds.) Interoperability and Open-Source Solutions for the Internet of Things: International Workshop, FP7 OpenIoT Project, Held in Conjunction with SoftCOM 2014, Split, Croatia, September 18, 2014, Invited Papers, pp. 13–25. Springer, Cham (2015)
51. The thing system. http://thethingsystem.com/dev/The-Thing-Philosophy.html
52. Xively. https://xively.com
53. Toulson, R., Wilmshurst, T.: Fast and Effective Embedded Systems Design: Applying the Arm Mbed. Elsevier (2012)
54. Hub of all things. http://hubofallthings.com
55. Arias Fernández, J., Bahers, Q., Blázquez Rodríguez, A., Blomberg, M., Carenvall, C., Ionescu, K., Kalra, S.S., Koutsoumpakis, G., Koutsoumpakis, G., Li, H.: IoT-framework (2014)
56. Calvin-base. https://github.com/EricssonResearch/calvin-base
57. OpenRemote. http://www.openremote.org/display/HOME/OpenRemote
58. ThingWorx. http://www.thingworx.com/platform/
59. Sense tecnic wotkit. http://sensetecnic.com
60. Derhamy, H., Eliasson, J., Delsing, J., Priller, P.: A survey of commercial frameworks for the Internet of Things, pp. 1–8 (2015)
61. Valera, A.J.J., Zamora, M.A., Skarmeta, A.F.G.: An architecture based on Internet of Things to support mobility and security in medical environments, pp. 1–5 (2010)
62. Liu, Y., Zhou, G.: Key technologies and applications of Internet of Things, pp. 197–200 (2012)
63. Heer, T., Garcia-Morchon, O., Hummen, R., Keoh, S.L., Kumar, S.S., Wehrle, K.: Security challenges in the IP-based Internet of Things. Wirel. Pers. Commun. **61**(3), 527–542 (2011)
64. Gan, G., Lu, Z., Jiang, J.: Internet of Things security analysis, pp. 1–4 (2011)
65. Katagi, M., Moriai, S.: Lightweight Cryptography for the Internet of Things, pp. 7–10. Sony Corp. (2008)
66. Babar, S., Stango, A., Prasad, N., Sen, J., Prasad, R.: Proposed embedded security framework for Internet of Things (IoT), pp. 1–5 (2011)
67. Borgohain, T., Kumar, U., Sanyal, S.: Survey of security and privacy issues of Internet of Things. CoRR **abs/1501.02211** (2015)
68. Roman, R., Najera, P., Lopez, J.: Securing the Internet of Things. Computer **44**(9), 51–58 (2011)
69. Mukherjee, A.: Physical-layer security in the Internet of Things: sensing and communication confidentiality under resource constraints. Proc. IEEE **103**(10), 1747–1761 (2015)
70. Ning, H., Liu, H., Yang, L.T.: Cyberentity security in the Internet of Things. Computer **46**(4), 46–53 (2013)
71. Kothmayr, T., Schmitt, C., Hu, W., Brunig, M., Carle, G.: A DTLS based end-to-end security architecture for the Internet of Things with two-way authentication, pp. 956–963 (2012)
72. Raza, S., Shafagh, H., Hewage, K., Hummen, R., Voigt, T.: Lithe: lightweight secure CoAP for the Internet of Things. Sens. J. IEEE **13**(10), 3711–3720 (2013)
73. Li, F., Xiong, P.: Practical secure communication for integrating wireless sensor networks into the Internet of Things. IEEE Sens. J. **13**(10), 3677–3684 (2013)
74. Sethi, M., Kortoci, P., Di Francesco, M., Aura, T.: Secure and low-power authentication for resource-constrained devices, pp. 30–36 (2015)

Part III
MCC and Big Data Paradigm in Smart Ambient Systems

The Art of Advanced Healthcare Applications in Big Data and IoT Systems

Claudia Ifrim, Andreea-Mihaela Pintilie, Elena Apostol,
Ciprian Dobre and Florin Pop

Abstract The goal of this chapter is to analyze existing solutions for self-aware Internet of Things. It will highlight, from a research perspective, the performance and limitations of existing architectures, services and applications specialized on healthcare. The chapter will offer to scientists from academia and designers from industry an overview of the current status of the evolution of applications based on Internet of Things and Big Data. It will also highlight the existing problems and benefits of the IoT for disabled people or people suffering from diseases and the research challenges found in this area.

Keywords Internet of Things · Big data · Analytics · Healthcare · Body sensor networks

The work has been funded by the Sectoral Operational Programme Human Resources Development 2007–2013 of the Ministry of European Funds through the Financial Agreement POSDRU/187/1.5/S/155536.

C. Ifrim (✉) · A.-M. Pintilie · E. Apostol · C. Dobre · F. Pop
Faculty of Automatic Control and Computers, Computer Science Department,
University Politehnica of Bucharest, Bucharest, Romania
e-mail: claudia.ifrim@hpc.pub.ro
URL: https://cs.pub.ro

A.-M. Pintilie
e-mail: andreea.pintilie@cti.pub.ro

E. Apostol
e-mail: elena.apostol@cs.pub.ro

C. Dobre
e-mail: ciprian.dobre@cs.pub.ro

F. Pop
e-mail: florin.pop@cs.pub.ro

1 Introduction

Nowadays, a massive amount of data is being generated and stored in the cloud. In this way its availability is increased, providing a boost of power to analytical and predictive tools. One of the main promises of analytics is data reduction with the primary function to support processing with the help of existing infrastructure. The motivation of this chapter direction comes from data being generated from a variety of sources, such as healthcare industry, communication, messaging and social networks, mobile sensors and many others, rather than a new storage mechanism. Data reduction techniques for Big Data have three perspectives: descriptive analytics, predictive analytics and prescriptive analytics. In the pay-as-you-go cloud environment, the storage can be very expensive.

The perspective of this chapter is represented by: monitoring, analysis and control of environments, based on collected data from a network of sensors.

The Internet of Things (IoT) is a paradigm where every object can be identified and has sensing, networking and processing capabilities. The objects can communicate with each other or with other devices or services available over the Internet. Whitmore et al. [1] Those objects will be ubiquitous and context-aware.

The objective of this chapter is to highlight the current status of the evolution, trends and research on Internet of Things applied in e-Health by examine the literature. In order to achieve our objective, a comprehensive review of the literature, that included conference papers, books and journal articles, was performed.

This chapter is organised as follows: in Sect. 2 we will cover the role of IoT in e-Health, followed by an overview of the Big Data systems prepared for the healthcare applications in Sect. 3. In Sect. 4 we will deal with the existing healthcare applications and in the last section we will expose our conclusions of this study.

2 IoT Solutions for e-Health Systems

2.1 Introduction

The exact meaning of e-Health term varies with the source. There is not a single consensus definition. Some benefits of e-Health extend from established telemedicine systems, others are only practical using a machine-to-machine (M2M) model and assume that patients have access to broadband service.

The World Health Organization defines e-Health as: e-health is the transfer of health resources and health care by electronic means. It encompasses three main areas:

- the delivery of health information, for health professionals and health customers through the Internet an telecommunications;

- the improvement of health services, e.g. through education and trainings for health workers;
- the use of e-commerce and e-business practices in health systems management.

E-health provides a new method for using health resources—such as information, money, and medicines—and in time should help to improve efficient use of these resources. The Internet also provides a new medium for information dissemination, and for interaction and collaboration among institutions, health professionals, health providers and the public [2].

Technology and Health are the two main coordinates when defining the e-Health term. The Internet and all the electronic devices are most frequently used to dissipate the information about health services or other detail regarding this area.

Based on the Internet of Things paradigm we have a lot of new opportunities that are reshaping the e-Health concept on a daily basis.

The applications for e-Health will thrive in the next future using IoT for home and assisted-living environment. There is a noteworthy interest for developing the monitoring systems dedicated to elderly or post trauma patients, including video and voice options. The systems allow the identification of falls and send notification to the medical personnel without any human intervention, by monitoring the automated movement. Since the traditional movement monitoring systems have known issues that trigger false alarms, a more accurate response can be achieved by a combination of voice and video verification in case of any notification or alarm sent by the system.

Another data stream that provides important information for physicians is the monitoring of blood pressure. Also, the remote monitoring of the patients to get readings for the blood glucose, pulse oximetry or hart monitoring is a strong source of important information, especially since the measurements reflects the patients condition on daily basis, under normal conditions, without the stress implied by the visit to the doctor.

The most important thing is that e-Health means also human interactions. This translates into a more dynamic situation where a technical support can be called, by phone, to solve some issues or, more important, the health-care services can now be moved from hospital environment to a patient's context.

The patient's health and fitness information can be remotely monitored by e-Health applications and when critical conditions are detected, alarms are triggered. Also, the e-Health applications can provide remote control of certain medical treatments or monitor of some parameters.

2.2 Context

The IoT can improve people's live and health through automation and augmentation. Its capabilities can save peoples life and improve decision making. It also can help people to adopt a healthy lifestyle based on the informations that personal devices can collect and expose in a simple graphic.

The main technical and managerial challenges in IoT development [3] highlighted by the reviewed literature are:

- Data management
- Data mining
- Privacy
- Security
- Chaos.

2.2.1 Data Management Challenge

All the devices and sensors are generating a big amount of data that must be stored, processed and analyzed.

Compared with general big data, the data generated by IoT has different characteristics, due to the variety of the devices generating it, due to the different types of collected data, of which the most classical characteristics include heterogeneity, variety, unstructured feature, noise, and high redundancy.

Although the current IoT data is not the dominant part of big data, by 2030, the quantity of sensors will reach one trillion and then the IoT data will be the most important part of big data, according to the forecast of HP. A report from Intel pointed out that big data in IoT has three features that conform to the big data paradigm:

- abundant terminals generating masses of data (Fig. 1);
- data generated by IoT is usually semi-structured or unstructured;
- data of IoT is useful only when it is analyzed.

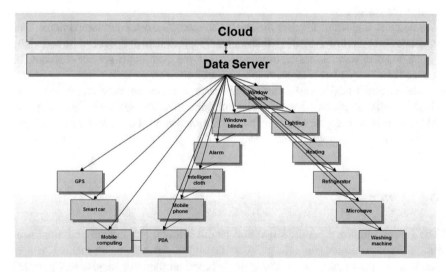

Fig. 1 Illustration of data acquisition equipment in IoT

The architecture of data centers are not prepared in this moment to deal with the heterogeneous nature and big volume of personal and enterprise data [4].

2.2.2 Data Mining Challenge

In order to use the big amount of data (discrete data or stream data) stored from all kind of devices, the use of data mining tools becomes a necessity.

Data needs to be tamed and understood using computers and mathematical models. Traditional data mining techniques are not directly applicable to unstructured data, images and video data. Considering the advanced knowledge necessary in analyzing and interpreting the big amount of data resulted nowadays from multiple sources, we can foresee a future shortage of competent data analysts. McKinsey Global Institute estimated that the United States needs 140,000–190,000 more workers with analytical skills and 1.5 million managers and analysts with analytical skills to make business decisions based on the analysis of big data [3, 5].

2.2.3 Privacy Challenge

Due to the increase number of personal devices (smart health equipment, smart cars emergency systems, all of which provides detailed informations on user location, health and purchasing preferences) that can be traced through the Internet of Things, personal privacy is an important issue to be considered.

The ownership of the data collected from smart objects must be clearly established and the smart objects and reading devices from the Internet of Things should each be equipped with privacy policies [6].

Protecting privacy is often counter-productive to service providers in this scenario, as data generated by the IoT is the key to improving the quality of people's lives and decreasing service provider's costs by streamlining operations. The IoT is likely to improve the quality of people's lives. According to the 2014 TRUSTe Internet of Things Privacy Index, only 22 % of Internet users agreed that the benefits of smart devices outweighed any privacy concerns [7].

2.2.4 Security Challenge

Because most of the Internet of Things devices are typically wireless and may be located in public places, the communication should be encrypted. However, many IoT devices are not powerful enough to support robust encryption [6, 8, 9].

In addition to an encrypted communication, identity management and unique identifiers are another important components of any security model.

Those are some essentials requirements for the IoT success and the security of a device and its generated data used in a monitoring system in the medical area represents a major problem that must be resolved.

2.2.5   Chaos Challenge

The evolution of IoT technologies (chips, sensors etc.) is in a hyper-accelerated innovation cycle, that is much faster than the typical consumer product innovation cycle.

Considering that the standards are still incomplete, there are privacy issues and insufficient security, that the communications is very complex and there is a high number of poorly tested devices, we should be aware that the base on which the IoT evolves is not strong enough to sustain its fast growth. If not carefully designed to consider all the implications, our lives can be easily turned to chaos by all the multiple devices and collaborative applications.

In an unconnected world, a small error or mistake does not bring down a system, but in a hyper-connected world, an error in one part of a system can cause disorder throughout. Smart home applications or medical monitoring and control systems consist of interconnected sensors and communication devices and controllers. If a sensor of a medical monitoring and control system malfunctions, the controller may receive an incorrect signal, which may prove fatal to the patient. It is not difficult to imagine smart home kits such as thermostats and residential power meters breaking down or being attacked by hackers, creating unexpected safety problems. The Internet bandwidth can get saturated with data traffic of proliferating devices, creating system-wide performance problems. A single device may have an insignificant problem, but for the system as a whole, the chain reactions of other connected devices can become disastrous.

To prevent chaos in the hyper-connected IoT world, businesses need to make every effort to reduce the complexity of connected systems, enhance the security and standardization of applications, and guarantee the safety and privacy of users anytime, anywhere, on any device.

2.3   IoT for e-Health Systems

"The Internet meets the physical world" is the new phase of the Internet evolution brought forth by the IoT. The today's few billions of endpoint will exponentially grow in numbers, and this will lead eventually to some scalability issues.

Starting from the traditional Internet who connected computers, the outcome will be now improved by the IoT powered e-Health solutions who will connect people, information, processes, devices and the context. The previously passive and not connected intelligent devices will be connected by the IoT and this connection will bring forth a huge amount of information, that can and will be used for algorithms bases actionable decisions. This new information will be based on strict evidence and will strongly impact the health-care services and the way these services are provided.

We all know that the health-care system of today is struggling to provide viable solutions for population. Through the many opportunities brought by the Internet of Things for e-Health, the wellness of the population will increase and the strain points of the health-care system will be reduced in time. The proactive monitoring of the

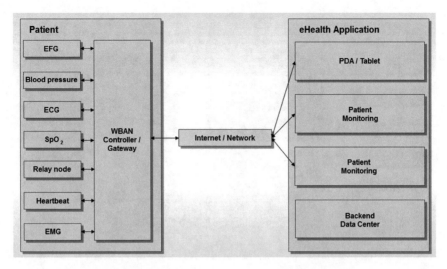

Fig. 2 Conceptual layout of the ASSET testbed—example of IoT architecture

patients, preventive medicine, follow-up care and permanent care disease management are only few of the most promising use of the connected e-Health cases.

The opportunities, changes and complexities of e-Health enabled by the IoT is significant and can be characterised by:

- number of devices interconnected
- various type of devices, applications and processes that are interacting
- number of devices that will generate information
- number of decision making points
- number of entry points into the system

The IoT devices could be categorized into two major classes:

- the current smart phones, tablets, and laptops (see Fig. 2)
- a set of interconnected sensors (e.g. Smart Cities, Manufacturing Automation, etc.) (see Fig. 3)

Based on how the devices are connected to the patient we can classify them into:

- implantable
- wearable

Those devices could be connected on a need basis, always connected or unconnected.

Based on how the device is connected to the network, the devices can be classified into:

- wired
- wireless
- non-connected

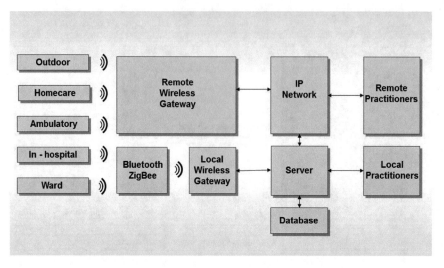

Fig. 3 Typical architecture of wireless sensor networks in health-care applications

Based on the timeframe when data is generated, the devices can be classified into:

- real-time streaming (e.g.: patient monitoring)
- discrete data sources (e.g.: oximeter—generates data at predefined intervals)
- one-time data source (e.g.: Magnetic resonance imaging scanning)

Considering the patient needs, a dedicate monitoring system could track data for a long term trend in medical condition or only for it's treatment period. The monitoring system could also send data to a processing element every x hours and even accept a delay of few seconds. The total loss of data for an entire measurement period where that timeframe is a very small fraction of the total collection time would be insignificant. In contrast, a device that actively monitors a serious, life-threatening condition that requires specific action to be taken within a given time period or where a single-occurrence is of importance, would impose tight requirements on the collection and dissemination of the data. In that case, it would not be acceptable to delay or lose a single packet of data.

Based on how the device is used, we can have devices used by:

- a single person—dedicated
- a limited group of people—shared with a limited group
- a group of people—shareable with a wider population.

As we can see, IoT is an important source of big data. Big data could come form medical care, traffic, agriculture, industry, etc.

According to the processes of data acquisition and transmission in IoT we can divide the IoT's network architecture into three layers:

- the sensing layer
- the network layer
- the application layer

In healthcare systems, IoT allows a flexible patient monitoring: the patient's status can be monitored from home by doctors and nurses. This has an impact on the average daily cost for patients and on the patient's comfort.

BodyGuardian Remote Monitoring System [10] can detect non-lethal arrhythmias in an ambulatory settings, by using a body-worn sensor, a mobile phone and the preventice care platform. The BodyGuardian's architecture is explained in [11]. The system uses a microprocessor, memory and a Bluetooth chip. After gathering the signals, the sensor pairs with a dedicated mobile device and the ECG, heart rate, respiratory rate and activity levels are transmitted via mobile phone through a cellular network and, ultimately, to the cloud. The Preventice Care Platform had the role to allow all the ECG, heart rate, respiratory rate and activity data to be displayed for physician evaluation, either on a web portal or an iPad application that is dedicated to this purpose.

Zebra Technologies and Healthcare Solutions [12] was concerned with improving the hospital's management and security. They defined a set of applications which facilitates resource planning, offers real-time visibility into patient's records, medications, equipments and provides a IoT platform for patient care, by integrating all devices connected to the hospital's network. They provided a patient identification schema based on barcoding that reduced medication errors by 65–86%, according the US Department of Veterans Affairs.

Even if both patients and providers benefit from the presence of IoT in healthcare domain, dangers such as security policies should be considered. Although security policies are well defined in a closed environment such as a hospital or medical care system, researchers are trying to find solutions for *BYOD* (Bring Your Own Device) [13]. In a BOYD system, nurses and doctors and other caregivers can use their own tablets, laptops, smartphones to have access to the hospital's resources. By using smartphones doctors can check their appointments, make new ones, access patients files, write prescriptions etc. Whether the use of external devices is supported or not by the hospital, BOYD represents a major security threat to the privacy and security of patients. In order to minimize the risks, many businesses are implementing BYOD policies that can further be adopted by healthcare systems.

Data protection is one of the main concerns in e-Health. Both Europe and US defined regulations concerning data protection in electronic health area: The e-Privacy Directive 2002/58/EC [14] in Europe and Health Insurance Portability and Accountability Act (HIPPA) [15] in USA. The EU Law definition for personal data is information "identified or identifiable natural person" and it states the rights and responsibilities of the "data controller" (the person/organization who has access and processes data) as well as the "data subject", the person whose data is being processed. HIPAA defines who can have access to personal health data, and it is a subset of a series of federal protection laws. In US data protection is reglemented in different states and by different federal laws.

Another danger is the ability to overwhelm doctors with information which will impact their main goal, taking care of the patient. ***OpenNotes patient engagement*** [16] project aims to provide patients access to their caregivers notes and possibility to contribute to them. According to John Mafi M.D. a member of the Open Notes program, the project's goal is to make "patients feel more in control of their own care and to correct factual errors in the record". Its purpose is also to improve communication with patients and other doctors. The main drawbacks of OpenNotes that doctors feared were:

- the struggled with a one-size-fits-all: the note had to include the doctor's opinions, suggestions, prescriptions, the billing information, previous patient's analysis, other doctor's prescriptions and the patient's input;
- addressing potentially sensitive issues, for example the doctors had to change "obesity" to "body mass index";
- unmanageability due to the number of actors (nurses, medical students, patients etc.) involved; the progress note documents have a predefined standard, while open notes can became hard to maintain when multiple actors review them.

Microsoft Kinect Sensor has also been tested in healthcare applications. It has been used in detection of stereotyped hand flapping movements in autistic children [17]. The system used the Dynamic Time Warping (DTW) in order to detect motor behaviours in children with Autism Spectrum Disorder (ASD). It was tested in a closed laboratory environment and in two schools with children having ASD. Multiple actors were involved: children, engineers, doctors, teachers. The system facilitated the identification of behavioural patterns when studying interaction skills in children with ASD.

Healthcare issues imposed the development directions of IoT. Drug management is one of the biggest challenges the industry faces today. According to Forbes,[1] the average cost to create and develop an approved drug is around 55 millions dollars. Electronics development leaded to major improvements in diagnostic applications in medical field. Such an example are "smart" pills. For example, in article [18] is presented an improved communication solution between an external device and an internal electronic pill while in [19] is presented a new system for local drug delivery. The concept of telemetry communication is based on master-slave model. The pill can be directed by the master in the organism and it can be assisted to record temperature, measure body pH, drug delivery etc. The pill's hardware can consists of a bit processor, memory and external peripheries [20]. IoT devices and process may have an impact on cost management in this area. WuXi PharmaTech and TruTag Technologies [21] are two companies which focused on such "smart" pills.

AdhereTech [22] is another healthcare technology company interested in developing smart healthcare devices. In order to improve medication adherence, the company focused on developing a smart pill bottle. The bottle has the role to verify whether patients are taking the medication, if the medication doses are correct and

---

[1]http://www.forbes.com/forbes/welcome/.

to generate alerts in case of a dose miss. Patients can also provide feedback via text or phone calls providing the reason why they missed their dose. For example, if a patient misses a dose due to side effects, AdhereTech can gather data in the first days of treatment, instead of having access to patient's feedback on his/her next visit to the doctor [23].

The main challenge of an e-Health system based on an IoT architecture is to support this wide range of device types in a variety of care needs and settings.

"Prevention must become a cornerstone of the health-care system rather than an afterthought. This shift requires a fundamental change in the way individuals perceive and access the system as well as the way care is delivered. The system must support clinical preventive services and community-based wellness approaches at the federal, state, and local levels. With a national culture of wellness, chronic disease and obesity will be better managed and, more importantly, reduced" [24].

3 Big Data Platforms for Healthcare Applications

Big Data services can be used to improve the quality of healthcare technologies. Nowadays e-Health data flows have generated up to 1000 peta bytes of data [25] and it is estimated to reach about 12 ZB by 2020. As presented above, data from various sources such as electronic medical records (EMR), mobilized health records (MHR), personal health records (PHR), external monitors, sensor, genetic sequencing, smart devices is integrated in order to enhance the medical research and the diagnostic process.

In [26] several technical requirements of providing easily accessible big data pools in healthcare domain were addressed, such as: semantic annotation, data digitalisation, sharing, privacy, security and quality. Based on their study, the availability of healthcare data mainly depends on the implementation of standards and platforms which can specify and provide secure and reliable data exchange.

Figure 4 depicts a platform that supports access from different actors (individual outlets, test facilities, government agencies) and provides access to data (EHR, MHR, EMR), as well as to data mining tools.

Ayasdi is an enterprise software company that sells big data analytics technology to organizations looking to analyze high-dimensional, high volume data sets. Organizations and companies have deployed Ayasdi's software across a variety of use cases, including the development of clinical pathways for hospitals, fraud detection, trading strategies, oil and gas well development, drug development and national security applications [27].

ClearDATA is a company which provides cloud computing for Healthcare.[2] The platform is HIPPA compliant and has more than 310,000 of healthcare professionals who are using it. The company identified the main concerns in healthcare industry,

---

[2]https://www.cleardata.com/.

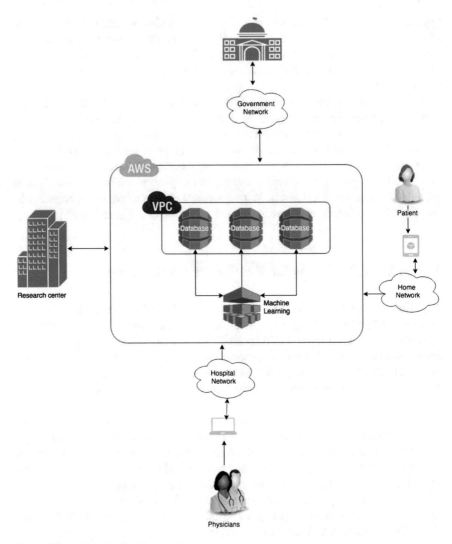

Fig. 4 Illustration of e-Health data platform

such as: (1) fragmented infrastructure, that is compounded by personnel and cost constraints, (2) fear of compromised data, (3) disparate systems, (4) security concerns and (5) compliance requirements. They provide:

- a Cloud Computing Platform dedicated to healthcare
- a Cloud Platform for AWS, backed by ClearDATA's BAA
- optimizations for critical healthcare workloads
- a Multi-Cloud Management

- security requirements based on HITRUST standards and the Common Security Framework
- HIPPA risk remediation services

Carestream[3] provides solutions for medical imaging. It also provides a Software-as-a-Service (SaaS) solution called Vue Cloud, which represents a secure, single access point to medical images. The portal grants customers permissions to create communities of approved radiologists, specialists or referring physicians. It also provides a virtual workflow to view exams, access tools, review a patient portfolio, collaborate on diagnoses and treatment, obtain second opinions, enable sub-specialty reading or view real-time department performance.

4 Healthcare Applications: A Brief Overview

Nowadays, the application areas based on an IoT architecture can be limited only by peoples imagination.

In the next future, applications based on IoT and sensors will increase exponentially due to the huge variety of the objects and sensors that can provide useful information regarding the environment. Another important fact is that those applications could improve people's lifestyle and wellness.

Healthcare and medical data are evolving and continuously growing complex data, based on a huge amount of various information values. The unlimited potential for storing of Big Data will lead to dramatically increased of saving, processing, querying, and analyzing medical data.

Aetna Life Insurance Company[4] took a sample comprising of 102 patients from a total of thousand patients in order to complete an experiment designed to predict the recovery in of patients diagnosed with metabolic syndrome. On a period of three consecutive years 600,000 laboratory test results was scanned during an independent experiment and 180,000 claims using a series of detections tests results of patients with metabolic syndrome.

Furthermore, the final results were compiled into a high personalized treatment plan assessing the risk factors and the recommended treatment plan for the patients. Based on this, doctors may decrease morbidity by 50 % in the next 10 years assisting and helping patients to lose weight by five pounds, or advising patients to reduce the total triglyceride in their bodies if the sugar content in their bodies is over 20 [28].

The Mount Sinai Medical Center in the U.S. utilizes technologies of Ayasdi, a big data company, to analyze all genetic sequences of Escherichia Coli, including over one million DNA variants, to investigate why bacterial strains resist antibiotics. Ayasdi's uses topological data analysis, a brand-new mathematic research method, to understand data characteristics [28].

---

[3]https://www.carestream.com/provides.

[4]https://www.aetna.com/.

HealthVault of Microsoft, launched in 2007, is an application of medical big data launched in 2007. HealthVault is available as a Web Service and as a mobile application for iOS and Windows Phone. It can tap into a huge array of medical-grade devices and applications, such as glucose monitors and blood pressure cuffs, to automatically import data into one location. Its goal is to manage individual health information in individual and family medical devices. In addition, it can be integrated with a third-party application with the software development kit (SDK) and open interface [28, 29].

The project presented in [30] used gene expression data with the goal of predicting a relapse, focusing on patients with early stages of colorectal cancer (CRC), by developing a gene expression classifier. In order to gather relevant data, the study was conducted in 3 different institutions for testing and even different countries for validation. The patients involved in the study were split into 2 groups: testing and validation. The study has been conducted from 1983 to 2002. The doctors managed to gather 33,834 gene probes. The results were used in predicting Relapse Free Survival (RFS)—the length of time that the patient survives without any signs of cancer—giving a hazard ratio (ratio of relapse rates between predicted-relapse and predicted-RFS patients) of 2.69, with a confidence interval at 95 %.

The Soprano project [31] has a medication alert which collects events from a medication dispenser about missed medication taking, selects the most appropriate communication channel and delivers the alert to the assisted person (AP). In case the AP does not acknowledge the alert, the process is escalated and a SMS is sent to an informal caregiver.

The Pip's project [32] objective is to include the entire set of business processes, professional practices, and products applied to the analysis and preservation of the citizen's wellbeing. Pips provides a Personalized Nutritional Support which offers: personalized nutritional advisor, shopping list, product information, product check according to personal profile, etc. The goal of this service is to inform and advise its users about a healthy lifestyle and to determine them to choose healthy nutritional habits by tailoring them to the user's needs.

Security in healthcare applications of sensor networks is a major concern. Since healthcare applications of sensor networks are almost similar to WSN application environment, most of the security issues are also similar and hence comparable [33].

The security issues can be related to system security and information security. Authors in [34] classified the threats and attacks into two major categories:

- **passive**—it may occur while routing the data packets in the system
- **active**—those type of threats are more harmful, criminal minded people may find the location of the user by eavesdropping. This may lead to life threatening situation.

5 Conclusions

The objective of this chapter is to highlight the current status of the evolution, trends and research on Internet of Things applied in e-Health by examine the literature.

IoT and Big data analytics have the power to transform the way healthcare performs nowadays. Using this power, healthcare organisations and industry can gain insight informations from data repositories and make informed decisions, not suppositions.

IoT requires real time data analysis services that can cope with huge amounts of data. These services raise new challenges from the Big Data point of view: we claim that fully distributed frameworks are required to achieve scalability.

Issues like security, privacy and standards for IoT and Big data analytics should also be considered. Applications in healthcare and data analytics have the potential to become more mainstream and accelerate their maturing process.

The normal trend of sensor device design is that they have little external security features and hence prone to physical tempering. This increases the vulnerability of the devices and poses tougher security challenges.

In the table below, we present the attacks which can occur in any health-care system using wireless sensor networks.

Attack assumptions	The risks to WBAN	Security requirements
Computational capabilities	Data modification impersonation	Data integrality authentication
Listening capabilities	Eavesdropping	Encryption
Broadcast capabilities	Replaying	Freshness protection

Acknowledgements The research presented in this chapter is supported by the following projects: "*KEYSTONE—semantic KEYword-based Search on sTructured data sOurcEs* (Cost Action IC1302)"; *CyberWater* grant of the Romanian National Authority for Scientific Research, CNDI-UEFISCDI, project number 47/2012; *clueFarm*: Information system based on cloud services accessible through mobile devices, to increase product quality and business development farms—PN-II-PT-PCCA-2013-4-0870; *MobiWay*: Mobility Beyond Individualism: an Integrated Platform for Intelligent Transportation Systems of Tomorrow—PN-II-PT-PCCA-2013-4-0321.
We would like to thank the reviewers for their time and expertise, constructive comments and valuable insight.

References

1. Whitmore, A., Agarwal, A., Xu, L.: The internet of things—a survey of topics and trends. Inf. Syst. Front. 239-256 (2014). Springer, US
2. World Health Organization. http://www.who.int/trade/glossary/story021/en/. Accessed Sept 2015
3. Lee, I., Lee, K.: The internet of things (IoT): applications, investments, and challenges for enterprises (2015)
4. Gartner: Gartner says the internet of things will transform the data center (2014). http://www.gartner.com/newsroom/id/2684616. Accessed Aug 2015
5. Manyika, J., Chui, M., Brown, B., Buhin, J., Dobbs, R., Roxburgh, C., Byers, A.H.: Big Data: The Next Frontier for Innovation, Competition, and Productivity. McKinsey Global Institute, USA (2011)
6. Bandyopadhyay, D., Sen, J.: Internet of things: applications and challenges in technology and standardization. Wirel. Pers. Commun. **58**(1), 4969 (2011)
7. TRUSTe: TRUSTe Internet of Things privacy index–US edition (2014). http://www.truste.com/resources/privacy-research/us-internet-of-things-index-2014/
8. Roman, R., Alcaraz, C., Lopez, J., Sklavos, N.: Key management systems for sensor networks in the context of the internet of things. Comput. Electr. Eng. **37**(2) 147–159 (2011)
9. Yan, T., Wen, Q.: A trust-third-party based key management protocol for secure mobile RFID service based on the Internet of things. In: Tan, H. (ed.) Knowledge Discovery and Data Mining, AISC 135, pp. 201–208. Springer, Berlin (2012)
10. Body Guardian. http://www.preventicesolutions.com/technologies/body-guardian-heart.html. Accessed Feb 2016
11. New applications in EP: the body guardian remote monitoring system, EP Lab Digest, vol. 12, issue 11 (2012)
12. Zebra Technologies and Healthcare Solutions. https://www.zebra.com/gb/en/solutions/healthcare-solutions.html. Accessed Feb 2016
13. Lennon, R.G.: Changing user attitudes to security in bring your own device (BYOD) and the cloud, Tier 2 Federation Grid, Cloud and High Performance Computing Science (RO-LCG), 2012 5th Romania, Cluj-Napoca
14. The e-Privacy Directive 2002/58/EC. http://eur-lex.europa.eu/LexUriServ/LexUriServ.do?uri=CELEX:32002L0058:en:HTML. Accessed Feb 2016
15. Health Insurance Portability and Accountability Act (HIPPA). http://www.hhs.gov/hipaa/. Accessed Feb 2016
16. Open Notes. http://www.opennotes.org/. (Accessed in February 2016)
17. Goncalves, N., Costa, S., Rodrigues, J.: Detection of stereotyped hand flapping movements in autistic children using the kinect sensor. In: 2014 IEEE International Conference on Autonomous Robot Systems and Competitions (ICARSC) May 14–15, Espinho, Portugal
18. Fawaz, N., Jansen, D.: Enhanced telemetry system using CP-QPSK band-pass modulation technique suitable for smart pill medical application. Wireless Days, 2008. WD '08. 1st IFIP, 1–5, 24–28 November 2008
19. Goffredo, R., Accoto, D., Santonico, M., Pennazza, G., Guglielmelli, E.: A smart pill for drug delivery with sensing capabilities. In: Engineering in Medicine and Biology Society (EMBC). 2015 37th Annual International Conference of the IEEE, 25-29 Aug 2015, pp. 1361–1364
20. Jansen, D., Fawaz, N., Bau, D., Durrenberger, M.: A small high performance microprocessor core SIRIUS for embedded low power designs, demonstrated in a medical mass application of an electronic pill. In: Embedded System Design Topic, Techniques and Trends, pp. 363–372. Springer (2007)
21. WuXi PharmaTech. http://www.wuxiapptec.com/press/detail/284/18.html. Accessed Feb 2016
22. AdhereTech. https://adheretech.com/. Accessed Feb 2016

23. DeMeo, D., Morena, M.: Medication adherence using a smart pill bottle. In: 2014 11th International Conference and Expo on Emerging Technologies for a Smarter World (CEWIT), pp. 1–4, 29–30 Oct 2014

24. BCC Research, Preventive Healthcare Technologies, Products and Markets. http://www.bccresearch.com/market-research/healthcare/preventive-healthcare-technologies-hlc070a.html. Accessed June 2015

25. Liu, W., Park, E.K.: Big data as an e-health service. In: 2014 International Conference on Computing, Networking and Communications (ICNC), pp. 982–988, 3–6 Feb 2014

26. Zillner, S., Oberkampf, H., Bretschneider, C., Zaveri, A., Faix, W., Neururer, S.: Towards a technology roadmap for big data applications in the healthcare domain. In: 2014 IEEE 15th International Conference on Information Reuse and Integration (IRI), pp. 291–296, 13–15 Aug 2014

27. Ayasdi. https://en.wikipedia.org/wiki/Ayasdi. Accessed Sept 2015

28. Chen, M., Mao, S., Liu, Y.: Big data: a survey. Mob. Netw. Appl. **19**(2), 171–209 (2014)

29. HealthVault. https://www.healthvault.com/ro/en. Accessed Sept 2015

30. Salazar, R., Roepman, P., Capella, G., Moreno, V., Simon, I., Dreezen, C., Lopez-Doriga, A., Santos, C., Marijnen, C., Westerga, J., Bruin, S., Kerr, D., Kuppen, P., van de Velde, C., Morreau, H., Van Velthuysen, L., Glas, A.M., Vant Veer, L.J., Tollenaar, R.: Gene expression signature to improve prognosis prediction of stage ii and iii colorectal cancer. J. Clin. Oncol. **29**(1), 1724 (2011)

31. Virone, G., Sixsmith, A.: Monitoring activity patterns and trends of older adults. In: EMBS 2008. 30th Annual International Conference of the IEEE, 20–25 Aug. 2008, pp. 2071–2075. Engineering in Medicine and Biology Society (2008)

32. PIPS personalised information platform for life and health services. http://www.hsr.it/wp-content/uploads/sc/schedaprogettopips.pdf. Accessed Feb 2016

33. Al Ameen, M., Liu, J., Kwak, K.: Security and privacy issues in wireless sensor networks for healthcare applications. J. Med. Syst. **36**(1), 93–101 (2012)

34. Ng, H.S., Sim, M.L., Tan, C.M.: Security issues of wireless sensor networks in healthcare applications. BT Technol. J. **24**(2), 138144 (2006)

A Smart Vision with the 5G Era and Big Data—Next Edge in Connecting World

Lokesh Kumar Gahlot, Pooja Khurana and Yasha Hasija

Abstract Last decade has seen a major transformation in the mobile computing and wireless technologies. And with the development in the Cloud computing, Mobile cloud computing (MCC) is likely to be the next phase of innovation in the telecommunication world. Mobile cloud computing (MCC) services are expected to be a source of quick development with huge capabilities of 5G mobile network. To revise the new opportunities developed by this major, the demand for intelligent, efficient and secure network will continue to increase. So the 5G technology is expected to be heart of next generation in the digital world. In this chapter we will cover a few aspects of 5G covering the challenges and advancements in the existing and emerging communication technologies; mobile cloud computing architecture and how 5G along with Big Data will lead to an era of smart world where everything will be connected to everything in smart and secure way.

Keywords 5G era · 5G architecture · 5G challenges · Mobile Cloud Computing · Device to Device Communication (D2D) · Network Function Virtualisation (NFV) · Software defined networks (SDN) · Massive MIMO

1 Introduction

Mobile cloud computing (MCC) is one of the most rapidly growing fields in the cloud technologies [1]. Mobile cloud computing (MCC) services are expected to be a source of quick development with huge capabilities of 5G mobile network. To deploy the new opportunities developed by this major, network must be intelligent, efficient and secure. So the 5G technology is expected to be heart of next generation in the digital world. To ensure higher levels of mobile user experiences, a huge intelligent secure network is needed.

L.K. Gahlot · P. Khurana · Y. Hasija (✉)
Delhi Technological University (Formerly Delhi College of Engineering),
Shahbad Daulatpur, Main Bawana Road, Delhi 110042, India
e-mail: yashahasija@gmail.com

© Springer International Publishing Switzerland 2017
C.X. Mavromoustakis et al. (eds.), *Advances in Mobile Cloud Computing and Big Data in the 5G Era*, Studies in Big Data 22,
DOI 10.1007/978-3-319-45145-9_7

5G will be an evolution in mobile communication technology with broadband and other different networks. It will have some unique capabilities with specific network and services to offer. It will be a sustainable network with unique technology. It will be a key factor in mobile communication technology. By 5G, mobile communication technology will have a tremendous growth platform in network communications technology [2]. 5G will have a capacity to connect a massive numbers of devices, clones or others. It will be a key impact for the Internet of things to provide networking platform. Some critical services that will require quick action associated to 5G include —reliability, low latency and scope at which network can be handled in a very secure manner using its native support. Cost reduction, automation and optimization will be the main stakeholders in 5G infrastructure.

The recent emergence of the cloud computing technology as a new field of development in information technology and network infrastructure development has transformed the way information is processed, stored and transmitted on a large-scale. In cloud computing, data is stored in the clouds on the Internet, and accessed or cached (viewed) by clients at desktops or portable devices etc. through the internet. Cloud computing is capable of providing highly scalable and reliable computation capability over the internet. Therefore the cloud computing technology is emerging as the key factor in IT infrastructure development, and also promises to achieve economies growth of deployment in operations and IT solutions.

5G systems will support a set of wireless interfaces in the air as:

1. Cellular and satellite solution
2. Current ongoing access schema
3. Ultradense network with small cells

Ultradense network with small cells will be seamless and will involve handling interface between wireless technologies [2]. It will be the main feature of 5G. Also the use of heterogeneous wireless technologies like Radio Access Technologies will further define its scope with increment in reliability and availability. Backhaul network techniques with mitigation and installation will be needed to deploy ultradense network with small cells.

5G will be based on software driven approach which majorly will rely on the following techniques:

1. Software defined Network (SDN)
2. Network Function Virtualization (NFV)
3. Mobile Edge Computing (MEC)
4. Device to Device Communication

It will be easy to use and fully optimize in terms of network management. The main challenges in optimization will be:

1. Complex Business Objective
2. End-to-End Energy Consumption
3. Quantitative and Qualitative experience with privacy and behaviour

Following are the main key drivers of 5G [3]:

1. Wider approach in terms of access technology—5G will utilize all its spectrum block and technologies to give the best optimized solution.
2. Massive Capacity
3. Heterogeneous network
4. Numerous number of connections
5. Ultra fast speed
6. Security
7. Mobility

1.1 Key Impacts for 5G

A very fast growing telecom sector and the current scenario of technical growth in mobile network have completely transformed the way communication and information transfer happens. Further development in technologies and their implementation have enabled us to redefine the entire landscape of the telecommunication sector which in turn has led to the global ICT growth and expansion with the facility to connect anytime—anywhere with people and communities. People can now communicate and share knowledge, information in a flexible, reliable and secure manner.

Figure 1 shows how the evolution has taken place in communication technology from 1G to 4G and now from here we will be heading towards the evolution of 5G.

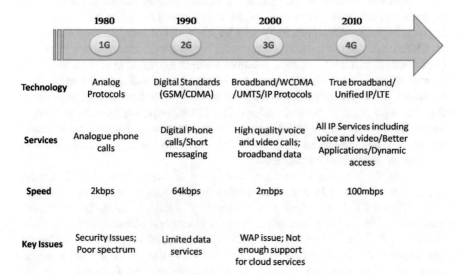

Fig. 1 Evolution of wireless technology from 1G to 4G

The evolution in the internet world will be driven by approach of 5G itself. This evolution will include:

1. The next generation ultra dense network infrastructure with reconstructed and re-designated mobile network technology
2. Massive network and cloud structure that will draw attention towards delivering vast services related to speed, data traffic, connectivity and security.

Essentially 5G will focus on two fundamental aspects:

1. massive capacity
2. massive network or connectivity

Massive capacity will allow users to use connection in a faster way between end users and the network to deliver services. It will be so fast such that user will not feel any distance between the client and the connected machine. More massive capacity will give more flexibility to manage connections in a better way and this will be made possible by the adoption of device to device communication or machine to machine (M2M) service innovations and interactions.

Mobile cloud computing (MCC) with 5G will revolutionize in such a way that it will provide global services in a heterogeneous manner with revolutionary aspects of scalability, availability, privacy, reliability, massive network and capacity. The mobile devices will also revolutionize along with the network revolution so that they can support voice calls as well as huge smart network connectivity. These devices can integrate the mobile network scenario to cloud computing technology.

According to some studies conducted by Cisco IBSG and IDC, average smartphone market grew by about 45 % in 2014, and a great percentage of worldwide population (80 %) has accessibility to the mobile world, which shows a huge impact to mobile world. Since the mobile phone market is growing and expanding rapidly and also the computation innovations in mobile devices continue to grow, huge network connectivity with large amount of computational space with speed, flexibility, accuracy and privacy is indispensable. All innovations in devices like smartphones, tablets, laptops, palmtops brought into mind to act as a host in the internet of world which needs massive network capacity and a service to handle it accurately. So cloud computing is the phenomenon that can provide an approach to manage huge capacity network safely. Cloud computing architecture provides a modulated structure to enable or access on demand network from a pool of shared resources. These resources may be in any form like storage services, network services, or application management.

The Fig. 2a, b shows hierarchical description of Technologies in 5G communication & layered architecture of communication using cloud resource respectively. Figure 2a, b also shows a mapping of technical hierarchy to layered architecture. Figure 2a depicts that in the present scenario different technologies are handled in different infrastructures. Also different infrastructures come under different levels of technological depth like cloud, network etc so all hardware devices come under client infrastructure as these are end user devices, as shown in Fig. 2a.

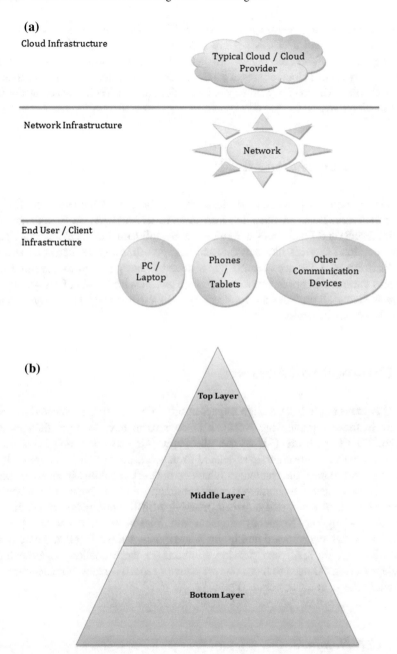

Fig. 2 **a** Hierarchical description of technologies in 5G communication. **b** Layered architecture of communication using cloud resources

Thus we can say the bottom most layer in Fig. 2b demonstrates end user level hierarchy. These end user devices are connected to network which is mid level hierarchy (Fig. 2b—middle layer). All network level operation comes under this level. Finally typical cloud comes at top level hierarchy which is the area of interest in this chapter. All cloud related aspect (Technical or Non Technical) comes in this level of hierarchy.

1.2 Cloud Computing Models

The cloud computing models are described in three base models according to service delivery. These are named as Infrastructure as a service (IaaS), Software as a Service (SaaS) and Platform as a Service (PaaS). Infrastructure as a Services (IaaS) offers the cloud services like network and storage infrastructure, hardware resources. Platform as a Services (PaaS) offers development platform to developers. It provides various APIs to developers. Software as a Service (SaaS) offers services to access the cloud. It provides the applications installed on cloud so that users can use through various browsers.

2 Challenges and Advances

As 5G is expected to have a huge impact in our daily life, and is expected to grow rapidly in the communicating world, it is obvious to have some challenges and blockers in the pathway [4]. As per the study [4], network traffic is going to increase rapidly, so the challenges related to network traffic will be obvious. It is, thus, expected to meet the challenges related to network capability in terms of traffic volume, transmission technologies and communication methods. The advancements in terms of small cell network, massive MIMO, millimetre wave communications, orthogonal frequency division multiplexed (OFDM) access passive optical networks, coordinated multipoint (CoMP) adopted small cell and macro cell BS (MBS) [5, 6, 7, 8] can provide solutions to these challenges. Essentially developments in 5G need to focus on two primary aspects—one is network capacity and other one is network connectivity.

2.1 Cell Numbers

The numbers of devices are increasing rapidly, so cell numbers should increase accordingly. The increment in cell number will be a logical way to improve the capacity of system with an approximate linear ratio as the signal to interference noise ratio is guaranteed. To increase the number of cells, some of the aspects need to be resolved because of the limitation of resources and cost. Cost, cell size, compatibility, cell management and interference are some of the aspects that need to be considered while increasing the number of cells. So now, new technological innovations like Small cell technology need to be adopted to increase the number of cells by order of magnitude.

2.2 Energy Consumption

Using the current resources like Base station systems, it is not possible to decrease the energy consumption in current technology which otherwise should be decreased in 5G. By using some of the technical frameworks, it has now become possible to decrease the energy consumption to some extent. After the use of bandwidth in current scenario by 1G, 2G or 3G (Spectrum Allocated) in a most efficient way, a new potential Spectrum will be required in the future. And also it is expected to be above 3 GHz or even higher. This provision is suitable from local aspect to increase the capacity.

2.3 Cost Factor

Cost reduction is a bit hard because the baseband requirements are based on the physical parameters such as high mobility and larger radius. The cost of Baseband and RF can however be reduced by the introduction of local-based IMT as the baseband and RF requirements can be relaxed because they need low mobility and small cell radius. In small cell technology both development cost and maintenance cost can also be reduced significantly.

As we discussed above, the main fundamental aspects necessary to develop 5G networks are as follows:

1. Massive capacity and massive connectivity
2. Platform capabilities to support capacity and connectivity
3. Support for diversity of users, application and services
4. Efficient use of available spectrum

The network capacity is expected to increase by 2020, thus it is important to keep the following aspects in mind:

1. Spectrum Band availability according to local law and Scenario there.
2. Use of new IMT band to achieve 10 Gb/s for individuals

To maximize the spectrum usability, in other words the efficient use of spectrum, it is important to use the entire spectrum possible and all technologies that can map these requirements. All spectrum access and air interface technologies will be required that are capable of mapping service requirements to the best suitable combinations of frequency and radio resources. To integrate, there will be need of cloud computing technologies with SDN. The integration of these two aspects (cloud computing technologies and SDN) will facilitate innovation of various new mobile network technologies with better quality of service (QoS) and less energy consumption.

2.4   Cloud Related Issues

After the comprehensive review of proposed architecture of Mobile Cloud Computing (MCC), the following issues and challenges come into picture:

1. Privacy and Security
2. Cloud Service Architecture related

Privacy and Security is the major issue when the data is moved to cloud. The movement of the data should be secure enough that it cannot be stolen. Privacy should be maintained between users so the privacy rights cannot be violated. There should be no data loss and data integrity should be maintained. It should be protected from any type of attacks like virus or malware or any misuse of access rights. So, there is a need for system security at server; network security, access and authentication, protection at storage level [9].

Cloud Service Architecture Related—Security related to mobile user, data access efficiency, inter-operability, cost and pricing issues are some of the issues in mobile cloud computing. Some model related issues that come under architecture are outlined below:

- IAAS model

  - virtual machine and repository security
  - network security

- PAAS model

  - API & language related security

- SAAS

  - Data security

2.5 Challenges in Technologies Provisioning the Internet of Things (IoT)

The connectivity through the internet is no longer restricted to the digital devices only. With the advent of Internet of Things (IoT), not only the devices but also other entities related to devices are now connected the internet. IoT has enabled humans and entities to have access and control over all the entities on the internet throughout the world anytime/anywhere.

However, there are certain limitations in the technologies provisioning IoT and are discussed as followed:

- It requires flexible architecture that maximizes the interactivity in distributed heterogeneous system of applications, environment and devices. It should support the following features:

- Multimodal centralized flexible architecture with effective routing and synchronisation
- Interoperability
- Scalability
- Interactivity

- IoT suffers from constraints of power and energy storage technologies that are important for energy efficient applications and services. So energy requirement should be key essential in IoT [10].
- The current network and communication technology for IoT is also limited in terms of design architecture, performance, reliability, privacy and security. Thus, IoT require new technological innovations like advance network management that supports secure, private and trusted interactions between things and the Internet.
- Providing globally unique identifiers (or unique digital names) to entities involved in the communication over the internet and specifying a digital unique identity (DUID) for each communication between these entities is yet another challenge [11].
- There is a need for more effective data storage and caching techniques to manage the rapidly growing data from the users and devices connected through the internet.
- Mining large amount of data generated for providing better services is not an easy task. Thus, a software-platform with high-performance is required.
- Supporting the design space of IoT using the conventional hardware systems is not sufficient. Thus, the adaption of new hardware technologies specifically for IoT is required.

2.6 Necessary Evolutionary Key Points in the Development of 5G

- Advancements in the waveform technologies along with advances in coding and modulation algorithms are essential for continuous improvements in spectral efficiency. These advancements will result in scalable feature for massive network connectivity and will also reduce the latency.
- Baseband and RF technological architecture will enable computational intensiveness. And it incorporates the adaptivity feature for the new air interfaces which are required in network evolution. Advanced RF domain processing will give benefits to the efficient and flexible usage of spectrum.
- A more advanced baseband computation will be required for such complex constraints of new technological solutions like mass-scale MIMO, small cell technology etc. The single-frequency full duplex radio technologies will be a massive contributor to the communication world that will result in increment in spectrum and energy efficiency along with cost reduction.
- Backhaul network designs with small cell system will enable the ultra dense networking of radio nodes. Self-organization will be available for spectrum blocks in backhaul networks. This technology will impact in high frequency spectrum radio accessing.

2.7 Software Defined Network

According to the Open Networking Foundation (ONF), the software defined network (SDN) is:

- A direct programmable approach as the network control is centralized and decoupled to the data. The decision making system that work as intelligence resides in controllers to maintain network globally.
- It simplifies its design and working on the basis of the open standards.
- Network operations are configured, managed and optimized dynamically. So it adds the feature of automatic dynamic adjustment of traffic flow according to need.

SDN is considered to be most useful approach for simplifying network management. SDN can distinguish the network service from the physical infrastructure by using the concept of open APIs and virtualization. It is envisioned to have directly programmable base stations and gateways in cellular SDN architectures similar to the programmable switches in wired SDN and can further be extended by network virtualization and flexible adaptation of air interfaces [12].

Several challenges are however expected in wireless Software Define Network (SDN) [13]. To use the SDN concept, huge evolution is required at infrastructure level because there are several questions and blockages like "how to configure programmable switch such that performance and flexibility should remain high". Also the unique way for the implementation of SDN infrastructure is required that should be globally unique as global standard which provide all network operation like computation, storage and resources environment. And the biggest challenge in SDN is security.

2.8 Massive MIMO

Massive Multiple-Input/Multiple-Output (MIMO) technology refers to the use of a large numbers of antenna elements to provide diversity, and compensate for path loss [14]. To provide high throughput at reduced energy consumption, Massive MIMO using large number of antenna elements allows for massive beam-forming and hence enables the miniaturization of antenna elements [15]. The miniaturization allows extremely low power consumption by each antenna element resulting in orders of magnitude improvement in the spectral and energy efficiency.

Before Massive MIMO can be used for 5G, there are several challenges that need to be addressed [4]. Beam-forming need a huge amount of channel state information that is a problematic impact especially for the downlink). The approach may be used in TDD, but is not effective in FDD. It also suffers from pilot contamination from other cells when power of transmission is high. It may also suffer from thermal noise if power transfer is low.

Another technique which is currently evolving (in next approach to 5G) is 3D MIMO, which also allows for 3D beam-forming. In normal beam-forming two dimensions beams are formed, while 3D MIMO also provide the control in both vertical and horizontal dimensions. 3D MIMO requires new channel models. And also it requires extended feedback mechanism which is a challenge in 5G. Thus, channel estimation and feedback mechanism are current challenges for 5G [4]. And also the very fast process algorithm will be required to deal with huge amount of data.

2.9 Machine to Machine (M2M)

There are several challenges specific to Machine to Machine (M2M) communications. Machine to Machine (M2M) devices influence the recent innovations towards satellite navigation devices that have facilitated for downloading traffic updates or feedback dynamically using integrated cellular modems [16]. Autonomous behavior, size and energy requirements are the main challenges of machine to machine (M2M). The nature and distribution of M2M traffic flows is not in accordance with the present network architectures. Thus, there is a need for some modifications in M2M devices for efficient use of in 5G. The challenges of network congestion and overload in M2M depend on applications where the technology is used. Recently, many models such as back-off management have been proposed in 3GPP that take care of network overload issues [4].

2.10 Big Data

Lots of opportunities and challenges in 5G era will also arise due to big data. Cellular networks will have to provide more efficient infrastructure with big data as vast amount of data will be generated in future by Internet of Things (IoT) or Machine to Machine (M2M) applications. New network architectures will therefore be crucial for big data applications.

3 Communication Networks in 5G and Their Architecture

New breakthroughs in advanced waveform technologies coupled with advancements in coding and modulation algorithms are essential for realizing continuous improvements in spectral efficiency [3]. So this section will provide a separate scenario of technologies for 5G wireless communication systems such as massive MIMO, radio network, and software defined cellular networks, device to device communication and future aspects of 5G networks.

3.1 Mobile Cloud Computing Architecture

It is quite clear that the progress from initial 3G networks to the mobile communication technology (broadband) has transformed our industry and society to a step ahead. And if 5G will come into picture, then it may be a huge generational gear shift to our society and surrounding. It will increase the level and standard of connectivity and deployment of radio access network (RAN) with vast change in performance and efficiency [17]. This means that 5G will have new radio interfaces, network topologies and business capabilities. As we are focusing on higher frequency radio spectrum for 5G, smaller cell radiuses due to higher frequency band may be a challenge to achieve wide spread coverage in a traditional way. So network topological model will need some evolution. The tracking of device will be tough as the beam itself has to track device because communication service is still differentiated from fixed line connection that is based on GSMA Intelligence. The beam should be directed to end user. So too many beams will be formed and each cell has to handle large number of beams at an instance of time.

Another alternative to this is MIMO (Multi-Input, Multi-Output) that is already introduced in the previous section. That is an array of antennae installed in a device and multiple radio connections are established between a device and a cell. Most of the technological aspects are used in 4G which may help towards the implementation of 5G.

3.2 Network Technologies Development Towards 5G

To develop more enhance mobile communication technology, continuous evolution and development towards advance technologies deployment is happening. Those advancements are discussed below:

- Software Defined Network (SDN)
- Network Function Virtualization (NFV)
- Advance Networks (HetNets, LPLT)

(Here HetNets stands for heterogeneous networks, LPLT stands for low power, low throughput).

3.3 Network Function Virtualisation (NFV) and Software Defined Networks (SDN)

Network Function Virtualisation (NFV) is a concept that enables differentiation of hardware from software or 'function', and acts as a real entity for the mobile world with increased performance of 'common, off-the-shelf' (COTS) IT platforms). And further Network Function Virtualisation (NFV) is extended to Software Defined Network (SDN) in which reconfiguration of an operator's network topology can be performed dynamically by software to adjust to load and demand, for example, additional increase in network capacity increase when data consumption is on its peak. These two technologies provides following advantages over other network architectures:

- Reduce CAPAX
- Cheap and simple architecture
- Ease of Upgrade
- Efficient use of capacity (reduce OPEX)

Figure 3 shows the basic architectures of mobile cloud computing (MCC). Figure 3 shows two networks—network A and network B, having some mobile communication devices. These devices are connected to network in a traditional way through wireless access point to base trans-receiver station (BTS) and base station subsystem (BSS) or satellite. Many types of services are running on the server available

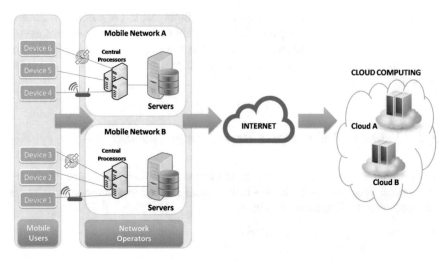

Fig. 3 Mobile cloud computing architecture

in each network like Database services, Agents (Home Agents—HA), Authentication, Authorization and Accounting etc. All requests are being processed by a central system that is connected to all the networks. So whenever user request comes, it is processed in the following way

1. Information extraction according to the home agent (HA) and databases
2. Based on the extracted information—Authentication, Authorization and Accounting (AAA) services are processed
3. Connect the user to cloud through internet
4. Relevant service request linking to the cloud

These requests are of any type depending on the cloud serves like database or storage services, computational services, API services, virtualization services etc.

3.4   Heterogeneous Small Cells Network

As it is known that the high data rates will be required to handle massive network capacity, one way is to reduce cell size so that the areal spectral efficiency increases by reusing the frequency [18]. And also we got the additional coverage by deployment of small cell. This is only possible by hardware miniaturization concept and will also result in cost-reduction [19].

Another approach by which the problem of highly dense network can be solved is Device to Device Communications [20, 21]. In this approach, each entity is able to communicate directly with other entities to exchange information thus reducing interference. Earlier all communications were routed through the base station and

Table 1 Comparative summary of 5G technologies

Technologies	Data rate	Network capacity	Device support	Energy efficiency	Latency	Cost	Interference	Coverage
Small cells	↑	↑		↑			↑	
Device-to-Device	↑			↑		↓	↑	↑
C-RAN		↑	↑	↑		↓	↓	↑
Millimeter wave		↑	↑	↑	↓			
Massive MIMO		↑	↑	↑	↓		↓	

*↑ High/Increased; ↓ Low

the gateway; however, this approach is not efficient for the devices which are very close to each other. Since in the future 5G era, the number of devices communicating with each other will be very large, it would be more efficient to allow direct communication between these devices.

So it is concluded that Massive MIMO technology, millimeter wave technology, small cell technology and some other approaches like Device to Device communication (D2D), Network Function Virtualisation (NFV), Software Defined Networks (SDN) etc. may give a rise to achieve high transmission rates in 5G networks. Table 1 summarizes the features of key 5G technologies.

4 Energy Efficiency

Currently, the area of interest for 5G is improvement in energy efficiency and throughput of 5G wireless backhaul network when we are using ultra dense small cell technology. Massive MIMO and millimeter wave technologies are the two technologies which are the most certain for system modeling in 5G with small cell technology [22, 23, 24]. Here we are discussing two scenarios in which energy efficiency and throughput for wireless backhaul network are compared [8].

1. A macrocell BS (MBS) is at the center of macrocell, and SBSs are uniformly distributed in the macrocell.
2. All traffic is maintained by SBSs only without MBS. And also these are uniformly distributed.

In the first case traffic is transmitted to MBS through milimeter wave link initially and then forwarded to main network by FTTC links. Here FTTC are distribution points, called fibre to the cell links or distribution links. Here two logical interfaces come into scenario as S1 and X2. S1 act as a feeder for user data from the advance gateway to the MBS. Here the advance gateway is considered as an entrance to the core network whereas X2 serves for exchange of mutual information among the small cells.

In the second case, as there is no MBS so backhaul traffic will relay on the adjacent SBS only by using milimeter wave technology. So, all traffic will be forwarded to the specified SBS which is connected to the core network. Here also two interfaces can be considered as S1 and X2 which are same as first case. The first case is considered as central solution and second case is considered as distributed solution [25].

An ideal wireless backhaul links are assumed between small cells and the MBS or the specified SBS. So the traffic is assumed to be only related to bandwidth and the average spectrum efficiency and the backhaul throughput of a small cell is considered as the product of bandwidth and average spectrum efficiency [8].

According to the scenario discussed in Xiaohu et al. [8]. The energy efficiency of the distribution solution is defined as:

$$\eta_{dist} = TH^{dist}_{sum} / E^{dist}_{system}$$

where E^{dist}_{system} is the system energy consumption for distributed solution and TH^{dist}_{sum} is the total backhaul throughput of a distribution solution (case 2). And the energy efficiency of the central solution is defined as:

$$\eta_{centra} = TH^{centra}_{sum} / E^{centra}_{system}.$$

where E^{centra}_{system} is the system energy consumption for central solution and TH^{centra}_{sum} is the total backhaul throughput of a central solution (case 1).

Xiaohu Ge et al. showed that there is a linear increment in the throughput with respect to number of cells in central solution but an exponential in the distribution solution. This is because the traffic is shared by small cells themselves in distributed solution. On the other hand, for energy efficiency logarithmic increment is observed with respect to number of cell in central solution but a linear increment in the distribution solution [8].

5   Latest 5G Activities Around the World

The evolution from 4G to 5G services will vary in requirements in terms of data rates, latency, network coverage, volume, security and privacy etc. This in turn necessitates the emergence of novel key technologies and network architecture along with the requirement of the system-level simulation methodologies and frameworks/standards for 5G systems integration, evaluation and deployment. Different research groups from academia, network infrastructure manufacturers and key mobile operators are engaging in individual and collaborative projects to drive the development of these technologies and standardizing the 5G framework for its realization by 2020.

The European Commission has funded various projects [26]. The key project that has the largest framework is the Mobile and Wireless Communications Enablers for the Twenty–twenty Information Society (METIS). The first phase of the project has been completed in 2015. METIS has developed various technologies that provide flexibility and can be easily configures to adapt to the large variations in the requirements across different scenarios [27]. Other main projects funded by EU Commission towards the development of 5G networks are 5G NOW, MCN, COMBO, MOTO and PHYLAWS, iJoin etc. Many key industrial players such as DOCOMO, Samsung, AT&T, Vodafone, Huawei, Nokia, Ericsson etc. are also involved in the development of the methodologies and testing technologies and architecture using various experiments. These players have successfully conducted various experiments and have developed various technologies/architectures such as femtocells, SBSs, C-RANs, OpenAirInterface etc. A more detail overview of the contribution of the various research groups and industry players can be found in the review article by Mitra and Agarwal [26] and Pirinen [28].

Clearly the emergence of the key technologies in 5G networks has also necessitated the formulation of heuristic guidelines and evaluation strategies to test the performance of these techniques and the network architecture. In this regard, different groups have used different simulation-based methods for performance evaluation. However, there are certain implicit challenges hidden for 5G system evaluation as described by Wang et al. [29]. One of the major projects that has defined various scenarios along with test cases for performance evaluation is the METIS project. In their final project report, METIS has presented evaluation results of already defined 5G KPIs (Key Performance Indicators) in 12 different scenarios. The detailed methodologies and evaluation result are available for different technologies under different scenarios are available as deliverable reports [30]. Beside, Wang et al. have provided a heterogeneous cloud-based framework of the system-level simulator facilitating the reliable and efficient evaluation of various technologies in varied scenarios [29].

6 Smart World Vision Under 5G and Big Data

In rapidly growing environment in mobile and cloud computing, the 5G networks will have vast capacity in telecommunication world. It will be the source of socio-economic growth in digital world through the wide variety of application developments in mobile world with more technical innovations and global collaboration in terms of the transformation of network infrastructure like the "zero-distance" connectivity between people and machine. 5G will provide a fundamental structure of smart cities. Smart sensors with smart technologies are application of smart city infrastructure. A new air interface with significantly improved mobile networks is on a race to the deadline of 2020.

5G is a development of a more sustainable model induced by previous generation technologies that have taught us something. So 5G will demonstrate the ways to Services that are initially expected to have an impact to become a popular technology in digital communication world. It is expected to be the 'next big thing'

Fig. 4 5G development steps

throughout the communication world. Figure 4 highlights some of the recent developments for realizing 5G network.

ITU Radio-communication Sector (ITU-R) plays a key role in the communication world through the radio-frequency spectrum. LTE and WiMAX which are included in the IMT-2000 technology group would be 3rd Generation, instead of 4th.

The communication network in 2020 will be more rich and complex than today. The network infrastructure will connect everything in a flexible and powerful manner by sensors, smart meters and smart gadgets. This will reinvent the roles and relationships between the players of mobile communication world. Communication networks will come with the network virtualisation and software based network that will provide more flexibility and creativity.

5G will bring new user experience with high quality experiences like HD video or tele-services available anywhere, regardless of area of user like a village or in a high speed train. It will provide access anywhere with heterogeneous technologies like WiFi, 4G and new radio interfaces. It will be a key enabler for the Internet of Things. It will also provide the platform to connect a massive number of objects anytime, anywhere to the Internet. Smart phones, wearable devices and other smart objects like drones, robots will have their own local networks. 5G will allow connectivity to all these and will be handled by specific network technologies.

7 Summary

As the 5G wireless technology is next wave to technical advancement in the field of communication world, 5G will have the following objectives:

- Massive network
- Massive capacity
- Huge diversity in services and application
- Efficient spectrum scenario

Adaptive network framework will become the necessity in the environment of 5G which can accommodate LTE and air interface. Cloud, Software Defined Network (SDN), Network Function Virtualization (NFV), millimeter wave technology,

Massive MIMO technology and D2D technology will be key platform features of 5G services.

To develop better technology in communication, some more studies should be carried out. So, more research and advancement is important in 5G wireless network technology to overcome the challenges. The following issues and challenges need to be addressed:

- Development of Cell architecture and distribution methodologies for wireless backhaul networks.
- Development of high throughput and energy efficient.
- Advancements in transmission technology to handle massive traffic for 5G wireless network.

The next decade will see the development in Radio Access Network technologies for implementing 5G network solutions which include:

- Service delivery architecture with Massive MIMO
- Advancements in multiple access waveform technologies
- System for interference control and Access protocols

References

1. Gupta, P., Gupta, S.: Mobile cloud computing: the future of cloud. Int. J. Adv. Res. Electr. Electron. Instrum Eng. **1**(3) (2012)
2. Ericsson: 5G Radio Access. Technical Report, Ericsson. (2014) www.ericsson.com/res/docs/whitepapers/wp-5g.pdf
3. Huawei white paper (2013) A 5G Technology Vision
4. Chin, W., Fan, Z., Haines, R.: Emerging technologies and research challenges for 5G wireless networks. Toshiba Research Europe Limited (2014)
5. Boccardi, B., et al.: Five disruptive technology directions for 5G. IEEE Commun. Mag. 74–80 (2014)
6. Andrews, J.G., et al.: What will 5G be? IEEE J. Sel Areas Commun. **32**(6), 1065–1082 (2014)
7. Wang, C.-X., Haider, F., Gao, X., et al.: Cellular architecture and key technologies for 5G wireless communication networks. IEEE Commun. Mag. **52**(2), 122–130 (2014)
8. Ge, X., Cheng, H., Guizani, M., Han, T.: 5G wireless backhaul networks: challenges and research advances. Netw. IEEE **28**(6), 6–11 (2014)
9. Talwar, S., Choudhury, D., Dimou, K., Aryafar, E., Bangerter, B., Stewart, K.: Enabling technologies and architectures for 5G wireless. In: Proceedings of IEEE MTT-S International Microwave Symposium (IMS), pp. 1–4, (2014)
10. Dabbagh, M., Hamdaoui, B., Guizani, M., Rayes, A.: Toward-Energy efficient cloud computing: prediction, consolidation, and overcommitment. IEEE Netw. 56–61 (2015)
11. Batalla, J., Krawiec, P.: Conception of ID layer performance at the network level for Internet of Things. Pers Ubiquitous Comput. **18**(2), 465–480 (2014)
12. Pop, F., Dobre, C., Comaneci, D., Kolodziej, J.: Adaptive scheduling algorithm for media-optimized traffic management in software defined networks. Computing. **98**(1–2), pp. 147–168 (2016)

13. Sezer, S., et al.: Are we ready for SDN? Implementation challenges for software-defined networks. IEEE Commun. Mag. **51**(7) 36–43 (2013)
14. Larsson, E., Edfors, O., Tufvesson, F., Marzetta, T.: Massive MIMO for next generation wireless systems. Commun Mag. IEEE **52**(2),186–195 (2014)
15. Choudhury, R.R.: A network overview of massive MIMO for 5G wireless cellular: system model and potentials. Int. J. Eng. Res. General Sci. **2**(4), 338–347 (2014)
16. Hasan, M., Hossain, E., Niyato, D.: Random access for machine-to-machine communication in LTE-Advanced networks: Issues and approaches. IEEE Commun. Mag. **51**(6) (2013). Special Issue on Smart Cities
17. Sabella, D., De Domenico, A., Katranaras, E., Imran, M.A., Di Girolamo, M., Salim, U., Lalam, M., Samdanis, K., Maeder, A.: Energy efficiency benefits of RAN-as-a-"Service concept for a cloud-based 5G mobile network infrastructure". Access IEEE **2**, 1586–1597 (2014)
18. Yunas, S.F., Valkama, M., Niemela, J.: Spectral and energy efficiency of ultra-dense networks under different deployment strategies. IEEE Commun. Mag. **53**, **1**, 90–100 (2015)
19. Chandrasekhar, V., Andrews, J.G., Gatherer, A.: Femtocell networks: a survey. IEEE Commun. Mag. **46**(9), 59–67 (2008)
20. Asadi, A., Wang, Q., Mancuso, V.: A survey on device-to-device communication in cellular networks. Commun. Surv. Tut. IEEE **24**, **16**(4), 1801–1819 (2014)
21. Lin, X., Andrews, J., Ghosh, A., Ratasuk, R.: An overview of 3GPP device-to-device proximity services. Commun. Mag. IEEE **52**(4), 8–40 (2014)
22. Bleicher, A.: Millimeter waves may be the future of 5G phones (2013). http://spectrum.ieee.org/telecom/wireless/millimeter-waves-may-be-the-future-of-5g-phones
23. Swindlehurst, A.L., Ayanoglu, E., Heydari, P., Capolino, F.: Millimeter-wave massive MIMO: the next wireless revolution? IEEE Commun. Mag. **1**, **52**(9), 56–62 (2014)
24. Rappaport, T.S., Shu, S., Mayzus, R., et al.: Millimeter wave mobile communications for 5G cellular: it will work! IEEE Access, **1**, 335–349 (2013)
25. Jungnickel, V., Manolakis, V., Jaeckel, S.: Backhaul requirements for inter-site cooperation in heterogeneous LTE- Advanced networks. In: IEEE International Conference on Communications Workshops (ICC), pp. 905–910 (2013)
26. Mitra, R.N., Agrawal, D.P.: 5G mobile technology: a survey. ICT Express. 2016 Jan 22
27. Osseiran, A., Boccardi, F., Braun, V., Kusume, K., Marsch, P., Maternia, M., Queseth, O., Schellmann, M., Schotten, H., Taoka, H., Tullberg, H.: Scenarios for 5G mobile and wireless communications: the vision of the METIS project. Commun. Mag. IEEE **52**(5), 26–35 (2014)
28. Pirinen, P.: A brief overview of 5G research activities. In: IEEE International Conference on 5G for Ubiquitous Connectivity, pp. 17–22, 4
29. Wang, Y., Xu, J., Jiang, L.: Challenges of system-level simulations and performance evaluation for 5G wireless networks. Access IEEE **2**, 1553–1561 (2014)
30. FP7 Integrating Project METIS (ICT 317669). Scenarios, requirements and KPIs for 5G mobile and wireless system. (2013) https://www.metis2020.com/documents/deliverables/

Towards a Cloud-Native Radio Access Network

Navid Nikaein, Eryk Schiller, Romain Favraud, Raymond Knopp,
Islam Alyafawi and Torsten Braun

Abstract Commoditization and virtualization of wireless networks are changing the economics of mobile networks to help network providers, e.g. Mobile Network Operator (MNO), Mobile Virtual Network Operator (MVNO), move from proprietary and bespoke hardware and software platforms towards an open, cost-effective, and flexible cellular ecosystem. In addition, rich and innovative local services can be efficiently materialized through cloudification by leveraging the existing infrastructure. In this work, we present a Radio Access Network as a Service (RANaaS), in which a Cloudified Centralized Radio Access Network (C-RAN) is delivered as a service. RANaaS describes the service life-cycle of an on-demand, elastic, and pay as you go RAN instantiated on top of the cloud infrastructure. Due to short deadlines in many examples of RAN, the fluctuations of processing time, introduced by the virtualization framework, have a deep impact on the C-RAN performance. While in typical cloud environments, the deadlines of processing time cannot be guaranteed, the cloudification of C-RAN, in which signal processing runs on general purpose processors inside Virtual Machines (VMs), is a challenging subject. We describe an example of real-time cloudified LTE network deployment using the OpenAirInter-

N. Nikaein · R. Favraud · R. Knopp
EURECOM, Biot, France
e-mail: navid.nikaein@eurecom.fr

R. Favraud
e-mail: romain.favraud@eurecom.fr

R. Knopp
e-mail: raymond.knopp@eurecom.fr

R. Favraud
DCNS Group, Paris, France

E. Schiller (✉) · I. Alyafawi · T. Braun
University of Bern, Bern, Switzerland
e-mail: Schiller@inf.unibe.ch

I. Alyafawi
e-mail: Alyafawi@inf.unibe.ch

T. Braun
e-mail: Braun@inf.unibe.ch

© Springer International Publishing Switzerland 2017
C.X. Mavromoustakis et al. (eds.), *Advances in Mobile Cloud Computing and Big Data in the 5G Era*, Studies in Big Data 22,
DOI 10.1007/978-3-319-45145-9_8

face (OAI) LTE implementation and OpenStack running on commodity hardware. We also show the flexibility and performance of the platform developed. Finally, we draw general conclusions on the RANaaS provisioning problem in future 5G networks.

1 Introduction

Every day, we encounter an increasing demand for wireless data use due to a growing number of broadband-capable devices, such as 3G and 4G mobile telephones. To satisfy a higher demand for data rates, service providers and mobile operators expect upgrades and expansion of the existing network, but the required Capital Expenditure (CAPEX) and Operational Expenditure (OPEX) are superior to the revenue growth [9]. The high upgrade and maintenance costs are mainly caused by the current architecture of mobile broadband networks, in which the Radio Access Network (RAN) is built upon the integrated Base Transceiver Station (BTS) architecture. Since mobile broadband providers operate on a large scale, the installation and maintenance of a large number of expensive BTSs over vast geographical areas increase the cost dramatically. Moreover, the new trend of smaller cells will more severely affect both the cost and maintenance problem in the future.

> A cost-effective RAN solution, which meets the ever-increasing amounts of mobile data traffic, has to fulfill a set of requirements. First, the new RAN has to quickly and automatically scale with the variable amount of mobile traffic. Second, it has to consume less power providing higher capacity and network coverage at the same time. Finally, it should allow mobile operators to frequently upgrade and operate the service over multiple/heterogeneous air-interfaces.

Only about 15–20 % of BTSs operating in the current RAN architecture are loaded more than 50 % (with respect to the maximum capacity), which makes the current RAN architecture energy inefficient [8]. An emerging solution to reduce upgrading costs and power consumption is the Centralized-RAN (C-RAN) [13, 17] with resource sharing and exploitation of load patterns at a given geographical area. Thus, C-RAN solution is able to adapt to user traffic variability and unpredictable mobility patterns than the current RAN. Moreover, it allows coordinated and joint signal processing to increase the spectral efficiency. Finally, the C-RAN represents a good match between the spatial-temporal traffic variations and available computational resources and hence power consumption.

Since C-RAN signal processing is centralized, it allows us to apply more sophisticated joint spatio-temporal processing of radio signals, which can increase the overall spectral efficiency. Cloud computing technologies based on virtualization allow us to lower the operational costs even more by running the RAN through (a) adoption of general purpose IT platforms instead of expensive specific hardware, (b) load balancing, and (c) fast deployment and resource provisioning. Running the RAN in the cloud environment is not new. The benefit of such an approach has demonstrated 71 % of power savings when compared to the existing system [7]. However, this approach comes at the cost of higher software complexity.

Recent works [27] have shown the feasibility of LTE RAN functions of software implementation over General Purpose Processors (GPPs), rather than the traditional implementation over Application-Specific Integrated Circuits (ASICs), Digital Signal Processors (DSPs), or Field-Programmable Gate Arrays (FPGAs). Different software implementations of the LTE base station, which is referred to as evolved Node B (eNB), already exist: (a) Amarisoft LTE solution, which is a pure-software featuring a fully-functional LTE eNB [2], (b) Intel solutions featuring energy efficiency and high computing performance using a hybrid GPP-accelerator architecture and load-balance algorithms among a flexible IT platform [25] and (c) OpenAirInterface (OAI) developed by EURECOM, which is an open-source Software Defined Radio (SDR) implementation of both the LTE RAN and the Evolved Packet Core (EPC) [12].

This chapter describes recent progress in the C-RAN cloudification (running software-based RAN in the cloud environment) based on the open source implementations and has the following organization. In Sect. 2, we introduce the concept, architecture, and benefits of centralized RAN in the LTE Network setup. Section 3 presents the critical issues of cloudified RAN focusing on fronthaul latencies, processing delays, and appropriate timing. Our performance evaluation of GPP-based RAN is provided in Sect. 4 and Base-Band Unit (BBU) processing time is modeled in Sect. 5. Possible architectures of cloudified RAN are described in Sect. 6. The description of the cloud datacenter supporting C-RAN resides in Sect. 7. Section 8 illustrates an example RANaaS with its life-cycle management. Finally, we conclude in Sect. 9.

2 Centralized RAN in the LTE Network

C-RAN based networks are characterized by the decomposition of a BTS into two entities namely Base-Band Unit (BBU) and, also known as Remote Radio Unit (RRU). In C-RAN, the RRH stays at the location of the BTS, while the BBU gets relocated into a central processing pool, which hosts a significant number of distinct BBUs [27]. In order to allow for signal processing at a remote BBU, a point-to-point high capacity interface of short delay is required to transport I/Q samples (i.e., digitized analog radio signals) between RRH and BBU. There are a few examples of link standards meeting the required connectivity expectations such as Open Radio Interface (ORI), Open Base Station Architecture Initiative (OBSAI), or Common Public Radio Interface (CPRI). Even though many recent works have shown the feasibility of C-RAN implementation and the C-RAN importance for the MNOs, there are still three open questions that has to be thoroughly investigated upon a C-RAN system design.

1. **Dimensioning of the fronthaul capacity**: a BBU pool has to support a high fronthaul capacity to transport I/Q samples for a typical set of 10–1000 base-stations working in the BBU–RRH configuration. Due to a low processing budget of RAN, the upper bound for the maximum one-way delay has to be estimated. Moreover, a very low jitter has to be maintained for the clock synchronization among BBUs and RRHs.
2. **Processing budget at the BBU**: in the LTE FDD setup, the Hybrid automatic repeat request (HARQ) mechanism with an 8 ms acknowledgment response time provides an upper bound for the total delay of both fronthaul latency and BBU processing time.
3. **The real-time requirements of the Operating-System and Virtualization-System**: to successfully provide frame/subframe timings, the execution environment of the BBU has to support strict deadlines of the code execution. Moreover, load variations in the cell (e.g., day/night load shifts) impose the requirement on the on-demand resource provisioning and load balancing of the BBU pool.

There are also many other challenges in this field [6], such as front-haul multiplexing, optimal clustering of BBUs and RRHs, BBU interconnection, cooperative radio resource management, energy optimization, and channel estimation techniques. The following subsections focus on the critical issues, and present C-RAN feasible architectures.

3 Critical Issues of C-RAN

In the following subsections, we evaluate the most important critical issues of the C-RAN. We concentrate on the fronthaul capacity problem, BBU signal processing, and real-time cloud infrastructure for signal processing [18].

3.1 Fronthaul Capacity

We start with the description of fronthaul requirements. A very fast link of low delay is necessary as the BBU processes the computationally most heavy physical (PHY) layer of the LTE standards. Many factors contribute to the data rate of the fronthaul, which depends on the cell and fronthaul configurations. Equation 1 calculates the required data rate based on such configurations:

$$C_{fronthaul} = \underbrace{2 \times N \times M \times F \times W \times C}_{\text{cell configuration}} \times \underbrace{O \times K}_{\text{fronthaul configuration}} , \qquad (1)$$

where N is the number of receiving/transmitting (Tx/Rx) antenna ports, M is the number of sectors, F represents the sampling rate, W is the bit width of an I/Q symbol, C number of carrier components, O is the ratio of transport protocol and coding overheads, and K is the compression factor. The following table shows the required fronthaul capacity for a simple set of configurations. An overall overhead is assumed to be 1.33, which is the result of the the protocol overhead ratio of 16/15 and the line coding of 10/8 (CPRI case). One can observe that the fronthaul capacity heavily depends on the cell configuration and rapidly grows with the increased sampling rate, number of antennas/sectors and carrier components (Table 1).

Figure 1 compares the fronthaul capacity between the RRH and the BBU pool for 20 MHz BW, SISO (max. 75 Mb/s on the radio interface). In the case without compression, the fronthaul has to provide at least 1.3 Gb/s; when the 1/3 compression ratio is used, the required fronthaul capacity drops to 0.45 Gb/s.

Further data rate reduction can be obtained by an RRH offloading the BBU functions. As shown in Fig. 2, the functional split can be provided by decoupling the L3/L2 from the L1 (labelled 4), or part of the user processing from the L1

Table 1 Fronthaul capacity for different configurations

BW (MHz)	N	M	F	W (bits)	O	C	K	$C_{fronthaul}$ (Mb/s)
1.4	1 × 1	1	1.92	16	1.33	1	1	81
5	1 × 1	1	7.68	16	1.33	1	1	326
5	2 × 2	1	7.68	16	1.33	1	1	653
10	4 × 4	1	15.36	16	1.33	1	1/2	1300
20	1 × 1	1	30.72	16	1.33	1	1	1300
20	2 × 2	3	30.72	16	1.33	1	1	7850
20	4 × 4	3	30.72	16	1.33	1	1	15600

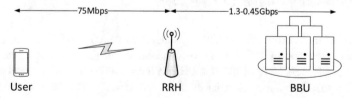

Fig. 1 Fronthaul capacity between the RRH and the BBU pool for 20 MHz BW, Single Input Single Output (SISO). Minimum required fronthaul capacity without compression is estimated at 1.3 Gb/s, the deployment of 1/3 compression ratio decreases the required capacity to 0.45 Gb/s

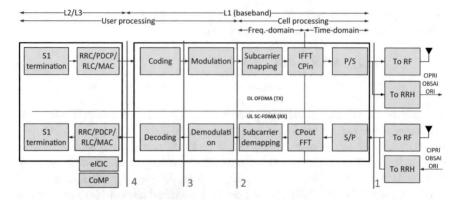

Fig. 2 Functional block diagram of downlink and uplink for LTE eNB

(labelled 3), or all user-specific from the cell processing (labelled 2), or antenna-specific from non-antenna processing (labelled 1), which is different for the Rx and Tx chain.

Trade-offs have to be performed among the available fronthaul capacity, complexity, and the resulted spectral efficiency. Regardless various possibilities in the BBU functional split, the fronthaul should still maintain the latency requirement to meet the HARQ deadlines. According to Chanclou et al. [5], the RTT between RRH and BBU equipped with a CPRI link cannot exceed 700 μs for LTE and 400 μs for LTE-Advanced. Jitter required by advanced CoMP schemes in the MIMO case is below 65 ns as specified in 3GPP 36.104. Next Generation Mobile Networks (NGMN) adopts the maximum fronthaul round-trip-time latency of 500 μs [16].[1] The propagation delay, corresponding to timing advance, between RRH and UE, affects only the UE processing time. The timing advance value can be up to 0.67 ms (equivalent to maximum cell radius of 100 km). Consequently, this leaves the BBU PHY layer only with around 2.3–2.6 ms for signal processing at a centralized processing pool. The next subsection elaborates on the BBU processing budget in the LTE FDD access method.

3.2 Processing Budget in LTE FDD

This subsection describes the processing budget problem of the Frequency-Division Long-Term Evolution (LTE-FDD). We concentrate on the Physical Layer (PHY) and Medium Access Control (MAC). PHY is responsible for symbol level processing, while MAC provides user scheduling and HARQ. The LTE FDD PHY is

---

[1]Different protocols have been standardized for the fronthaul, namely CPRI representing 4/5 of the market, OBSAI representing 1/5 of the market, and more recently the Open Radio Interface (ORI) initiated by NGMN and now by the European Telecommunications Standards Institute (ETSI) Industry Specification Group (ISG).

implemented in the asymmetric way using Orthogonal Frequency-Division Multiple Access (OFDMA) and Single-Carrier Frequency-Division Multiple Access (SC-FDMA) on the downlink and uplink respectively. To control the goodput of the air interface, the PHY uses various Modulation and Coding Schemes (MCSs). 3GPP defines 28 MCSs with indexes between 0 and 27. A distinct MCS is characterized by a specific modulation (i.e., QPSK, 16-QAM, 64-QAM having a varying number of data bits per modulation symbol carried) and the so called code rate, which measures the information redundancy in a symbol for error correction purposes [3, 9]. The smallest chunk of data transmitted by an eNB through the LTE FDD PHY is referred to as Physical Resource Block (PRB) and spans 12 sub-carriers (180 kHz) and 7 modulation symbols (0.5 ms), which gives 84 modulation symbols in total. In LTE FDD, we are provided with channels of 1.4 MHz, 3 MHz, 5 MHz, 10 MHz, 15 MHz, and 20 MHz bandwidth, which can simultaneously carry 6, 15, 25, 50, 75, and 100 PRBs respectively. Therefore, the workload of signal processing in software is heavily influenced by the MCS index, number of allocated PRBs, and the channel bandwidth. Moreover, Hybrid Automatic Repeat Request protocol (HARQ) on the MAC layer introduces short deadline for signal processing on the PHY. Due to HARQ, every transmitted chunk of information has to be acknowledged back at the transmitter to allow for retransmissions. In LTE-FDD, the retransmission time is equal to T_{HARQ} of 8 ms. Let us briefly explain the retransmission mechanism. Every LTE FDD subframe (subframe is later referred to as SF) lasts for 1 ms and contains information chunks carried within PRBs. The HARQ protocol states that the Acknowledgment (ACK) or Negative Acknowledgment (NACK) for a data chunk received at subframe N has to be issued upon a subframe $N + 4$ and decoded at the transmitter before subframe $N + 8$, which either sends new data or again negatively acknowledged chunks. In the following subsection, we briefly summarize the BBU functions.

3.3   BBU Functions

Figure 2 illustrates the main RAN functions in both TX and RX spanning all the layers, which has to be evaluated to characterize the BBU processing time and assess the feasibility of a full GPP RAN. Since the main processing bottleneck resides in the physical layer, the scope of the analysis in this chapter is limited to the BBU functions. From the figure, it can be observed that the overall processing is the sum of cell- and user-specific processing. The former only depends on the channel bandwidth and thus imposes a constant base processing load on the system, whereas the latter depends on the MCS and resource blocks allocated to users as well as the Signal-to-Noise Ratio (SNR) and channel conditions. The figure also shows the interfaces where the functional split could happen to offload the processing either to an accelerator or to an RRH.

   To meet the timing and protocol requirements, the BBU must finish processing before the deadline previously discussed at the beginning of Sect. 3.2. Each MAC

PDU sent at subframe N is acquired in subframe $N + 1$, and must be processed in both RX and TX chains before subframe $N + 3$ allowing ACK/NACK to be transmitted in subframe $N + 4$. On the receiver side, the transmitted ACK or NACK will be acquired in subframe $N + 5$, and must be processed before subframe $N + 7$, allowing the transmitter to retransmit or clear the MAC PDU sent in subframe N. Figure 3a, b show an example of timing deadlines required to process each subframe on the downlink and uplink respectively. Figure 3c graphically represents the communication between the UE, RRH, and BBU. It can be observed that the total processing time is 3 ms. The available processing time for a BBU to perform the reception and transmission is upper-bounded by HARQ round trip time (T_{HARQ}), propagation time ($T_{Prop.}$), acquisition time ($T_{Acq.}$), and fronthaul transport time ($T_{Trans.}$) as follows:

$$T_{rx} + T_{tx} \leq T_{HARQ}/2 - (T_{Prop.} + T_{Acq.} + T_{Trans.} + T_{Offset}) , \qquad (2)$$

where $T_{HARQ} = 8$ ms, $T_{Prop.}$ is compensated by the timing advance of the UE: $T_{Prop.} = 0$, $T_{Acq.}$ is equal to the duration of the subframe: $T_{Acq.} = 1$ ms, and there is no BBU offset on the downlink: $T_{Offset} = 0$. Depending on the implementation, the maximum tolerated transport latency depends on the BBU processing time and HARQ period. The LTE FDD access method puts a particular focus on perfect timing of (sub-) frame processing. To accomplish this goal, the processing system has to fulfill real-time requirements. The next subsection focuses on the real-time cloud system design capable of C-RAN provisioning.

3.4   Real-Time Operating System and Virtualization Environment

A typical general purpose operating systems (GPOS) is not designed to support real-time applications with hard deadline. Hard real-time applications have strict timing requirements to meet deadlines. Otherwise unexpected behaviors can occur compromising performance. For instance, Linux is not a hard real-time operating system as the kernel can suspend any task when a desired runtime has expired. As a result, the task can remain suspended for an arbitrarily long period of time. The kernel uses a scheduling policy that decides on the allocation of processing time to tasks. A scheduler that always guarantees the worst case performance (or better if possible) and also provides a deterministic behavior (with short interrupt-response delay of 100 μs) for the real-time applications is required. Recently, a new scheduler, named SCHED_DEADLINE, is introduced in the Linux mainstream kernel that allows each application to set a triple of (*runtime*[*ns*], *deadline*[*ns*], *period*[*ns*]), where *runtime* ≤ *deadline* ≤ *period*.[2] The scheduler is able to preempts the kernel code to meet the deadline and allocates the required runtime (i.e., CPU time) to each task period.

---

[2]http://www.kernel.org/doc/Documentation/scheduler/sched-deadline.txt.

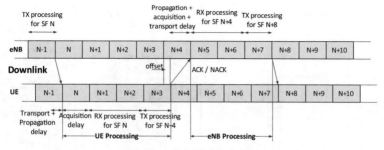

(a) *Downlink HARQ timing.*

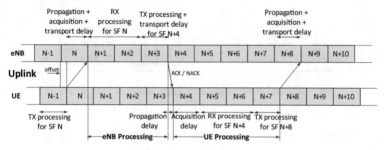

(b) *Uplink HARQ timing.*

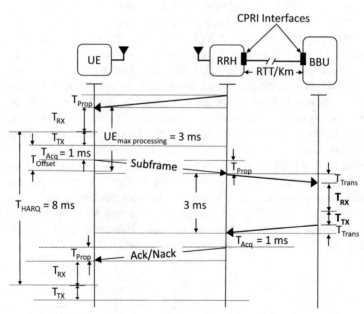

(c) *HARQ process timing requirement.*

Fig. 3 FDD LTE timing

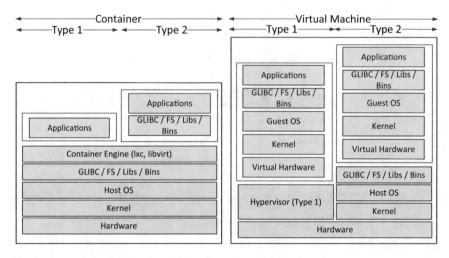

Fig. 4 Comparison of a virtual machine and container virtualized environment

A good deadline scheduler can simplify C-RAN deployment, because Software-based Radio providing RAN in software is a real-time application that requires hard deadlines to maintain frame and subframe timing. In the C-RAN setting, the software radio application runs on a virtualized environment, where the hardware is either fully, partially, or not virtualized. Two main approaches exist to virtualization: virtual machines (e.g., Linux KVM[3] and Xen[4]) or containers (e.g., LinuX Container LXC[5] and Docker[6]) as shown in Fig. 4. In a virtual machine (VM), a complete operating system (guest OS) is used with the associated overhead due to emulating virtual hardware, whereas containers use and share the OS and device drivers of the host. While VMs rely on the hypervisor to requests for CPU, memory, hard disk, network and other hardware resources, containers exploit the OS-level capabilities. Similar to VMs, containers preserve the advantage of virtualization in terms of flexibility (containerize a system or an application), resource provisioning, decoupling, management and scaling. Thus, containers are lightweight as they do not emulate any hardware layer (share the same kernel and thus application is native with respect to the host) and therefore have a smaller footprint than VMs, start up much faster, and offer near bare metal runtime performance. This comes at the expense of less isolation and greater dependency on the host kernel.

---

[3]http://www.linux-kvm.org.

[4]http://www.xenserver.org.

[5]http://linuxcontainers.org.

[6]http://www.docker.com.

Two other important aspects when targeting RAN virtualization are:

- **I/O Virtualization**: I/O access is a key for a fast access to the fronthaul interface and to the hardware accelerators that might be shared among BBUs. In hypervisor approach to virtualization (i.e., VM), IO virtualization is done through the hardware emulation layer under the control of hypervisor, where as in a container this is materialized through device mapping. Thus, direct access to hardware is easier in containers than in VMs as they operate at the host OS level. In a VM, additional techniques might be needed (e.g., para-virtualization or CPU-assisted virtualization) to provide a direct or fast access to the hardware. When it comes to sharing I/O resources among multiple physical/virtual servers, and in particular that of radio front-end hardware, new techniques such as single/multi root I/O virtualization (SR/MR-IOV) are required.
- **Service composition of the software radio application**: A radio application can be defined as a composition of three types of service [15], atomic service that executes a single business or technical function and is not subject to further decomposition, composed service that aggregates and combines atomic services together with orchestration logic, and support service that provides specific (often common) functions available to all types of service. An atomic service in RAN can be defined on per carrier, per layer, per function basis. For instance, a radio application could be defined as a composition of layer 1 and layer 2/3 services supported by a monitoring as a service.

4 OpenAirInterface Based Evaluation of the Cloud Execution Environment

Section 1 gives a brief insight into various software-based implementations of BBU. This section, provides an overview of the OpenAirInterface (OAI), which is a key software component in our studies. The main advantage of OAI is that it an open-source project that implements the LTE 3GPP Release-10 standard. It includes a fully functional wireless stack with PHY, MAC, Radio Link Control (RLC), Packet Data Convergence Protocol (PDCP) and Radio Resource Control (RRC) layers as well as Non-Access-Stratum (NAS) drivers for IPv4/IPv6 interconnection with other network services [20]. Regarding the LTE FDD, OAI provides both the uplink and downlink processing chains with SC-FDMA and OFDMA respectively (c.f., Sect. 3.2). For efficient numerical computing on the PHY, OAI uses specially optimized SIMD Intel instruction sets (i.e., MMX/SSE3/SSE4). Figure 5 presents the OAI multi-threaded signal processing at the subframe level. As an example, the mobile air-interface of a client terminal started transmitting subframe $N-1$ at time (a). The decoder thread of the OAI lte-softmodem starts processing the subframe $N-1$ at (1) after the subframe is entirely received at time instance (b). Due to the fact that the encoding thread starting at (2) has to get input from the decoding thread to comply with HARQ retransmission scheme, the decoding thread gets at

Fig. 5 Processing orders in OAI

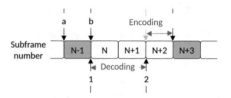

most 2 ms to finish signal processing. Again, HARQ requires data to be acknowledged at subframe $N + 3$, therefore the encoding thread has to finish before (c) and receives at most 1 ms for processing. This description, however, does not include RRH-BBU propagation delays, which shorten the computing budget (both decoding and encoding) by a few hundred microseconds. Summing up, the OAI decoder gets twice as much time as the encoder; roughly 2 ms are allocated for decoding and 1 ms for encoding.[7]

In the following subsections, we evaluate the OAI execution performance on different platforms.

4.1 Experiment Setup

Four set of different experiments are performed. The first experiment (c.f., Sect. 4.2) analyses the impact of different x86 CPU architecture on BBU processing time, namely Intel Xeon E5-2690 v2 3 GHz (same architecture as IvyBridge), Intel Sandy-Bridge i7-3930K at 3.20 GHz, and Intel Haswell i7-4770 3.40 GHz. The second experiment (c.f., Sect. 4.3) shows how the BBU processing time scales with the CPU frequency. The third experiment (c.f., Sect. 4.4) benchmarks the BBU processing time in different virtualization environments including LXC, Docker, and KVM against a physical machine (GPP). The last experiment (c.f., Sect. 4.5) measures the I/O performance of virtual Ethernet interface through the guest-to-host round-trip time (RTT).

All the experiments are performed using the OAI *DLSCH and ULSCH simulators* designed to perform all the baseband functions of an eNB for downlink and uplink as in a real system. All the machines (hosts or guests) operate Ubuntu 14.04 with the low latency (LL) Linux kernel version 3.17, x86-64 architecture and GCC 4.7.3. To have a fair comparison, only one core is used across all the experiments with the CPU frequency scaling deactivated except for the second experiment.

The benchmarking results are obtained as a function of allocated PRBs, modulation and coding scheme (MCS), and the minimum SNR for the allocated MCS for 75 % reliability across 4 rounds of HARQ. Note that the processing time of the turbo decoder depends on the number of iterations, which is channel-dependant.

---

[7]This rule was established empirically, because in full load conditions (i.e., all PRBs allocated in the subframe; the same MCS for all PRBs) the OAI LTE FDD TX requires 2 times less processing time than the OAI LTE FDD RX.

The choice of minimum SNR for an MCS represents the realistic behavior, and may increase number of turbo iterations and consequently causing high processing variation. Additionally, the experiments are performed at full data rate (from 0.6 Mb/s for MCS 0 to 64 Mb/s for MCS 28 in both directions) using a single user with no mobility, Single-Input and Single-Output (SISO) mode with Additive White Gaussian Noise (AWGN) channel, and 8-bit log-likelihood ratios turbo decoder. Note that if multiple users are scheduled within the same subframe on the downlink or uplink, the total processing depends on the allocated PRB and MCS, which is lower than a single user case with all PRBs and highest MCS. Thus, the single user case represents the worst case scenario.

The processing time of each signal processing module is calculated using timestamps at the beginning and at the end of each BBU function. OAI uses the `rdtsc` instruction implemented on all x86 and x64 processors to get very precise timestamps, which counts the number of CPU tics since the reset. Therefore the processing time is measured as a number of CPU tics between the beginning and end of a particular processing function divided by the CPU frequency.[8]

To allow for a rigorous analysis, the total and per function BBU processing time are measured. For statistical analysis, a large number of processing_time samples (10000) are collected for each BBU function to calculate the average, median, first quantile, third quantile, minimum and maximum processing time for all the subframes on the uplink and downlink.

4.2 CPU Architecture Analysis

Figure 6 depicts the BBU processing budget in both directions for the considered Intel x86 CPU architecture. It can be observed that processing load grows with the increase of PRB and MCS for all CPU architectures, and that it is mainly dominated by the uplink. Furthermore, the ratio and variation of downlink processing load to that of uplink also grows with the increase of PRB and MCS. Higher performance (lower processing time) is achieved by the Haswell architecture followed by SandyBridge and Xeon. This is primarily due to the respective clock frequency (c.f., Sect. 4.3), but also due to a better vector processing and faster single threaded performance of the Haswell architecture.[9] For the Haswell architecture, the performance can be further increased by approximately a factor of two if AVX2 (256-bit Single instruction multiple data (SIMD) compared to 128-bit SIMD) instructions are used to optimize the turbo decoding and FFT processing.

---

[8]https://svn.eurecom.fr/openair4G/trunk/openair1/PHY/TOOLS/time_meas.h.

[9]http://en.wikipedia.org/wiki/Haswell_(microarchitecture).

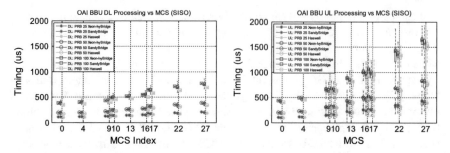

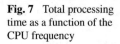

Fig. 6 BBU processing budget on the downlink (*left*) and uplink (*right*) for different CPU architecture

Fig. 7 Total processing time as a function of the CPU frequency

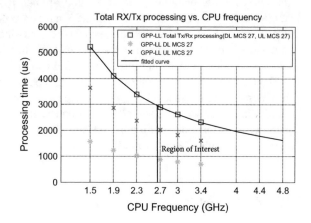

4.3 CPU Frequency Analysis

Figure 7 illustrates the total BBU processing time as a function of different CPU frequencies (1.5, 1.9, 2.3, 2.7, 3.0, and 3.4 GHz) on the Haswell architecture. The most time consuming scenario is considered with 100 PRBs and MCS 27 on both downlink and uplink. In order to perform experiments with different CPU frequencies, Linux ACPI interface and *cpufreq* tool are used to limit the CPU clock. It can be observed that the BBU processing time scales down with the increasing CPU frequency. The figure also reflects that the minimum required frequency for 1 CPU core to meet the HARQ deadline is 2.7 GHz.

Based on the above figure, the total processing time per subframe, T_{subframe}, can be modeled as a function of the CPU frequency [1]:

$$T_{\text{subframe}}(x) \, [\mu s] = \alpha/x \,, \tag{3}$$

where $\alpha = 7810 \pm 15$ for the MCS of 27 in both directions, and x is CPU frequency measured in GHz.

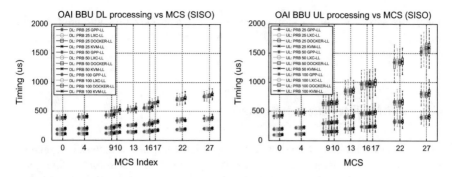

Fig. 8 BBU processing budget on the downlink (*left*) and uplink (*right*) for different virtualized environments

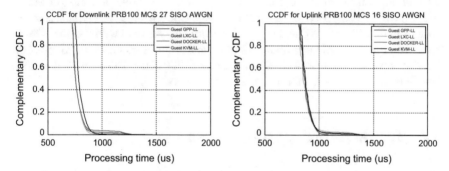

Fig. 9 BBU processing time distribution for downlink MCS 27 and uplink MCS 16 with 100 PRB

4.4 Virtualization Technique Analysis

Figure 8 compares the BBU processing budget of a GPP platform with different virtualized environments, namely LXC, Docker, and KVM, on the SandyBridge architecture (3.2 GHz). While on average the processing times are very close for all the considered virtualization environments, it can be observed that GPP and LXC have slightly lower processing time variations than that of DOCKER and KVM, especially when PRB and MCS increase.

Figure 9 depicts the Complementary Cumulative Distribution Function (CCDF) of the overall processing time for downlink MCS 27 and uplink MCS 16 with 100 PRB.[10] It can be observed that the execution time is stable for all the platforms on the uplink and downlink. The processing time for the KVM (hypervisor-based) has a longer tail and is mostly skewed to longer runs due to higher variations in the non-native execution environments (caused by the host and guest OS sched-

---

[10]The CCDF plot for a given processing time value displays the fraction of subframes with execution times exceeding this value.

uler). Higher processing variability is observed on a public cloud with unpredictable behaviors, suggesting that cares have to be taken when targeting a shared cloud infrastructure [1].

4.5 I/O Performance Analysis

Generally, the fronthaul one-way-delay depends on the physical medium, technology, and the deployment scenario. However in the cloud environment, the guest-to-host interface delay (usually Ethernet) has to be also considered to minimize the access to the RRH interface. To assess such a delay, bidirectional traffics are generated for different set of packet sizes (64, 768, 2048, 4096, 8092) and inter-departure time (1, 0.8, 0.4, 0.2) between the host and LXC, Docker, and KVM guests. It can be seen from Fig. 10 that LXC and Docker are extremely efficient with 4–5 times lower round trip times. KVM has a high variations, and requires optimization to lower the interrupt response delay as well as host OS scheduling delay. The results validate the benefit of containerization for high performance networking.

4.6 New Trends in C-RAN Signal Processing

This chapter is an attempt to analyze three critical issues in processing radio access network functions in the cloud through modeling and measurements. The results reveal new directions to enable a cloud-native radio access network that are outlined below.

New functional split between BBU and RRH: To reduce the fronthaul data rate requirements, optimal functional split is required between BBU and RRH. This depends on the deployment on the cell load, spatial multiplexing (number of UEs/RE/RRH, e.g., MU detection and CoMP), and scenario and can be dynamically assigned between RRH and BBU. In addition some non-time critical function may be performed at a remote cloud. Three principles must be considered while retain-

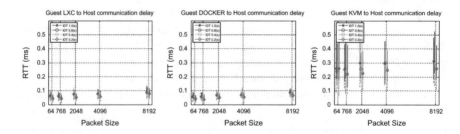

Fig. 10 Round trip time between the host and LXC, Docker, and KVM guests

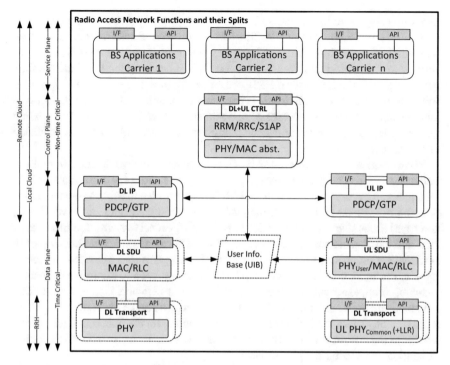

Fig. 11 Functional split between BBU and RRH

ing the benefit of coordinated signal processing and transmission, namely (1) minimize the FH data rate, (2) minimize the split on the time-critical path, (3) no split of the deterministic functions. The proposed split is shown in Fig. 11. In TX chain, full PHY layer can be moved from BBU to RRH (c.f., label 4 in Fig. 2) in order to minimize the fronthaul capacity requirements as the operation of PHY layer remain deterministic as long as the L2/MAC layer provides transport blocks for all channels with the required pre-coding information. When it comes to RX chain, moving cell processing to RRH seems promising as it halves the fronthaul capacity requirements. Additional fronthaul capacity reduction can be obtained if part of user processing can be dynamically assigned to RRH (i.e., log-likelihood ratio) depending on the number of UEs scheduled per resource elements and per RRH. The control plane protocols may be moved to a remote cloud as they are not time-critical functions.

Number of CPU cores per BBU: In LTE-FDD, the total RX (Uplink) + TX (Downlink) processing should take less than 3 ms to comply with HARQ RTT, leaving 2 ms for RX and 1 ms for TX. Due to that TX requires the output of RX to proceed, the number of concurrent threads/cores per eNB subframe is limited to 3 even, if each subframe is processed in parallel. By analyzing processing time for a 1 ms LTE sub-frame, 2 cores at 3 GHz are needed to handle the total BBU processing of an eNB. One processor core for the receiver, assuming 16-QAM on the uplink, and approximately 1 core for the transmitter processing with 64-QAM on the downlink, are required to meet the HARQ deadlines for a fully loaded system. Processing load is mainly dominated by uplink and increases with growing PRBs and MCSs [1, 4]. Furthermore, the ratio and variation of downlink processing load to that of uplink also grows with the increase of PRB and MCS. With the AVX2/AVX3 optimizations, the computational efficiency is expected to double and thus a full software solution would fit with an average of one x86 core per eNB. Additional processing gain is achievable if certain time consuming functions are offloaded to a dedicated hardware accelerator.

Virtualization environment for BBU: When comparing results for different virtualization environments, the average processing times are very close making both container and hypervisor approach to RAN virtualization a feasible approach. However, the bare metal and LXC virtualization execution environments have slightly lower variations than that of DOCKER and KVM, especially with the increase of PRB and MCS increase. In addition, the I/O performance of container approach to virtualization proved to be very efficient. This suggests that fast packet processing (e.g. through DPDK) is required in hypervisor approach to minimize the packet switching time, especially for the fronthaul transport network. Due to the fact that containers are built upon modern kernel features such as cgroups, namespace, chroot, they share the host kernel and can benefit from the host scheduler, which is a key to meet real-time deadlines. This makes containers a cost-effective solution without compromising the performance.

5   Modeling BBU Processing Time

We confirm with the results from Sect. 4 that the total processing time increases with PRB and MCS, and that uplink processing time dominates the downlink. A remaining analysis to study the contribution of each BBU function to the overall processing time is to be done together with an accurate model, which includes the PRB and MCS as input parameters. In this study, three main BBU signal processing modules are considered as main contributors to the total processing including (de-)coding, (de-)modulation, and iFFT/FFT. For each module, the evaluate processing time is measured for different PRB, MCS, and virtualization environment on the Intel SandyBridge architecture with CPU frequency of 3.2 GHz (c.f., Fig. 12).

The plots in Fig. 12 reveals that processing time for iFFT and FFT increase only with the PRB while (de-)coding and (de-)modulation are are increasing as a function of both PRB and MCS. Moreover, the underlying processing platform adds a processing offset to each function. It can be seen from different plots in Fig. 12 that coding and decoding functions represent most of processing time on the uplink and downlink chains, and that decoding is the dominant factor. The QPSK, 16-QAM, and 64-QAM modulation schemes correspond to MCS 9, 16, and 27. The OAI implementation speeds up the processing by including highly optimized SIMD integer DSP instructions for encoding and decoding functions, such as 64-bit MMX, 128-

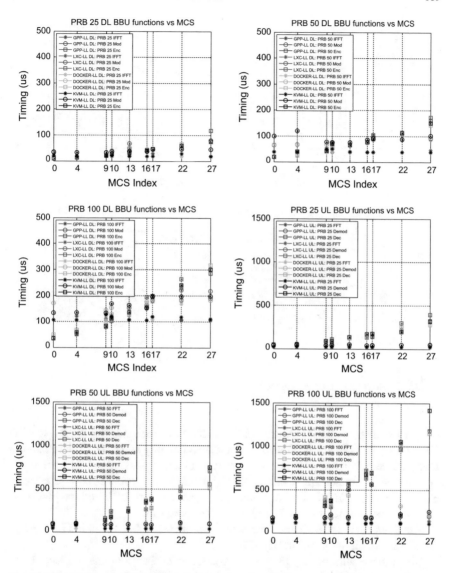

Fig. 12 Contribution of (i-)FFT, (de-)modulation, and (de-)coding to the total BBU processing for different PRB, MCS, and platforms

bit SSE2/3/4. However, when operating the OAI in a hypervisor-based virtualization, some extra delay could be added if these instructions are not supported by the hardware emulation layer (c.f., Fig. 4).

From Fig. 12, we observe that the downlink and uplink processing curves have two components: dynamic processing load added to a base processing load. The dynamic processing load includes user parameters, such as (de-)coding and

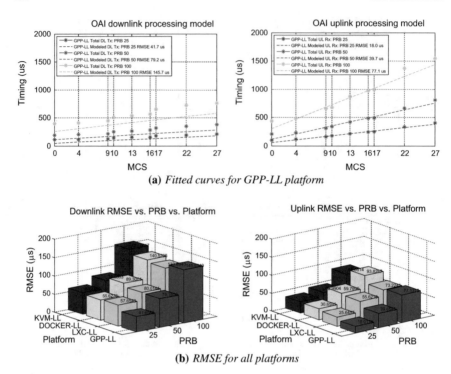

(a) *Fitted curves for GPP-LL platform*

(b) *RMSE for all platforms*

Fig. 13 Modeling BBU processing time

(de-)modulation, which is in linear relation to the allocated MCS and PRBs. Note the (de-)coding functions depend also on the channel quality and SNR. The remaining user parameters, namely DCO coding, PDCCH coding, and scrambling, are modelled as the root mean square error (RMSE) for each platform. The base processing load includes iFFT/FFT cell-processing parameter for each PRB and the platform-specific parameter relative to the reference GPP platform.

The fitted curve of the total processing time for GPP is illustrated in Fig. 13a and the RMSE for all platforms in Fig. 13b.

Given the results in Fig. 13, we propose a model that compute the total BBU downlink and uplink processing time for different MCS, PRB, and underlying platform, as indicated by the following formula.

$$T_{\text{subframe}}(x, y, w) \, [\mu s] = \underbrace{c[x] + p[w]}_{\text{base processing}} + \underbrace{u_r[x]}_{\text{RMSE}} + \underbrace{u_s(x, y)}_{\text{dynamic processing}} \quad , \qquad (4)$$

PRB, MCS, and underlying platform are represented by the triple (x, y, w). The $p[w]$ and $c[x]$ are the base offsets for the platform and cell processing, $u_r[x]$ is the reminder of user processing, and $u_s(x, y)$ is the specific user processing that depends on the allocated PRB and MCS. We fit $u_s(x, y)$ part linearly to $a(x)y + b(x)$, where y is the

Table 2 Downlink processing model parameters in us

x	c	p				$u_s(x, y)$		u_c			
		GPP	LCX	DOCKER	KVM	a	b	GPP	LCX	DOCKER	KVM
25	23.81	0	5.2	2.6	3.5	4.9	24.4	41.6	57.6	55.6	59.4
50	41.98	0	5.7	9.7	13	6.3	70	79.2	80	89.3	79.7
100	111.4	0	7.4	13	21.6	12	147	145.7	133.7	140.5	153

Table 3 Uplink processing model parameters in us

x	c	p				$u_s(x, y)$		u_c			
		GPP	LCX	DOCKER	KVM	a	b	GPP	LCX	DOCKER	KVM
25	20.3	0	5.4	4.8	8.8	11.9	39.6	18	25.6	30.6	32
50	40.1	0	6	9.2	15.8	23.5	75.7	39.6	55.6	59.8	42.9
100	108.8	0	13.2	31.6	26.6	41.9	196.8	77.1	73.2	93.8	80

input MCS, a and b are the coefficients. Tables 2 and 3 provide the uplink and downlink modelling parameters of Eq. 4 for a SandyBridge Intel-based architecture with the CPU frequency set to 3.2 GHz. For different BBU configuration (e.g., Carrier aggregation or Multiple-Input and Multiple-Output (MIMO)), CPU architecture and frequency (c.f., Figs. 6 and 7), a and b has to be adjusted. In our setup, the achieved accuracy using our model is illustrated given an example. Given that PRB equals to 100, Downlink MCS to 27, Uplink MCS to 16, and performing within LXC platform, the estimated total processing time is 723.5 μs ($111.4 + 7.4 + 12 \times 27 + 147 + 133.7$) against 755.9 μs on the downlink, and 1062.4 μs ($108.8 + 13.2 + 41.9 \times 16 + 196.8 + 73.2$) against 984.9 μs on the uplink.

6 Potential Architectures of C-RAN

While from the operators' perspective such an architecture has to meet the scalability, reliability/resiliency, cost-effectiveness requirements, from the software-defined RAN, two key requirements have to be satisfied: (1) realtime deadline to maintain both protocol, frame and subframe timing, and (2) efficient and elastic computational and I/O resources (e.g. CPU, memory, networking) to perform intensive digital signal processing required, especially for different transmission schemes (beam-forming, MIMO, CoMP, and Massive MIMO).

Broadly, three main choices are possible to design a RAN, each of which provide a different cost, power, performance, and flexibility trade-offs.

- **Full GPP**: where all the processing (L1/L2/L3) functions are software-defined. According to China Mobile, the power consumption of the OAI full GPP LTE modem is around 70 W per carrier [7].

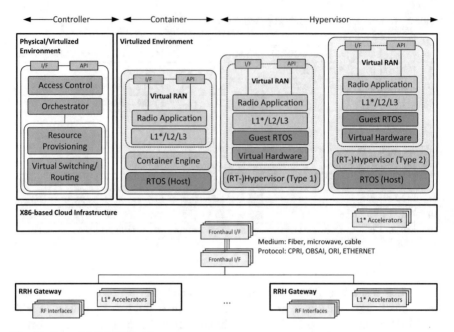

Fig. 14 Potential C-RAN architectures

- **Accelerated**: where certain computationally-intensive functions, such as turbo decoding and encryption/decryption, are offloaded to a dedicated hardware such as an FPGA, GPU, and/or DSP. The remaining functions are software-defined and performed on the host/guest OS. In this case, the power consumption can be reduced to around 13–18 W per carrier.
- **System-on-Chip**: where the entire Layer 1 is performed in a dedicated hardware (e.g. a SoC), and the layer 2 functions are run on the host/guest OS. This can reduce the power consumption to around 8 W per carrier.

As shown in Fig. 14, the hardware platform can either be a full GPP or a hybrid. In the later case, all or part of the L1 functions might be split and placed either locally at the BBU cloud infrastructure or remotely at the RRH unit. In either cases, some of the L1 functions might be offloaded to dedicated accelerator. It can be seen that a pool of base station (BS) can be virtualized inside the same (or different) cloud infrastructure and mapped to RF interface within the RRH gateway. A virtualized RAN (vRAN) can communicate with core networks (CN) through a dedicated interface (e.g. S1 in LTE), and with each other directly through another interface (e.g. X2 in LTE). In addition, vRAN can rely on the same cloud infrastructure to provide localized edge service such as content caching and positioning, and network APIs to interact with the access and core networks [23]. Different service compositions and chaining can be considered, ranging from all-in-one software radio application virtualization to per carrier, per layer or per function virtualization [10]. The virtualization technology can either be based on container or a hypervisor, under the control of a cloud OS, managing the life-cycle of a composite service (orchestrator logic) as well as provisioning the required resources dynamically.

Nevertheless, a full GPP approach to RAN brings the required flexibility in splitting, chaining, and placement of RAN functions while meeting the realtime deadlines along with the following principles [11, 19, 26]:

- **NFV and Micro service Architecture**: breaks down the network into a set of horizontal functions that can be bundled together, assigned with target performance parameters, mapped onto the infrastructure resources (physical or virtual), and finally delivered as a service. This implies that micro virtualized network functions (VNF) are stateless (services should be designed to maximize statelessness even if that means deferring state management elsewhere) and composable (services may compose others, allowing logic to be represented at different levels of granularity; this allows for re-usability and the creation of service abstraction layers and/or platforms). In addition, they can be autonomous (the logic governed by a service resides within an explicit boundary), loosely coupled (dependencies between the underlying logic of a service and its consumers are limited to conformance of the service contract), reusable (whether immediate reuse opportunities exist, services are designed to support potential reuse), and discoverable (services should allow their descriptions to be discovered and understood by (possibly) humans, service requesters, and service discovery that may be able to make use of their logic).[11]
- **Scalability**: monitors the RAN events (e.g. workload variations, optimization, relocation, or upgrade) and automatically provision resources without any degradations in the required/agreed network performance (scale out/in).
- **Reliability**: shares the RAN contexts across multiple replicated RAN services to keep the required redundancy, and distributes the loads among them.
- **Placement**: optimizes the cost and/or performance by locating the RAN services at the specific area subjected to performance, cost, and availability of the RF front-end and cloud resources.
- **Multi-tenancy**: shares the available spectrum, radio, and/or infrastructure resources across multiple tenants (MNOs, MVNOs) of the same cloud provider,
- **Real-time Service**: allows to open the RAN edge service environment to authorized third-parties to rapidly deploy innovative application and service endpoints. It provides a direct access to real-time radio information for low-latency and high-bandwidth service deployed at the network edge [23]. The Real-time Service shall be automatically configurable to rapidly adjust to varying requirements and utilization of the cloud environment (c.f., Sect. 7).

Table 4 compares the requirements of general-purpose cloud application against the cloud RAN.

---

[11]Micro-service architecture is in opposition to the so-called "monolithic" architecture where all functionality is offered by a single logical executable, see http://martinfowler.com/articles/microservices.html. It has to be noted that the micro-service architecture supports the ETSI NFV architecture [10], where each VNF can be seen as a service.

Table 4 General purpose cloud applications versus C-RAN

Application	GP-Cloud computing	Cloud-RAN
Data rate	Mb/s, bursty	Gb/s, stream
Latency/Jitter	Tens of ms	<1 ms, jitter in ns
Lifetime of data	Long	Extremely short
Number of clients	Millions	Thousands–Millions (IoT)
Scalability	High (Micro-service and stateless)	Low
Reliability	Redundancy/load balancing	Redundancy, offloading, load balancing
Placement	Depends on the cost and performance	Specific areas with Radio Frontend. Depends on the cost and performance.
Timescale (operation, recovery)	Non-realtime	Realtime

7   Cloud Architecture for the LTE RAN

In cloudified C-RAN, the BBU becomes software-based, hence the concept of C-RAN cloudification, in which the BBU life-cycle is managed through a cloud operating system and run over the cloud infrastructure, is sound and may become an important business connection between mobile telephony operators and cloud providers. Generally, a cloud provider delivers their (publicly available) service in the form of three different flavors, namely Infrastructure-as-a-Service (IaaS), Platform-as-a-Service (PaaS) and Software-as-a-Service (SaaS) [22], however, in the scope of this work, we put a particular focus on the IaaS-based systems. In the IaaS mode, a cloud operator delivers a resource as a so called Virtual Machines (VM), which comes with processing power, RAM, and storage (optionally other services too) accessible through the Internet. A user operating a VM system has the experience of remote access to an ordinary computer, which is accomplished through a virtualization procedure. Virtualization, which is enabled through a special software layer called a hypervisor, allows us to simultaneously run many VMs (instances) on a single physical cloud server.

When a Cloud-RAN is deployed on a public cloud, then multiple instances compete for the same infrastructure (e.g., computing power, storage, RAM). Hence, in an ordinary setup, we cannot be provided with deadlines for required real-time computing. It is therefore necessary to work out new organizational models of publicly available data centers as currently cloud providers do not offer real-time support in their virtual environment. Here, we briefly present our efforts to allow for real-time support in IaaS clouds. We start with an OpenStack installation of a well established cloud orchestration system. OpenStack looks after computing power, storage, and networking resources of the cloud infrastructure (server pools) and orchestrates the execution of VMs including (re-) configuration upon initialization or a user request.

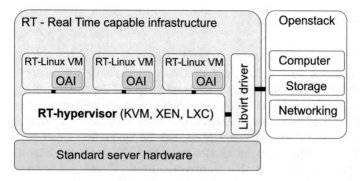

Fig. 15 OpenStack management architecture

The response time of the full-virtualization KVM-based OpenStack system did not fully satisfy our requirements due to unpredictable processing delays. We therefore decided to modify the host system (cloud compute servers) by installing a low latency kernel and replace the default virtualization technique with the Linux Containers (LXC) plugin of OpenStack (c.f., Fig. 15). LXC is Operating System Level virtualization providing high performance as all CPU instructions are natively executed. Moreover, it allows us for the real-time process prioritization on the guest operating system (VM). In our case, the lte-softmodem OAI application is prioritized real-time within the LXC container using the SCHED_DEADLINE or SCHED_FIFO schedulers provided by the low latency Linux kernel. Good performance of RAN satisfied through LXCs could have a big impact on the security of cloud infrastructure as LXCs do not provide a good separation of VMs from physical servers and should be avoided in the case of less time-critical applications such as EPC, HSS, etc. Therefore a heterogeneous cloud infrastructure maintaining both real-time (e.g., LXC-based) and general-purpose (e.g., KVM-based) computing regions[12] can properly serve purposed of the MNO. The region's workload is not know in advance, therefore the cloud provider has to be provided with flexibility to on-demand re-program the infrastructure when required, e.g., to activate a larger number of real-time compute nodes for RAN if the current workload exceeds the capacity of the real-time infrastructure, but the overall cloud-global capacity can still withhold the workload when reconfigured (i.e., adapting the size of real-time and non-real-time regions). To this end, we can employ JUJU[13] and Metal As a Service (MAAS)[14] to program physical cloud compute nodes and provide the concept of *programmable cloud* that dynamically adjusts the cloud region size.

---

[12]A cloud region is an organizational unit of the cloud containing a pool of cloud workers with specific properties such as the same configuration or geographical location.

[13]http://www.ubuntu.com/cloud/tools/juju.

[14]http://www.ubuntu.com/cloud/tools/maas.

8   C-RAN Prototype

In this section, we demonstrate a RANaaS proof-of-concept (PoC) (c.f., the architecture presented in Fig. 16). Our cloud infrastructure consists of the OpenStack orchestrating software with appropriately designed compute servers. Normally, OpenStack manages large pools of resources, but in our example, it controls a local nano datacenter developed to execute RANaaS. Our compute node is deployed on a commodity computer running Ubuntu 14.04 with the low latency Linux kernel version 3.17, while the OpenStack installation uses the LXC plugin on compute nodes to support LXC virtualization. For cloud orchestration OpenStack developed a Heat module that provides a human- and machine-accessible service for the management of the entire life-cycle of a virtual infrastructure and applications. This orchestration engine relies on text-based templates, called Heat Orchestration Templates (HoTs), to manage multiple composite cloud applications and organize them as a stack of virtualized entities (e.g. network, LXCs) called the Heat stack.

Following the LTE protocol stack,[15] our demonstration has to instantiate an E-UTRAN part, evolved packet core (EPC), and home subscriber server (HSS). The EPC consists of a mobility management entity (MME) as well as a Serving and Packet data network Gateway (S/P-GW). Mobile Operators (e.g., MNO, MVNO) use the User Interface (UI) to manage the life-cycle of RANaaS. The Service Manager (SM) component receives user queries from the UI and manages the cloud execution through the Service Orchestrator (SO) component, which leverages the use of the Heat API for cloud orchestration.

In the demonstrated scenario, a HoT file describes the whole virtual infrastructure including the LTE network elements as well as the required network setup tailored to a specific business case. Using the HoT template, Heat manages the service instantiation of every required LTE network function implemented in OAI spread among multiple VMs. As we previously explained, RANaaS has strict latency and timing requirements to achieve a required LTE frame/subframe timing. To this end, we use the SCHED_DEALINE Linux scheduler to allocates the requested runtime (i.e., CPU time) upon every sub-frame to meet the deadline.

Listing 1 presents an example HoT file, which instantiates the RAN as a Service (RANaaS) stack. The template is provided to Heat, which automatically spawns a VM using an arbitrary image previously uploaded to OpenStack (enb-1 provides the installation of the OAI lte-softmodem), attaches the network (e.g., PUBLIC_NET defined in OpenStack), and pre-configures the VM through a bash script provided as user-data. Other VMs illustrated in Fig. 16 could be instantiated in a similar way. Heat allows us to use previously defined resource attributes. For instance, if an eNB requires the address of an HSS, one can reference to it through the *get_attr* Heat function, i.e., get_attr: [EPC, first_address], where the EPC is a previously defined resource and first_address is the attribute of the resource (an IP address of the first interface). Consequently, the whole LTE as a Service (LTEaaS) containing an HSS, EPC, and eNBs can be instantiated from a single HoT file with one request to Heat.

---

[15] Here, the work stack does not refer to Heat and should be understood as a protocol stack.

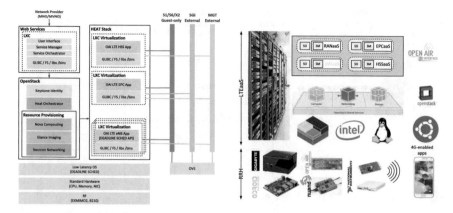

Fig. 16 RANaaS prototype (*left*) and hardware setup (*right*)

Listing 1 The LTEaaS HoT file

```
heat_template_version: 2013-05-23
description: LTEaaS
parameters:
  key_name:
    type: string
    description: >
      Name of a KeyPair to enable
      SSH access to the instance
    default : cloudkey
resources:
  HSS: ...
  MME: ...
  S+P-GW: ...
  eNB:
    type: OS::Nova::Server
    properties:
      image: enb-1
      flavor: eNB.large
      key_name: cloudkey
      networks: [{ network: PUBLIC_NET }]
      user_data:
        str_replace:
          template: |
            #!/bin/bash
            MY_IP=`ip addr show dev eth0 | \
            awk -F'[ /]*' '/inet /{print $3}'`
            sed -i 's#MY_IP_ADDRESS_REPLACE#'$MY_IP'#g' \
            enb.band7.tm1.usrpb210.conf
            sed -i 's#MME_IP_ADDRESS_REPLACE#'$MME_IP'#g' \
            enb.band7.tm1.usrpb210.conf
            ./build_oai.bash --eNB -w USRP > /tmp/oai.log
            ./lte-softmodem -O \
            enb.band7.tm1.usrpb210.conf
          params:
            $MME_IP: { get_param: mme_ip }
```

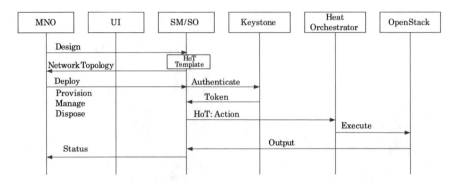

Fig. 17 The RANaaS life-cycle management

LTEaaS describes the service life-cycle of an on-demand, elastic, pay as you go RAN that is running on top of the cloud infrastructure. We believe that life-cycle management is a key for successful adoption and deployment of C-RAN and related services (e.g. MVNO as a Service). It is a process of network design, deployment, resource provisioning, operation and runtime management, and disposal as shown in Fig. 17. In this figure, SM/SO indicates Service Manager/Service Orchestrator, while Keystone and Heat Orchestrator are OpenStack services; the box OpenStack refers to other OpenStack services such as Compute, Storage, Networking, etc. With the help of the UI, the MNO first designs the HoT and spawns other actions such as Deploy, Provision, Manage, and Disposal, which are then managed by the SM/SO that directly communicates with the Heat Orchestrator.

8.1   LTEaaS: eNB Resource Provisioning

This section presents the performance study of the time-critical eNB application running in the LTE as a Service (LTEaaS) architecture. We conducted several experiments particularly relying on the LTE eNB and UE implementation [19, 21] using the OAI platform that implements standard compliant 3GPP protocol stack. We deploy the LTEaaS on the cloud center as shown in Fig. 16 and as described above. The parameters of the real-time OAI eNB are the following: FDD 10 MHz channel bandwidth (50 PRBs) in SISO mode over band 7. MCS are fixed to 26 in downlink and 16 in uplink to produce high processing load. The eNB sends grants to the UE for UL transmission only in downlink SF #0, 1, 2 and 3. Useful UL SFs are then SF # 4, 5, 6, 7. The others UL SFs can possibly be used for HARQ retransmissions.

We compare the feasibility and performance of the proposed LTEaaS architecture using two different linux OS schedulers: namely SCHED_FIFO (not SCHED_ OTHER) or SCHED_DEADLINE (low-latency policy) while running the eNB in LXC containers. Linux cgroups and cpu-sets are used to control the CPUs cores accessible to the container. Bandrich C500, a commercial LTE UE dongle is connected to the instantiated eNB using the classical LTE over-the-air attachment procedure. We measure the uplink goodput (data-rate over a period of a second) for each scheduler applying to the eNB and for different numbers of available CPU cores (CPU is i7-3930k 3.2 GHz with hyper-threading and turbo mode disabled). The measurement lasts 120 s while iperf is generating UDP traffic between the UE and a local server connected to the EPC.

Figures 18 and 19 present the complementary cumulative distribution function of the running time of each RX thread at the eNB, when using SCHED_FIFO or SCHED_DEADLINE with 3 or 2 CPU cores available. Each of these threads corresponds to a specific UL SF. It should be noted that those threads are not the only ones running, as there are also a management thread and a TX thread for each DL SF. In Fig. 18, the value *(1)* of 0.65 ms indicates the BBU and protocol processing time of a fully loaded SF (most of the time for SFs #4, 5, 6 and 7 shown as solid lines in the figure, from time to time corresponding to HARQ retransmission for the other subframes), while the increase *(2)* of 0.2 ms is related to the RLC packet reassembly event that also triggers the PDCP integrity check.

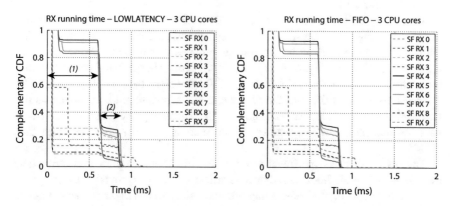

Fig. 18 OAI LTE soft-modem running on 3 CPU cores

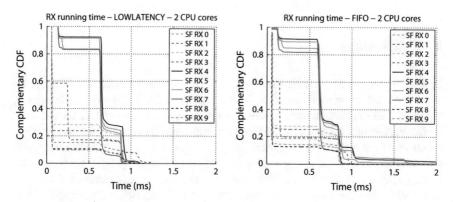

Fig. 19 OAI LTE soft-modem running on 2 CPU cores

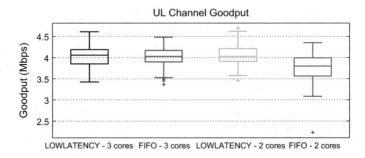

Fig. 20   Impact of the execution environment on the LTE soft-modem uplink performance

Both schedulers behave similarly in this scenario when 3 CPU cores are available as shown on Fig. 18. There is no missed deadline in either case, meaning that the processing power is sufficient to directly execute the required threads in their constrained time (2 ms after receiving RF samples for RX, and 1 ms for generating the RF samples for TX).

When only 2 CPU cores are available, the results change for the FIFO scheduler as shown in Fig. 19. Using the low-latency scheduler, the results are similar than with 3 CPU cores and there is no missed subframes. But using the FIFO scheduler, it can be seen that the SF processing time is sometimes larger than 2 ms as indicated by the tails of the curves and during the 120 s transfer, there are 708 missed SFs. It represents a loss of 0.6 % of the SFs due to late scheduling. Figure 20 shows that while this loss might seem small, it impacts the average uplink goodput with a more than 6 % decrease. The DL channel should present a similar behavior when fully loaded.

The results of this experiment are in line with what was presented throughout this chapter and underlines that adequate hardware resources provisioning (programmable cloud concept) and scheduling are mandatory to achieve high performances in cloud architectures.

9   Conclusions

In this chapter, we have studied and analyzed several important aspects of the radio access network cloudification. First, we have presented C-RAN as a cost effective, scalable, energy efficient, and flexible service for MNOs and MVNOs. Second, current requirements of the LTE standard were translated in terms of various requirements for C-RAN including fronthaul properties, processing software latencies, and real-time capabilities of the operating system. Third, by using OAI, we have evaluated C-RAN in various execution environments such as dedicated Linux, LXC, and KVM. We drew new conclusions on the RRH-based BBU offloading and virtualization environment for C-RAN; we highlighted advantages of containerization over virtualization in C-RAN provisioning. Fourth, we described the properties of RANaaS focusing on the radio-processing organization and micro-service, multitenant architecture; we pointed out main differences between RANaaS and general purpose cloud computing. Finally, we described the cloud architecture for LTE RAN and focused on the C-RAN prototype and its life-cycle management.

References

1. Alyafawi, I., Schiller, E., Braun, T., Dimitrova, D., Gomes, A., Nikaein, N.: Critical issues of centralized and cloudified lte-fdd radio access networks. In: 2015 IEEE International Conference on Communications (ICC), pp. 5523–5528 (2015). doi:10.1109/ICC.2015.7249202
2. Amari LTE 100, a Software LTE Base Station on a PC. http://www.amarisoft.com/
3. Berardinelli, G., Ruiz de Temino, L., Frattasi, S., Rahman, M., Mogensen, P.: Ofdma vs. sc-fdma: performance comparison in local area imt-a scenarios. IEEE Wirel. Commun. **15**(5), 64–72 (2008). doi:10.1109/MWC.2008.4653134
4. Bhaumik, S., Chandrabose, S.P., Jataprolu, M.K., Kumar, G., Muralidhar, A., Polakos, P., Srinivasan, V., Woo, T.: Cloudiq: a framework for processing base stations in a data center. In: Proceedings of the 18th Annual International Conference on Mobile Computing and Networking, Mobicom '12, pp. 125–136. ACM, New York, NY, USA (2012). doi:10.1145/2348543.2348561
5. Chanclou, P., Pizzinat, A., Le Clech, F., Reedeker, T.L., Lagadec, Y., Saliou, F., Le Guyader, B., Guillo, L., Deniel, Q., Gosselin, S., Le, S., Diallo, T., Brenot, R., Lelarge, F., Marazzi, L., Parolari, P., Martinelli, M., O'Dull, S., Gebrewold, S., Hillerkuss, D., Leuthold, J., Gavioli, G., Galli, P.: Optical fiber solution for mobile fronthaul to achieve cloud radio access network. In: Future Network and Mobile Summit (FutureNetworkSummit) 2013, pp. 1–11 (2013)
6. Checko, A., Christiansen, H., Yan, Y., Scolari, L., Kardaras, G., Berger, M., Dittmann, L.: Cloud ran for mobile networks—a technology overview. IEEE Commun. Surv. Tutorials **17**(1), 405–426 (2015). doi:10.1109/COMST.2014.2355255
7. China Mobile Research Institute: C-RAN White Paper: The Road Towards Green RAN. http://labs.chinamobile.com/cran (2013)
8. Costa-Perez, X., Swetina, J., Guo, T., Mahindra, R., Rangarajan, S.: Radio access network virtualization for future mobile carrier networks. IEEE Commun. Mag. **51**(7), 27–35 (2013). doi:10.1109/MCOM.2013.6553675
9. Dahlman, E., Parkvall, S., Skold, J.: 4G LTE/LTE-Advanced for Mobile Broadband, 1st edn. Academic Press (2011)
10. ETSI: Network Functions Virtualisation (NFV), White paper. Technical report, ETSI (2014)
11. EU FP7 Mobile Cloud Networking public deliverable D2.5: Final Overall Architecture Definition. Technical report, EU (2015)
12. EURECOM: Open Air Interface. http://www.openairinterface.org/
13. Haberland, B., Derakhshan, F., Grob-Lipski, H., Klotsche, R., Rehm, W., Schefczik, P., Soellner, M.: Radio base stations in the cloud. Bell Labs Tech. J. **18**(1), 129–152 (2013). doi:10.1002/bltj.21596
14. Interworking and JOINt Design of an Open Access and Backhaul Network Architecture for Small Cells based on Cloud Networks (iJOIN): an FP7 STREP project co-funded by the European Commission. http://www.ict-ijoin.eu/
15. Mobile Cloud Networking (MCN): an FP7 IP project co-funded by the European Commission. http://www.mobile-cloud-networking.eu
16. NGMN: Further Study on Critical C-RAN Technologies (v0.6). Technical report, The Next Generation Mobile Networks (NGMN) Alliance (2013)

17. NGMN: Suggestions on Potential Solutions to C-RAN by NGMN Alliance. Technical report, The Next Generation Mobile Networks (NGMN) Alliance (2013)

18. Nikaein, N.: Processing radio access network functions in the cloud: critical issues and modeling. In: Proceedings of the 6th International Workshop on Mobile Cloud Computing and Services, MCS'15, pp. 36–43. ACM, New York, NY, USA (2015). doi:10.1145/2802130.2802136

19. Nikaein, N., Knopp, R., Gauthier, L., Schiller, E., Braun, T., Pichon, D., Bonnet, C., Kaltenberger, F., Nussbaum, D.: Demo—closer to Cloud-RAN: RAN as a service. In: Proceedings of ACM MOBICOM Demonstrations (2015). doi:10.1145/2789168.2789178

20. Nikaein, N., Knopp, R., Kaltenberger, F., Gauthier, L., Bonnet, C., Nussbaum, D., Ghaddab, R.: OpenAirInterface 4G: an open LTE network in a PC. In: Proceedings of the 20th Annual International Conference on Mobile Computing and Networking, MobiCom '14, pp. 305–308. ACM, New York, NY, USA (2014). doi:10.1145/2639108.2641745

21. Nikaein, N., Schiller, E., Favraud, R., Katsalis, K., Stavropoulos, D., Alyafawi, I., Zhao, Z., Braun, T., Korakis, T.: Network store: exploring slicing in future 5g networks. In: Proceedings of the 10th International Workshop on Mobility in the Evolving Internet Architecture, MobiArch'15, pp. 8–13. ACM, New York, NY, USA (2015). doi:10.1145/2795381.2795390

22. Oracle: Oracle Cloud, Enterprise-Grade Cloud Solutions: SaaS, PaaS, and IaaS. https://cloud.oracle.com/home

23. Patel, M., Joubert, J., Ramos, J.R., Sprecher, N., Abeta, S., Neal, A.: Mobile-Edge Computing. Technical report, ETSI, white paper (2014)

24. Scalable and Adaptive Internet Solutions (SAIL): an FP7 IP project co-funded by the European Commission. http://www.sail-project.eu

25. Schooler, R.: Transforming Networks with NFV & SDN. Intel Architecture Group (2013)

26. Wilder, B.: Cloud Architecture Patterns. O'Reilly (2012)

27. Wubben, D., Rost, P., Bartelt, J., Lalam, M., Savin, V., Gorgoglione, M., Dekorsy, A., Fettweis, G.: Benefits and impact of cloud computing on 5g signal processing: flexible centralization through cloud-ran. IEEE Signal Process. Mag. **31**(6), 35–44 (2014). doi:10.1109/MSP.2014.2334952

Smart City Surveillance in Fog Computing

Ning Chen, Yu Chen, Xinyue Ye, Haibin Ling, Sejun Song
and Chin-Tser Huang

Abstract The Internet and Internet of Things (IoT) make the Smart City concept an achievable and attractive proposition. Efficient information abstraction and quick decision making, the most essential parts of situational awareness (SAW), are still complex due to the overwhelming amount of dynamic data and the tight constraints on processing time. In many urban surveillance tasks, powerful Cloud technology cannot satisfy the tight latency tolerance as the servers are allocated far from the sensing platform; in other words there is no guaranteed connection in the emergency situations. Therefore, data processing, information fusion and decision making are required to be executed on-site (i.e., near the data collection locations). Fog Computing, a recently proposed extension of Cloud Computing, enables on-site computing without migrating jobs to a remote Cloud. In this chapter, we firstly introduce the motivations and definition of smart cities as well as the existing challenges. Then the

N. Chen (✉) · Y. Chen
Department of Electrical and Computer Engineering, Binghamton University,
Binghamton, NY, USA
e-mail: nchen14@binghamton.edu

Y. Chen
e-mail: ychen@binghamton.edu

X. Ye
Department of Geography, Kent State University, Kent, OH, USA
e-mail: xye5@kent.edu

H. Ling
Department of Computer and Information Sciences, Temple University,
Philadelphia, PA, USA
e-mail: hbling@temple.edu

S. Song
School of Computing and Engineering, University of Missouri-Kansas City,
Kansas City, MO, USA
e-mail: songsej@umkc.edu

C.-T. Huang
Department of Computer Science and Engineering, University of South Carolina,
Columbia, SC, USA
e-mail: HUANGCT@cse.sc.edu

© Springer International Publishing Switzerland 2017
C.X. Mavromoustakis et al. (eds.), *Advances in Mobile Cloud Computing
and Big Data in the 5G Era*, Studies in Big Data 22,
DOI 10.1007/978-3-319-45145-9_9

concepts and advantages of Fog Computing are discussed. Additionally, we investigate the feasibility of Fog Computing for real-time urban surveillance using speeding traffic detection as a case study. Adopting a drone to monitor the moving vehicles, a Fog Computing prototype is developed. The results validate the effectiveness of our Fog Computing based approach for on-site, online, uninterrupted urban surveillance tasks.

1 Smart City

The world's population has been increasingly concentrated in urban areas at an unprecedented scale and speed [1–3]. This rapid process of urbanization has brought profound influence on the daily life of urban citizens, but at the same time, as the byproducts of urbanization, a series of issues have emerged, such as environmental pollution, energy consumption, urban crimes, and traffic congestion. In addition, urbanization has led to social, environmental, economic, political transformations. All these issues resulted from urbanization process are challenging the city governments, urban planners and stakeholders. In the following sections, the motivations and concepts of smart cities are presented, and the challenges are discussed.

1.1 Motivations and Concepts

The United Nations World Urbanization Prospects reported that the urban population in the world has grown up rapidly from 746 million, 30 % of the world's population in 1950, to 3.9 billion which is 54 % of the total population of the world in 2014 [4]. In addition, the United Nations estimated that the percentage could reach as high as 66 % in 2050 and the urbanization progress will be much faster in some developing counties such as China. This rapid pace has improved the living quality of urban residents through developing physical infrastructure, transportation system, as well as education and health facilities. However, the negative effects have also emerged. A variety of serious ecological and social issues are listed below.

- **Air pollution**: Along with the growing population in urban areas, vehicles are necessary transportation tools providing convenience for citizens. However, as more vehicles running in the cities, huge volumes of vehicle exhaust can be produced every day which is hazardous and causes health issues. Due to rapid economic growth and industrialization in many developing countries, it is noteworthy that industry gases emissions has also played an important role in contributing to air pollution.
- **Traffic congestion**: Traffic congestion in urban areas caused by the huge numbers of vehicles. It leads to long traffic delays, especially during the rush hours. Besides wasting our time in traffic, there are some underlying but more significant impacts.

For example, more fuel is consumed due to the higher burning rate during the traffic jams, which is one of the main sources of air pollution leaving alone the loss of money.

- **Car accidents**: The growing number of vehicles in urban areas could give rise to fatal car accidents as well. A report from Texas Department of Transportation [5] shows that in 2014, 344 people in Texas were killed in crashes involving speed-over limit of which 210 people were the drivers and the other 134 persons were passengers or pedestrians. Therefore, a smart, on-line urban speeding vehicle surveillance system would be helpful to reduce the number of car accidents.

The above urban issues are not exclusive. There could be more concerns relating to economical, religions and political issues. Cities are not the places people simply live together anymore. The urbanization are reflecting the people's pursuing for better life and the big transforms of culture, economic, society and politics. Considering the issues and problems of urbanization, people are seeking for technical solutions to make sustainable developing and environment friendly cities.

The evolution of information and communication technology has provided people the opportunities to solve the urbanization issues. The Internet has been an important part of daily life and various kinds of city dynamics are combined tightly with digital sensors and networking systems. With the ubiquitously deployed sensors and pervasively available computing devices, the cities are largely digitalized. Human mobility, energy consumption, air quality, traffic conditions and other index of city dynamics can be recorded.

In recent years, new networking system and computing architectures have been proposed to deal with the rapid process of urbanization and to provide Internet-based services such as Big Data, Internet of Things (IoT), Cloud Computing, Fog Computing etc. They are different technologies, but also relating to each other. The Internet of Things connects not only digital sensors and devices, but also physical infrastructure together by which people can get real-time data from a remote location and information can be exchanged between devices [6]. The large scale of data collected from ubiquitous digital sensors are valuable to analyze the city and the answers for urban issues can be explored in massive urban data. In big data [7] technologies, the hidden relationship and reasons for urban issues will be exploited.

Grid computing, utility computing, Cloud Computing and Fog Computing are different computing architectures for different applications. The appearance of Cloud Computing has solved the problems such as how to store the massive volume of urban dynamic data and how to analyze the urban big data with powerful computing tools. The data centers in Cloud Computing provide the users more flexibility without the need for capital outlay which reduce the maintenance cost and potential risks. Fog Computing, as the extension of Cloud Computing, enables the computation tasks accomplished at the edge of network which would be ideal for latency sensitive services and in some extreme conditions. The combination of all the developing information and communication technologies can help cities collect data, deliver data to data centers and analyze data for living patterns and city dynamics which takes lots of advantages for urban planning and governance.

Considering the urbanization issues to be solved and the existing technologies, the concept of 'Smart Cities' emerges. Smartness means understanding and learning: understanding the new patterns and learning how to deal with the new patterns just as in smart cities. People leverage the information and communication technologies to collect the complex urban dynamic data from which underlying dynamic patterns would be extracted by data mining techniques to understand the city. From understanding and analyzing of complex urban data, urban planners can make the cities more intelligent and the city governance more efficient. There is not a consensus about smart city definitions. From different perspectives of distinct relative stakeholders, the definitions are differing from each other.

While there are many different definitions for smart city, below is one that provides a clear vision [8]:

> A smart city is a system integration of technological infrastructure that relies on advanced data processing with the goals of making city governance more efficient, citizens happier, businesses more prosperous and the environment more sustainable.

According to the above definition of smart city, the main goal of smart cities is to make citizens happier by utilizing information technologies. Smart cities use data sensing and acquisition technologies to collect data regarding every aspect of cities, data transmission technologies to send the urban data to analysis centers and data mining technologies to fuse and analyze the urban data to extract valuable information.

Figure 1 illustrates a four-layer hierarchical smart city architecture, which consists of: data sensing and acquisition layer, data storage and management layer, data analysis layer and smart applications layer. The first layer at the bottom is the data sensing and acquisition layer. In smart city, the data sensing and acquisition is the fundamental part of the entire system. In this layer, heterogeneous networked sensors are deployed ubiquitously in smart cities to collect dynamic urban data and the advanced communication equipments are utilized to transmit the unstructured urban data to the second layer. The second layer is the data storage and management layer. The dynamic urban data is characterized as massive volume at both spatial and temporal scale. In addition, the data are collected by different sensors in different formats. Therefore, we need a layer to store the big urban data effectively and also are capable of filtering useful data efficiently from heterogeneous data sources.

The third layer is the data analysis layer which is composed of computing centers. The valuable data from data storage layer will be sent to the upper layer for analysis. In the data analysis layer there can be different computing tools. Cloud data centers can be used for long-term analysis and batch processing jobs with powerful computation capability. Fog Computing can be utilized for on-site processing and instant decision making. The top layer is the application layer. Once the computation results are obtained in the data analysis layer, they will be transferred to the appli-

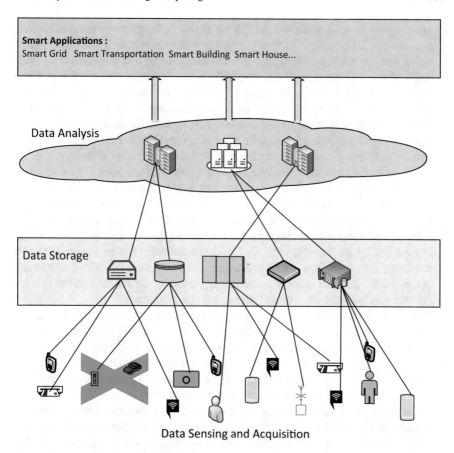

Smart Applications :
Smart Grid   Smart Transportation Smart Building Smart House...

Data Analysis

Data Storage

Data Sensing and Acquisition

Fig. 1 A four-layer smart city architecture

cation layer. This layer consists of a wide range of smart applications such as smart grid, smart house and smart building, which make the smart city more intelligent.

1.2 Challenges

Smart city is a powerful strategy to deal with the severe problems along with the process of urbanization. Information and communication technologies are adopted to obtain the essential understanding to urban issues. Although the innovative technologies are helpful to improve the situations, the smart city is still facing a lot of challenges. How to obtain effective and correct data sets for certain applications and how to analyze heterogeneous data need to be further discussed. In this section, several major challenges in smart city are presented.

- **Data sensing and acquisition** As discussed above, the data sensing and acquisition are the fundamentals to smart cities since data are depicting the dynamic of cities, but currently problems in city wide data sensing and acquisition do exist. Lack of enough sensors for some specific mission critical missions is still one major constraint, for example, urban surveillance for over-speed traffic monitoring. Installing new digital sensors at a city wide range is a kind of solution but would take the additional power consumption as well. Different to traditionally digital sensors deployed for specific purposes, recently citizens are seen as sensors as well. People always record what they see and hear which would be published in social networks. Associating with spatial and temporal tags, these data sets are ideal for smart city applications such as city regional semantics recognition, smart transportation, trace analysis and so on. However, the negative part of this strategy seeing citizens as sensors are quite obvious. First of all, data collected from personal digitally social networks can be noisy, how to extract useful data is a big challenge. Secondly, this kind of data sometimes can be misleading since the initial purpose of these data are not for the urban analysis. The data literacy is a problem. The last issue is that accessing personal data always cause privacy debates which has to been taken into serious consideration.
- **Data management** One goal of smart cities is to reveal the underlying relationship between complex urban dynamic data and urban phenomenons. The massive urban dynamic data are collected by pervasively deployed digital sensors or citizens as sensors. The wide deployment of sensors is the foundation for the mission critical data-driven tasks and urban analysis. However, sensors widely deployed in smart cities are heterogeneous, especially between digital sensors and citizen as sensors, the urban data collected could be in different forms and unstructured as numerical forms and word forms. It is a big challenge to obtain the valuable data from such a big data base. What's more noteworthy, the data collected by digital sensors are always quantitative, but the data from citizens could be qualitative. For same set of quantitative data, combining with different qualitative data, they could have significantly different meanings. How to combine heterogeneous data from a layered sensor environments is non-trivial.
- **Data processing** Data processing is the kernel part for urban applications. Mining the data, building the model, extracting new patterns are critical in smart city. Although a variety of data mining algorithms have been developed, more novel algorithms are in need. Some of the currently available machine learning algorithms can only solve the problems under certain context or certain distribution. In many research areas, the models established are under certain conditions, which are not suitable for many real-time, or mission critical urban applications.

2   Urban Surveillance

Human caused disasters are one of most concerning issues in smart city, for example, car accidents caused by intentionally overspeed driving. Such kind of disasters not only lead to the loss of money, but also could be fatal to innocent people. Urban surveillance, utilizing the ubiquitous sensors and data processing technologies, is an essential part for quick detection and prompt response to urgent situations in smart city. In this section, the motivations and the existing problems of urban surveillance is discussed.

2.1 Urban Surveillance: Motivations

As the rapid urbanization and pervasively increasing of urban dynamic data, smarter management strategies are expected for administrators and planners. Efficient urban surveillance is important for situational awareness (SAW), which is essential for many critical and dynamic data-driven missions [9–11]. Considering the wide dimensions of cities, urban surveillance is becoming an indispensable part in urban planning. Along with the prosperity, the urbanization has also witnessed the increase of crimes and violations. Considering the limited law enforcement resources, the digitalized urban surveillance can provide urban residents a safer residence environment.

Cameras, smart phones, transportation cards or any other digital devices can be utilized for data collection, which enable timely tracking, analysis and decision making. Beside situational monitoring for crimes and violations, urban surveillance is also a powerful tool in some special environments that may be dangerous to human operators. For example, the chemical products storage and the surrounding environment where risk of explosion exists. In addition, trace data of individuals or communities are also very useful for epidemics dissemination control, abnormal illegal events detection and even early alarm for terrorism activity. In urban surveillance, un-interrupted dynamic data sensing, real-time massive data analysis and instant accurate decision making for sudden disasters are quite critical and significant. Understanding and analyzing the large-scale complex urban dynamic data is of great importance to enhance both human lives and urban environments.

Target detection and potential danger recognition from surveillance data is achieved through an exploitation of a layered sensor environment, and real time detection is ideal [9]. However, this mission is challenging due to the lack of powerful computing infrastructures at the surveillance site that is able to process the big dynamic data. Outsourcing all the data to remote data center may not be able to meet the tight latency constraints [12]. Therefore, a smart surveillance strategy is expected to leverage the computing power close to the job site, such that it is feasible to achieve the goal of instant decision making.

2.2 Urban Surveillance: Open Problems

Urban surveillance is a significant part for situational awareness and mission critical tasks. However, there are still some open problems in building a real-time urban surveillance system in practice. Here we illustrate two of the major issues, data sensing and computing architecture.

- **Surveillance data sensing**: Widely deployed sensors and pervasively used smart devices enable urban surveillance. Meanwhile, surveillance sensors are not enough in certain specific tasks. Let's consider the urban traffic surveillance as an example. In the United States, the traditional way to catch over speeding drivers is that the policemen patrolling on the road. Now with the help of cameras installed along the roads or at the intersections, the over speeding surveillance can be more efficient. However, due to the limited resource of police department and the number of installed cameras at fixed locations, overspeed drivers can still drive at whatever speed they like as long as they are aware of the police officer and remember to slow down near the cameras. Obviously, new sensors and detecting strategies are needed. Ideally the urban surveillance data-driven system should be capable to recognize the potential danger efficiently, to infer the reasons accurately and to make decisions quickly.
- **Computing architecture**: It is ideal that the urban surveillance system can fuse and process the dynamic urban data from a heterogeneous layered sensors in a real-time manner, especially for latency-sensitive applications. However, the existing computing architecture cannot serve this kind of applications well. At the current stage of urban surveillance, Cloud Computing is recognized as a promising solution for large scale data processing. The real-time, dynamic data are collected from sensors deployed for critical surveillance tasks. It implies that huge volume of data need to be transferred to Cloud center for pattern recognition and decision making. However, Cloud cannot satisfy the requirements of all the mission critical applications, especially those requesting tight response time. As various surveillance missions are emerging, current computing architecture failed to satisfy the requirements.

3 Fog Computing

Dynamic data fusion is highly desired for urban surveillance, especially in the context of natural or human caused disasters. Cloud Computing provides cost efficient solutions for large scale, batch data processing jobs. As mentioned above, however, the fast development of ubiquitously deployed sensors for data collection and mobile computing techniques is pushing systems to the boundary where Cloud Computing is not able to satisfy users' requirements [13]. Not only because powerful Cloud cannot meet the tight latency tolerance as being allocated far away from the sensing area, but also there is often no guaranteed connection during the emergency. Therefore, to meet the requirements of mission critical situations, it is crucial to provide the functionalities of data processing, information fusion, and decision-making in an on-site manner. Fog Computing [13–15], a recently proposed extension and complement for Cloud Computing, enables computing at the network edge without outsourcing jobs to remote Cloud centers [15]. In this section, the concepts of Fog Computing is introduced and then an architecture including both Cloud Computing and Fog Computing paradigm is given.

3.1 Fog Computing: Concepts

The Internet of Things are connecting physical infrastructure together providing us the opportunity to access the remote data, sense the situation and control the physical systems for efficient resource management and accurately customized services. Cloud Computing is a promising computing paradigm providing users hardware or software infrastructure with flexibility [16]. Furtherly, Cloud Computing free the users with no need for capital outlay and maintenance cost.

Besides the economy and flexibility, Cloud Computing is charactered as powerful computation capability and batch processing for which it is ideal for large scale urban dynamic data analysis [17]. However, the developing of Internet of Things are leading to a wide variety of innovative applications and surveillance missions. The ubiquitous smart devices and advanced networking technologies are giving rise the new location awareness services which require on-site processing and low latency quality of service. Based on the tight requirements proposed by some applications, the Cloud Computing cannot be feasible for all the tasks anymore.

Fog Computing, as the extension of and complement architecture to Cloud Computing, is a promising computing paradigm to meet the requirements proposed by rapidly developing Internet of Things. The definition of Fog Computing can be given as below:

> Fog Computing is a distributed computation paradigm that leverages the huge number of heterogeneous devices deployed at the edge of the network, which are connected with each other and collaborate with each other by sharing computation, storage and communication functionalities.

As depicted in definition, the ubiquitously deployed digital devices serve as Fog Computing nodes. Cloudlet, personal laptops, smart phones, tablets and even routers could be Fog nodes. Some of the advantages of Fog Computing are listed as below:

- **High availability**: With the prosperity of the Internet of Things, more and more digital devices can serve as Fog Computing nodes as long as they are capable of processing and storing data.
- **Location awareness**: In Fog Computing paradigm, the fog nodes are deployed close to the data source and the computing results are often used locally. The location awareness is highly desired for many smart city surveillance applications.
- **Low latency** Comparing to the remote Cloud data center, Fog Computing paradigm enables the computation directly on site, at the edge of network, it removed

the round trip time from job site to Cloud centers. This characteristic is essential to many latency sensitive applications.

- **Networking efficiency:** In Cloud Computing scenario, the data needs to be transmitted from end users to remote data center. It relies on the network conditions and actually causes unnecessary network traffic since in many cases the data are only locally significant.
- **Security and privacy**: Processing the collected data on-site can also reduce the risk of being intercepted or compromised by attackers.

3.2 Fog Computing Architecture

Figure 2 shows a Fog Computing architecture. The layer at the bottom consists of end users with arbitrary locations. There would be a wide variety of processing tasks considering the large amount of end users in the Internet of Things. The amount of real-time, dynamic user data could be tremendous and the users request based processing applications are heterogeneous. It is certain that the large scale data are difficult to be stored at the end user side because of the limited storage capability. Furthermore, some mission critical tasks are latency sensitive, the large round trip

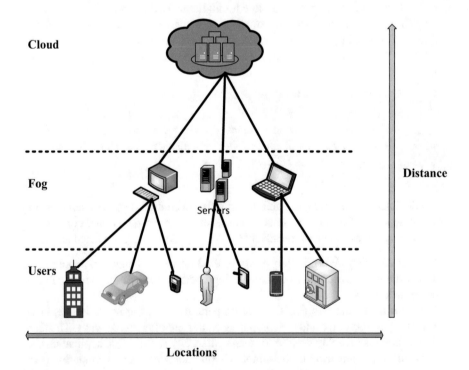

Fig. 2 The Fog Computing architecture

Table 1 Comparison between Fog and Cloud

Computing paradigm	Fog Computing	Cloud Computing
Resource allocation	Distributed	Cluster
Real-time interaction	Yes	No
Latency	Low	High
Devices	Heterogeneous	Virtual machines
Computation capability	Normal	Powerful

time between end user and Cloud center will not be tolerated. Therefore, the Fog Computing layer is allocated between traditional Cloud layer and end users layer.

The collected data will be transferred to Fog nodes for processing. Once the job is done, the Fog nodes send the results back to the user applications and upload meta data to the Cloud center for further analysis.

The top layer is still the Cloud layer which will provide powerful computing infrastructure to end users. Tremendous amount of data after being processed by Fog layer are stored in the Cloud centers and advanced data mining technologies are utilized to obtain the thorough and long-term analysis which is an indispensable part in smart cities.

A comparison between Fog Computing and Cloud Computing is given in Table 1 below.

The main difference between Fog paradigm and Cloud paradigm shown in the table actually verifies that they are complementary to each other. Fog and Cloud tackle different requirements for distinct tasks in smart city. Cloud Computing with powerful hardware and software infrastructure is more suitable for large scale data analysis for deeper insight. In contrast, Fog Computing is not as powerful as Cloud Computing. But with the property of low latency and real-time processing, Fog Computing is more desirable in situational awareness tasks of urban surveillance.

4 A Case Study

Traffic surveillance enables traffic monitoring and traffic light adjustment by utilizing the data collected from the widely deployed cameras and sensors [18–20, 20–22]. Real-time traffic data helps the police department optimize the resources to deal with accidents at particular locations in a more efficient and active way. In traffic surveillance, how to effectively monitor speeding vehicles is always a big concern. Overspeed driving not only hurts drivers, but also may cause fatal consequences to innocent people like pedestrians or people in non-overspeed vehicles. A report from Texas Department of Transportation [5] shows that in 2014, 344 people in Texas were killed in crashes involving speed-over limit of which 210 people were the drivers and the other 134 persons were passengers or pedestrians. Therefore, a smart, on-line urban speeding vehicle surveillance system would be very helpful to reduce the number of car accidents.

We took real-time traffic monitoring as a case study to evaluate the feasibility of a Fog Computing based urban surveillance solution [10, 23]. In our scheme, a drone in the sky monitors vehicles on the ground and a raw video stream is collected and sent back to the controller on ground, which implements the real-time vehicle surveillance. And the raw video stream is sent to Fog Computing nodes nearby, such as a laptop, like the one in police cars. The Fog node tracks the moving target in the video and calculates the speed. Our scheme not only can monitor the traffic in much wider area compared to cameras on ground, but also process the video data and get the speed outcome in real time, which is critical for immediate response to the speeding violation.

This section is organized as follows. First we introduce the Fog Computing based urban surveillance system for speeding detection. Then the target tracking and speed calculation algorithms are presented. Finally, the experimental results are reported with some discussions.

4.1 System Architecture

Figure 3 illustrates the three-layer system architecture of the Fog Computing based urban surveillance system. It consists of the remote Cloud center, on-site/near-site computing Fog, and data collection units, such as sensors and cameras. The kernel is the Fog layer, which is formed by multiple computing units, including drones, computers carried by the vehicles, and computing devices of first responders. They monitor the area concurrently from different positions. When they are collecting real-time data streams, each of them also needs processed information for instant decision making. Although it is ideal that all the collected data are sent back to central Cloud facility for thorough global analysis, there is no guarantee that a reliable communication network to remote Cloud center is always available. In addition, not all data is globally significant. It does not need to create unnecessary traffic in the networks. Instant on-site decision making also reduces the risk of exposing the data to eavesdroppers in transmission channels. Therefore, the Fog, which consists of the computing devices carried by the units in or near the monitoring area, can fulfill the requirements very well.

In the existing traffic monitoring systems, police officers and cameras along the road are not sufficient to effectively keep the drivers from driving above the speed limit, especially because of the limited monitoring range of police cars and the number of cameras. In the proposed Fog based surveillance system, drones flying in the sky generate real-time raw video stream, and send it back to the ground station and display it on the screen. Thus the human operator can monitor the traffic of a much wider range, detect fast moving vehicles on the screen, and get the speed information when they find someone is suspicious.

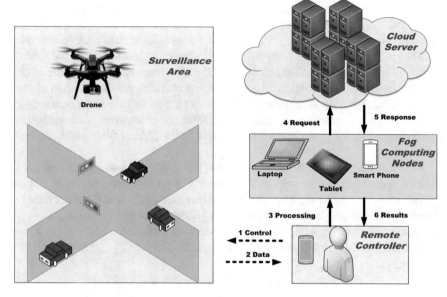

Fig. 3 System architecture

It is worth to note that the nodes in Fog layer provide both computation power and storage space. The raw video data can be pre-processed and stored in the Fog nodes at first and be delivered to the remote Cloud center for a long-term analysis to build a historical record of the traffic condition. For example, with the collection of traffic events in a long period, a big picture may be constructed, which is very valuable for city planners and policy makers.

4.2 Algorithms

In order to catch speeding vehicles, two algorithms are needed: one is the target tracking algorithm and the other is the vehicle speed calculation algorithm.

A. Target Tracking Algorithm

When a fast moving vehicle appears in the real-time video, it needs to be locked immediately and be tracked effectively frame by frame. In practical scenarios, not only the suspicious vehicle itself appears in the video frame, but also occlusion, background clutter, the change of illumination and the noise. It becomes more challenging if the user needs to conduct the job in the night, when the road is dark. Other than the requirements of robustness and effectiveness of the tracking algorithm, the processing time in the Fog nodes could be another big concern. When a user locks one moving vehicle in the video, the speed information of that vehicle is needed instantly.

Considering the specific requirements, a real time robust L1 tracker using accelerated proximal gradient approach [24] is selected. This accelerated real time L1 tracker is casted by using the sparse representation in the particle filter framework.

The particle filter implements the recursive Bayes estimation using the method of non-parametric Monte Carlo simulation. It uses a large number of particles that are transferred in the state space to estimate the probability density function of state variables. Particle filter is an efficient tool to solve the problem in non-linear system. In addition, the distribution of random variables are unnecessary to be Gaussian distribution. Two steps are essentially involved in the particle filter: prediction and update.

We denote x_t to represent the state variable describing the motion of the target in frame t. y_t denotes the observation of the moving target in frame t. In target tracking applications, we assume state variable x_t is only related to x_{t-1} and the observation at frame t is only related to x_t, which means observations among $y_{1:t} = \{y_1, y_2, \cdots, y_t\}$ are independent of each other, given $x_{1:t}$. It is assumed that at frame $t-1$, the probability density distribution is $p(x_{t-1}|y_{t-1})$. In prediction step, $p(x_t|y_{t-1})$ is derived from $p(x_{t-1}|y_{t-1})$:

$$p(x_t, x_{t-1}|y_{t-1}) = p(x_t|x_{t-1}, y_{t-1})p(x_{t-1}|y_{t-1}) \tag{1}$$

In Eq. (1), given x_{t-1}, x_t and y_{t-1} are independent, so Eq. (1) becomes:

$$p(x_t, x_{t-1}|y_{t-1}) = p(x_t|x_{t-1})p(x_{t-1}|y_{t-1}) \tag{2}$$

Then compute the integration of Eq. (2) over x_{t-1}:

$$p(x_t|y_{t_1}) = \int p(x_t|x_{t-1})p(x_{t-1}|y_{t-1})dx_{t-1} \tag{3}$$

With Eq. (3), we can move forward to the update step by using Bayes rules:

$$p(x_t|z_t) = \frac{p(x_t|x_{t-1})p(x_t|y_{t-1})}{p(y_t|y_{t-1})} \tag{4}$$

$p(y_t|x_t)$ is the observation likelihood. In the particle filter, the posterior probability above is estimated by N samples, denoted by $S_t = \{x_t^1, x_t^2, x_t^3, \cdots, x_t^N\}$ with different weights. Due to the lack of knowledge about variable distribution, sequential important distribution $q(x_t^{(i)}|y_t)$ is used to generate the samples. The weight is:

$$W_t^{(i)} \propto \frac{p(x_t^{(i)}|y_t)}{q(x_t^{(i)}|y_t)} \tag{5}$$

and the weight can be updated as follows:

$$W_t^i = w_{t-1}^i \frac{p(y_t|x_t^i)p(x_t^i|x_{t-1}^i)}{q(x_t|x_{1:t-1}, y_{1:t})} \tag{6}$$

The observation likelihood depicts the similarity between the target candidate and the target templates [25].

In sparse approximation, the signal y can be linearly represented by the atoms of the over-complete dictionary D.

$$y = D \cdot x \tag{7}$$

where x is the coefficient of each atom in the dictionary D. In moving target tracking algorithm, over-complete dictionary consists of target templates denoted by $T = t_1, t_2, t_3, \cdots, t_n$. With the target templates, a target candidate can be represented as follows:

$$y \approx T \cdot x = x_1 t_1 + x_2 t_2 + \cdots + x_n t_n \tag{8}$$

Because of the sparsity in sparse approximation, for a good target candidate, most coefficients of the target templates should be zero, which means a good target candidate can be nearly represented by several target templates. In other words, the coefficients of a bad target candidate can be relatively equally with smaller number.

In the real scenarios of our monitoring videos, other than the target object, occlusion, noise, shadows, sometimes even darkness would appear. So we have to consider the error. Therefore, trivial templates denoted by $I = i_1, i_2, i_3, \cdots, i_n$ are introduced in this algorithm and the Eq. (8) is rewritten as follow:

$$y = \begin{bmatrix} T & I \end{bmatrix} \begin{bmatrix} x \\ e \end{bmatrix} \tag{9}$$

where e represents the coefficients of trivial templates. In a further consideration, it is reasonable to assume that the coefficients of a good candidate should be positive, which can also be considered as the non-negative constraints. Hence, in this scenario, positive and negative trivial templates should be involved. Then Eq. (9) is rewritten as:

$$y = \begin{bmatrix} T & I & -I \end{bmatrix} \begin{bmatrix} x \\ e_+ \\ e_- \end{bmatrix} = D \cdot m \tag{10}$$

Here, $D = (T\ I\ -I)$ and $m^T = (x\ e_+\ e_-)$. What we want to know is the coefficients m of the target templates and trivial templates, but in the over-complete dictionary $D^{m \times n}$, m is much smaller than n, which means the solution of Eq. 10 is not unique. Some constraints are indispensable in order to get a unique solution in the sparse representation. Fortunately, we can solve this problem as an $L1$ norm least squares problem.

$$min\|Dm - y\|_2^2 + \lambda\|m\|_1 \tag{11}$$

$\| \cdot \|_1$ denotes l_1 norm and $\| \cdot \|_2$ denotes l_2 norm here respectively. As mentioned above, trivial templates are brought into the dictionary to deal with the noise and occlusion. But what if there is no occlusion? The target object should be well approximated by the target templates from previous frames. Additionally, in case of no occlusion in the frame, the trivial templates would impact the detection accuracy otherwise and bring some computation complexity. So in this accelerated l_1 norm tracking algorithm, another coefficient μ_t is introduced to improve the constraint (11). The revised constraint is as below:

$$min\frac{1}{2}\|y_t - Dm\|_2^2 + \lambda\|m\|_1 + \frac{\mu_t}{2}\|m_I\|_2^2 \tag{12}$$

where m_I is the coefficients of trivial templates in this target tracking sparse approximation problem: $m = [m_T \ m_I]$. If occlusion is detected in a video frame, μ_t is zero. Otherwise, μ_t is supposed to be some specific value.

In practical experiments, solving such kind of modified l_1 norm minimization could be pretty time consuming. A fast numerical method called accelerated proximal gradient [26] is applied to solve this problem. This approach is designed for solving the optimization problem as below:

$$min \ F(a) + G(a) \tag{13}$$

and the accented proximal gradient is fast for some specific types of function G.

After solving the l_1 least squares minimization problem and obtaining the sparse coefficients m, the observation likelihood of state variable x_t^i can be expressed as:

$$p(y_t|x_t^i) = \frac{1}{\Gamma}exp\{-\alpha\|y_t^i - T_t m_T^i\|_2^2\} \tag{14}$$

where α is used to control the shape of a Gaussian Distribution and Γ is a normalized factor. m_T denoted the coefficients of target templates. The optimal state x_t^i satisfies:

$$x_t^i = \arg\max_{x_t^i \in S_t} \ p(y_t|x_t^i) \tag{15}$$

B. Speed Calculation Algorithm
Leveraging the moving target tracking algorithm discussed above, the position of a vehicle in consecutive frames can be identified using a bounding box to lock the moving vehicle on the road of interests. If the real distance that the vehicle has traveled is available, the speed can be calculated easily.

As shown in Fig. 4, the pink rectangle represents one video frame whose resolution is $m \times n$, and the dark blue circle represents actual field taken by the drone camera. α is the field of view of the drone camera and the dash line h is the height from the camera to the ground. Knowing height h and the field of view α, the longest distance of the field monitored by the drone camera, which is the diameter of the dark blue circle as well, can be:

Fig. 4 Field of view

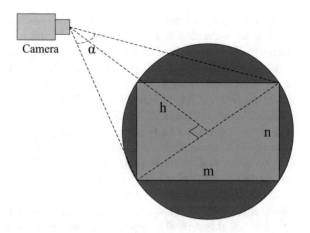

$$d = 2h \times tan\frac{\alpha}{2} \tag{16}$$

According to the way a camera works, the image plane is also a circle that is circumscribed of the CCD (Charge Coupled Device) plate in the camera. We assume that each pixel represents nearly equally length, so the actual length that one pixel represents in an image can be obtained by:

$$l = \frac{d}{\sqrt{m^2 + n^2}} \tag{17}$$

The l_1 tracker algorithm indicates that the position of the moving vehicle frame by frame. Assume in two consecutive video frames, the vehicle position changes from (x_1, y_1) to (x_2, y_2). The pixel number that the vehicle moves across in the frame is:

$$n = \sqrt{(x_2 - x_1)^2 + (y_2 - y_1)^2} \tag{18}$$

Then, $R = l \times n$ is the actual range that the vehicle moves between two consecutive video frames and combined with the time interval, we can calculate the speed v:

$$v = \frac{R}{t} \tag{19}$$

4.3 Experimental Results

We have evaluated the proposed Fog Computing based speeding vehicle detection scheme. The experimental results are reported in this section.

A prototype has been built to validate the feasibility of Fog Computing based smart urban surveillance. Two DJI Phantom 3 Professional drones are used to monitor the traffic on road. Two Google Nexus 9 tablets are connected to the remote controller of the drones such that the collected video streams are played to the operators in real-time. One laptop functions as a Fog Computing node, which configuration is Intel core i7-2720QM 2.20 GHz and the RAM memory is 8.00 GB. The given FOV (Field of View) of the camera on the drone is 94°. However, in real camera engineering, manufacturers would always make the image plane not perfectly circumscribed with the CCD plate. Therefore, the diameter would be a little larger than the actual length represented by the diagonal of the image. We have taken a photo of an object whose dimensions are known to calibrate FOV. The calibrated FOV is 89.39°.

It is a challenge to track an object in a noisy environment, such as the shadows on the road, multiple similar targets on the road or even dark video frames because of the night. It is critical to ensure that we will not lose the track of the vehicle of interest.

Figure 5 shows the results of target tracking in day time, with tree occlusions on the road from the roadside. The test video stream was taken above a road on Binghamton University campus. The vertical height from the camera on our drone to the ground is 140.0 m. The black car is the target and a white bounding box is on its body in the image for locking. The tracking algorithm has successfully tracked that black car all the time and never lost it.

Then some more challenging scenarios were considered. Figure 6 shows the tracking results with multiple vehicles moving on the road. This video stream was taken above a local freeway and the vertical height is 262.5 m. The results verified that the proposed Fog based surveillance scheme is able to track the target coexisting with multiple similar objects.

Fig. 5 Tracking test sequence 1

Fig. 6 Tracking results with multiply similar objects

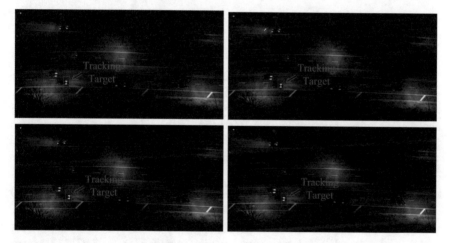

Fig. 7 Tracking results in dark night

Also, the capability of tracking in dark is highly desired by city administration since there is less police force on road in evening times and drivers are more likely to drive faster. At night it is too dark to distinguish the body of cars from the background. Our strategy is to lock the headlights of a moving vehicle. The results shown in Fig. 7 verified the effectiveness of the tracking scheme in dark night.

Fig. 8 Speed calculation test scenario

5 Speeding Vehicle Detection

The first experiment examined the accuracy of speed evaluation. The video is taken above a campus road of our university, and the vertical height of the drone is 140.0 m. As shown in Fig. 8, a black Toyota Camry is moving on the road with the constant speed of 27 mile per hour. There are 160 frames in the video stream processed with the time interval of 100 *ms* between two consecutive frames. The speed of tested vehicle was calculated every ten frames, and the speed estimation error is defined as follows:

$$error = \frac{|estimation - actual|}{actual} \tag{20}$$

Figure 9 presents the estimated driving speed of the vehicle. It is close to the actual driving speed, 27 mile per hour, varying from 26.2358 to 28.7736. The estimation errors are shown in Fig. 10, which is small (lower than 10 %). This experiment has proved that the proposed scheme can efficiently estimate the speed of vehicles driving on road with an acceptable accuracy.

Then, we have applied our system to monitor real traffic on highway and caught speeding vehicles successfully. The video is taken on the highway I81-N, where the speed limit is 65 mile per hour. Figure 11 shows that a red freight truck at the left lane is speeding up to pass a white freight truck in the right lane.

The speed estimation results are depicted in Fig. 12. The red line represents the speed of the vehicle on the left lane and the blue line represents the slower one on the right lane. The red line is above the blue line almost all the time, which is consistent with the observation in the video. The ranges of the speed over the limit of these two vehicles are depicted in Fig. 13. The faster vehicle is speeding and its speed was

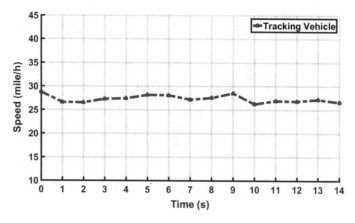

Fig. 9 Speed calculation results

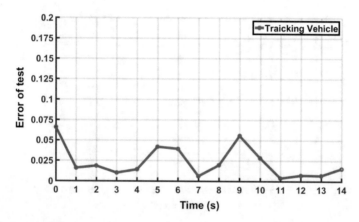

Fig. 10 Speed estimation errors

Fig. 11 Test sequence on highway

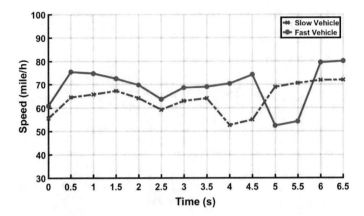

Fig. 12 Highway speed calculation

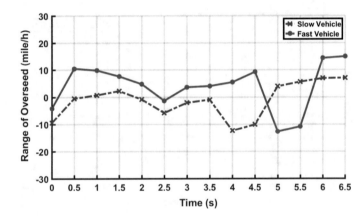

Fig. 13 Range of overspeed

10 miles higher than the limit. In particular, after 6 s, it became 15 miles faster than the speed limit. In contrast, the blue line shows that the speed of the slower vehicle keeps around the speed limit pretty well.

6 Summary

The emergence of Internet of Things and advanced computing architecture are motivating us to implement the concept of smart city. The focus of smart city is to make citizens happier, management more efficient and business more prosperous by utilizing innovative information and communication technologies. However, implementing a smart city still faces some challenges. The information fusion from heterogeneous layered sensors, efficiently and correctively analysis of large scale dynamic data are difficult. Situational awareness, which is an essential part of urban surveillance, requires short processing delay and quick decision making which would be hard for existing methods.

Cloud Computing is recognized as a promising solution for smart city. But it cannot meet the requirements from mission critical applications. The latency is the main obstacle to implement Cloud Computing based urban surveillance. Fog Computing can enable computation directly at the edge of network which is close to end users providing storage and processing schemes. Such that Fog Computing is potentially the ideal tool to be utilized in situational awareness tasks.

In this chapter, we proposed to apply Fog Computing in smart urban surveillance and validated the feasibility using speeding vehicle detection as a case study. Leveraging the Fog Computing paradigm, a conceptual proof prototype has been built, in which DJI Phantom 3 Professional drones are used for real-time surveillance video collection, tablets and laptops serve as Fog Computing nodes. It allows users monitor the traffic and track the speeding vehicles effectively. Field experimental studies have been conducted and the results show that the proposed scheme can efficiently track the target and catch speeding vehicles in real-time. This work has validated the feasibility of applying the Fog Computing paradigm to make the city smarter.

References

1. Arribas-Bel, D.: Accidental, open and everywhere: emerging data sources for the understanding of cities. Appl. Geogr. **49**, 45–53 (2014)
2. Batty, M.: Smart cities, big data. Environ. Plann. Part B **39**(2), 191 (2012)
3. Zheng, Y., Capra, L., Wolfson, O., Yang, H.: Urban computing: concepts, methodologies, and applications. ACM Trans. Intell. Syst. Technol. (TIST) **5**(3), 38 (2014)
4. U. Nations, World Urbanization Prospects 2014: Highlights. United Nations Publications (2014)
5. T. D. of Transportation, Texas motor vehicle crash statistics (2014). http://www.txdot.gov/government/enforcement/annual-summary.html. Accessed 01 Nov 2015
6. Zanella, A., Bui, N., Castellani, A., Vangelista, L., Zorzi, M.: Internet of things for smart cities. IEEE Internet of Things J. **1**(1), 22–32 (2014)
7. Chen, M., Mao, S., Liu, Y.: Big data: a survey. Mob. Netw. Appl. **19**(2), 171–209 (2014)
8. Yin, C., Xiong, Z., Chen, H., Wang, J., Cooper, D., David, B.: A literature survey on smart cities. Sci. China Inf. Sci. **58**(10), 1–18 (2015)
9. Blasch, E., Seetharaman, G., Suddarth, S., Palaniappan, K., Chen, G., Ling, H., Basharat, A.: Summary of methods in wide-area motion imagery (wami). In: SPIE Defense + Security. International Society for Optics and Photonics, pp. 90 890C (2014)
10. Chen, Y., Blasch, E., Chen, N., Deng, A., Ling, H., Chen, G.: Real-time wami streaming target tracking in fog. In: the 2016 SPIE Defense, Security, and Sensing (DSS) (2016)
11. Wu, R., Chen, Y., Blasch, E., Liu, B., Chen, G., Shen, D.: A container-based elastic cloud architecture for real-time full-motion video (fmv) target tracking. In: Applied Imagery Pattern Recognition Workshop (AIPR), 2014 IEEE, pp. 1–8. IEEE (2014)
12. Wu, R., Liu, B., Chen, Y., Blasch, E., Ling, H., Chen, G.: Pseudo-real-time wide area motion imagery (wami) processing for dynamic feature detection. In: 2015 18th International Conference on Information Fusion (Fusion), pp. 1962–1969. IEEE (2015)

13. Bonomi, F., Milito, R., Natarajan, P., Zhu, J.: Fog computing: a platform for internet of things and analytics. In: Big Data and Internet of Things: A Roadmap for Smart Environments, pp. 169–186. Springer (2014)
14. Stojmenovic, I., Wen, S.: The fog computing paradigm: scenarios and security issues. In: 2014 Federated Conference on Computer Science and Information Systems (FedCSIS), pp. 1–8. IEEE (2014)
15. Yi, S., Li, C., Li, Q.: A survey of fog computing: Concepts, applications and issues (2015)
16. Armbrust, M., Fox, A., Griffith, R., Joseph, A.D., Katz, R., Konwinski, A., Lee, G., Patterson, D., Rabkin, A., Stoica, I., et al.: A view of cloud computing. Commun. ACM **53**(4), 50–58 (2010)
17. Stantchev, V., Barnawi, A., Ghulam, S., Schubert, J., Tamm, G.: Smart items, fog and cloud computing as enablers of servitization in healthcare. Sensors Transducers (1726-5479) **185**(2) (2015)
18. Buch, N., Velastin, S., Orwell, J., et al.: A review of computer vision techniques for the analysis of urban traffic. IEEE Trans. Intell. Transp. Syst. **12**(3), 920–939 (2011)
19. Kitchin, R.: The real-time city? big data and smart urbanism. GeoJournal **79**(1), 1–14 (2014)
20. Megalingam, R.K., Mohan, V., Leons, P., Shooja, R., Ajay, M.: Smart traffic controller using wireless sensor network for dynamic traffic routing and over speed detection. In: Global Humanitarian Technology Conference (GHTC), 2011 IEEE, pp. 528–533. IEEE (2011)
21. Sarowar, S.S., Shende, S.M.: Overspeed vehicular monitoring and control by using zigbee
22. Srinivasan, S., Latchman, H., Shea, J., Wong, T., McNair, J.: Airborne traffic surveillance systems: video surveillance of highway traffic. In: Proceedings of the ACM 2nd international workshop on Video surveillance & sensor networks, pp. 131–135. ACM (2004)
23. N. Chen, Y. Chen, Y. You, H. Ling, and R. Zimmermann, "Dynamic urban surveillance video stream processing using fog computing," in *the 2nd IEEE International Conference on Multimedia Big Data (BigMM 2016)*
24. Bao, C., Wu, Y., Ling, H., Ji, H.: Real time robust l1 tracker using accelerated proximal gradient approach. In: 2012 IEEE Conference on Computer Vision and Pattern Recognition (CVPR), pp. 1830–1837. IEEE (2012)
25. Mei, X., Ling, H.: Robust visual tracking using l_1 minimization. In: 2009 IEEE 12th International Conference on Computer Vision, pp. 1436–1443. IEEE (2009)
26. Tseng, P.: On accelerated proximal gradient methods for convex-concave optimization. SIAM J. Optim. (2008)

Part IV
MCC and Big Data Control and Data Management

Secure Opportunistic Vehicle-to-Vehicle Communication

Alexandra-Elena Mihaita, Ciprian Dobre, Florin Pop,
Constandinos X. Mavromoustakis and George Mastorakis

Abstract How much time do you spend stuck in traffic? Well, the average person spends around 43 hours a year stuck just because of the over-populated streets. The road infrastructure is something that cannot be easily improved, which is why the field of Intelligent Transportation Systems (ITS) has emerged. The most common example of such systems is the navigators which integrate the location monitoring of drivers with the services that help predict faster(or at least, most pleasant) alternative route(s). But, since everyone tries to find the optimal route, conflict of interest between the drivers can appear: one driver can choose to send bad data in order to give a false image of the map and gain advantages. The present chapter describes a solution to create a security mechanism in the context of ITS. The solution is a heterogeneous solution in which both symmetrical and asymmetrical encryption are used. Section 1 makes a short introduction into the field of ITS with its main challenges. The related work is then presented in Sect. 2. A theoretical approach (see Sect. 3) over the security mechanism proposed in the chapter is then made, followed by the practical description of the implementation (Sect. 4) and the constructive details (Sect. 5). Section 6 presents the experimental evaluation and the results of

A.-E. Mihaita (✉) · C. Dobre · F. Pop
Faculty of Automatic Control and Computers, University Politehnica of Bucharest,
313, Splaiul Independentei, 060042 Bucharest, Romania
e-mail: alexa.mihaita@gmail.com

C. Dobre
e-mail: cipran.dobre@cs.pub.ro

F. Pop
e-mail: florin.pop@cs.pub.ro

C.X. Mavromoustakis
Department of Computer Science, University of Nicosia,
46 Makedonitissas Ave., 2414 En-gomi, Nicosia, Cyprus
e-mail: mavromoustakis.c@unic.ac.cy

G. Mastorakis
Department of Informatics Engineering, Technological Educational Institute of Crete,
Estavromenos, 71500 Heraklion, Crete, Greece
e-mail: mastorakis@gmail.com

© Springer International Publishing Switzerland 2017
C.X. Mavromoustakis et al. (eds.), *Advances in Mobile Cloud Computing
and Big Data in the 5G Era*, Studies in Big Data 22,
DOI 10.1007/978-3-319-45145-9_10

the security proposal. Finally, the conclusion are presented in the final Sect. 7 and some future work is mentioned.

1 Introduction

Given the growth in number of vehicles on road, we all had at least one episode of frustration while being stuck in traffic, getting late for a meeting or desperate driving to arrive only late at work. The over-populated road infrastructure cannot simply be upgraded so easily (not that it helps much, as we show in Sect. 6), which is why people turned their attention to "smart technologies". Intelligent Transportation Systems (ITS), in particular, are receiving increasing attention lately, due to the benefits wireless devices, combined with sensing technologies and ICT smart services, would bring. Navigators are among most common examples of such systems, that integrate monitoring of a drivers position with services designed to offer alternative route(s) to make, in theory at least, his voyage to the destination faster (or at least, more pleasant). Most of us used probably applications on our smartphone or car computer such as Google Traffic, or Waze, or services from TomTom or Garmin, just to give examples of such solutions. Besides this examples, ITS include *Advanced Traffic Management Systems* (ATMS), integrating traffic control strategies based on the real traffic information, *Advanced Traveler Information Systems* (ATIS), centered on the user experience and on giving the drivers all the information that they require in a easy-to-understand manner, *Advanced Vehicle Control and Safety Systems* (AVCSS), focusing on making the roads a safer place with a lower accidents rate, *Advanced Public Transportation Systems* (APTS), translating technologies used in ATMS, ATIS and AVCSS in the public transportation field to improve the quality of service and the efficiency, and/or *Commercial Vehicle Operation systems* (CVO), designed to translate ideas from the ATMS, ATIS and AVCSS into the field of commercial operation vehicles such as buses, ambulances or even police cars.

Vehicular communications are a part of any such ITS, and represent a particular set of networks in which the vehicles and roadsides are considered to be the peers of the networks between which messages are exchanged, in order to form a safer and smarter way to drive. We assume vehicles are equipped with wireless communication capabilities—a reality with all major car manufacturers nowadays. Communication can happen between a vehicle and wireless units offered by the road infrastructure (vehicle-to-infrastructure, or V2I, communication), and between vehicles as they pass each others (vehicle-to-vehicle, or V2V, communication). V2I communications allow the exchange of critical information for the safety and operation of the roads in order to avoid or, at least, minimize the effects of a motor vehicle crash. It is an important piece in the need of anticipating the status of the roads in order to create alternative routes or warnings for the users. V2V communication, on the other hand, can be quite useful for allowing a vehicle to "sense threats and hazards with a 360 degree awareness of the position of other vehicles and the threat or hazard they present; calculate risk; issue driver advisories or warnings; or take preemptive actions to avoid and mitigate crashes" [1]. In practice, both technologies should work together, as vehicle can receive updates on a global scale from a Data Center through V2I communication, and ad-hoc updates from nearby locations from other participants in traffic through V2V communication.

The MIT Technology Review named Vehicle-to-Vehicle (V2V) one of the biggest tech breakthroughs of 2015, predicting that it will become widely available as soon as next year. General Motors was the first major car company to commit, announcing in September that it would release a V2V-equipped Cadillac by 2017 [2]. World Congress on Intelligent Transport Systems in 2014 led to this conclusion: connected cars will be the ultimate Internet of Things. They will collect and make sense of massive amounts of data from a huge array of sources. Cars will talk to other cars, exchanging data and alerting drivers to potential collisions. They will talk to sensors on signs on stoplights, bus stops, even ones embedded in the roads to get traffic updates and rerouting alerts. And they will communicate with your house, office, and smart devices, acting as an digital assistant, gathering information you need to go about your day [3].

V2V communication determines a vehicle to have a preemptive behavior in order to prevent car damages (e.g. crashing into a poll), accidents (e.g. generated from high speed or lack of assurance when transcending another car) and even traffic jams. Preemptive behavior assumes the fact that the vehicle need to be well informed about the status of the network and about its position in the network in order to compute a series of probabilities to determine the best course of action. V2I communications allow the exchange of critical information for the safety and operation of the roads in order to avoid or, at least, minimize the effects of a motor vehicle crash. It is an important piece in the need of anticipating the status of the roads in order to create alternative routes or warnings for the users.

Both V2V and V2I have been developed in order to create a safer, more efficient way to drive. The problem is when a selfish vehicle driver appears that wants to know all the information about the traffic status, to receive all the warnings if there are any, but it does not want to pass them along or, worse, to generate false data in order to maximize his experience. This kind of behavior is considered malicious and needs to be prevented and that can be done using a security mechanism (see illustration in Fig. 1).

Imagine an intersection where all the vehicles can be uniquely identifiable and each car has the ability to give information about the traffic it experienced until that

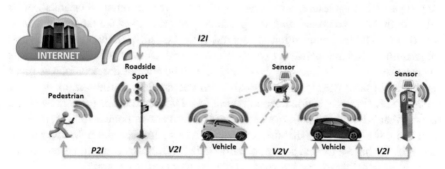

Fig. 1 Intelligent traffic example

point through the GPS position, traffic alerts, pollution or economic driving. Then, all the vehicles could have an image of the traffic status and could decide further paths in order to avoid traffic jams and even accidents caused by the correlation of fast speed driving with little space between the vehicles. But if a car says that a certain path was crowded then all the cars that wanted to pass on that street will reconsider and try using a different path, which leads to traffic jams on the alternative route. This motivates a vehicle to listen to the information about traffic jams and not pass it forward, so that he will be able to circulate faster than the others. In order to prevent such behaviors, we need to be able to distinguish the messages each car sends and protect their content. This preventive behavior is part of a bigger mechanism called the security mechanism.

In this chapter, a security mechanism for vehicle communication is proposed, one which guarantees properties such as message authentication and non-repudiation, content protection and location validation. The mechanism is designed to work with both V2V and V2I communication technologies.

Up until now, research in securing the VANET networks has been focused on three leading ideas: cryptography-based, radio signal-based and resource-based, each idea presenting advantages and disadvantages.

The cryptography-based approach grants position reliability and identity with non-repudiation having as main disadvantage the high overhead added to a system. The main focus has been using the Public Key Infrastructure (PKI) which has introduced problems about the public key distribution and validation.

The radio signal-based approach verifies the accuracy in the positions of all neighbors by the power signal of a transmitted message. The disadvantage is derived from the advantage in the sense that the approach of position detection based of signal power is not always accurate because of the signal properties of bouncing off from different surfaces.

The resource-based method verifies the resources of a vehicle in order to determine if it respects the minimum requirements of the network (e.g. radio resources needed for send/receive messages, computational or identification resources test). The main disadvantage was demonstrated by Newsome et al. [4] that claimed that the method can not by applied successfully to ad-hoc networks.

The presented security mechanism is cryptography oriented, and implies a public key type of infrastructure, where messages are signed in order to establish one's identity and are encrypted with the public key in order to make them secret from the others. The communication is secured with the Diffie Hellman key exchange algorithm for the link between two vehicles and with .X509 certificates for the link between the infrastructure and the vehicles. The novelty of the approach is given by the fact that it tries to protect nodes from one another, in the sense that if one node is malicious, its action need to be limited. Translated to the real life events, it means that all vehicles will provide secured information about their GPS position and about the traffic jams that they experienced to the Infrastructure. In this way, all the vehicles have a trusted communication with the Infrastructure and each car can be held responsible for the information that it brings to the network.

Another advantage introduced by the security mechanism presented in this chapter is that all the processing data is made on the infrastructure side, which eases the complexity of the application existing on vehicles. All the vehicles generate raw data about their location and traffic alerts when they remain in a range for more then a quantum of time, and then send that data to the infrastructure. The infrastructure then correlates the data to the data existing in the network to that point in order to determine the rightfulness of the data and to estimate the risks that it involves. If traffic alerts have been spotted for a certain path, the infrastructure has the duty of finding alternative paths and to split the vehicles so that those alternative routes do not become also crowded.

Another idea brought by the proposed mechanism is the security of the communication when using third parties to forward messages without allowing them to alter the messages. This idea was based on the Cooperative MAC protocol that allows helpers to pass along messages from a sender to a receiver in order to help the communication between the two communication parties but the novelty was the fact that all the senders sign the messages they send and then add a hash to the messages, the helper takes the message from the sender and also signs the message. Thus, in case of a difference between the received hash of the message and the computed hash of the message, the receiver can determine where things went wrong and signal it to the infrastructure.

An adaptation of the Diffie Hellman key exchange algorithm was thought of for vehicle to vehicle communication securisation which added the benefit of never using the same key twice, which leads to less vulnerability for an attacker to take advantage of.

The chapter is structured as follows: First, a short related work is presented in Sect. 2. In Sect. 3 the theoretical concepts and ideas leading to the development of the secure mechanism are discussed. This is followed in Sect. 5 by details for design and implementation. An evaluation of the solution is presented in Sect. 6. The chapter ends with conclusions.

2 Related Work

Intelligent Traffic Systems (ITS) are advanced applications that imply multiple technologies of electronics, control and sensing, computer and others along with innovative services in order to allow users to be better informed about the road status, for safety and efficiency increase on the streets, to relieve traffic congestion and reduce air pollution.

In order to do so, ITS get information from road sensors and vehicles in traffic that run dedicated mobile applications. Considering that each car has limited resources and that each driver pursues his interest of reaching the destination fastest, traffic participants could start acting malicious by providing false information about their location or about the traffic status.

M. Raya discussed in Ref. [5] about the vehicular security evolution, starting from tachometers to GPS tracking devices and dedicated mobile applications. In order to better determine the requirement for a VANET, the author proposed in paper [5] a distribution of the VANET application into two main categories based on their focus:

- safety related applications are vital to traffic situations and the most dangerous. This kind of applications are focused on embedding information from sensors, other traffic participant and the infrastructure in order to avoid life-endangering situations. Their greatest challenge is processing all the data in the given time slot.
- other applications focus on the management of the network like payment services, location inquires or improving quality of service (QoS). These applications need security but their time constraints are looser.

The CARAVAN scheme presented in [6] is a second type application so the security mechanism chosen is not time dependent which makes possible dividing vehicles into groups, selecting a leader for each group as a proxy and forwarding the data. This example minimizes the vulnerabilities of eavesdropping.

Safety related vehicular applications tend to be more complicated because of the sensitive data and limited time for communication. Gerlach et al. [7] describe the main principles for vehicular communication security. Their work is considered to be a reference in the field. Hubaux et al. [8] discuss about privacy and position forgeries in VANETs. They introduce the concept of Electronic License Plate (ELP) as a unique identifier for vehicles in ITS. Authors of [9] present ideas for securing VANETs with digital signatures, and in [10] authors introduce a security architecture that uses PKI for digital tachometers.

Authors in [11] discuss open problems in security over VANETs, resuming them as follows:

- secure positioning. GPS signal is weak within cities which makes it vulnerable to spoofing or jamming attack. This kind of data is necessary in the authentication phase but also for accountability about the shared data.
- data verification. Correlation and verification of all the data in the network are needed in order to prevent forging attacks.
- DoS resilience. Wireless communication can be easily jammed. A possible solution is the implementation of multiple transceivers with disjoint frequency bands.

In Ref. [11] authors also discuss about the characteristics of attacks and split them at a superficial view, into five categories:

- bogus information. A malicious attacker sends false information in order to confuse the other users and create a distorted image of the network.
- cheating with positioning information. This type of attack can be used when running from responsibility in case of accidents.
- ID disclosure. The disclosure of one's neighbor can be done in order to rebuild their traces.
- denial of service attacks. These attacks are used in order to make the network unavailable.

Table 1 Attacker categories

Attacker	Description
Insider versus outsider	The insider is an authenticated member of the network
	The outsider is an intruder and his actions are therefor limited
Malicious versus rational	The malicious does not seek personal retribution but to create damage to the network
	The rational is more predictable since it seeks personal advantages
Active versus passive	The active can generate messages or signals
	The passive can not generate messages but only to eavesdropping

- masquerade or impersonating attacks. These attacks are used for sending bogus information (Tables 1 and 2).

A description of different attacks on VANET is presented in papers [12–14] as follows:

- denial of service or distributed denial of service are meant to make the network unavailable and thus the users uninformed.
- message suppression attacks happen when a user, willingly, does not forward critical messages in order to release them at later times and thus create a wrong image of the traffic.
- fabrication attacks happen when a user creates false information in order to obtain certain privileges.
- alteration attacks occur when the messages exchanged are different from the receiver to the destination, thus creating a false image of the traffic.
- replay attacks are the ones that collect data over a time slot only to reuse that data at later times in order obtain certain privileges.
- sybil attacks are part of the impersonating attacks and happen when an attacker uses a different set of identification at the same time.
- ID disclosure attack adds prejudice to users privacy, revealing their secret data.
- location tracking is an attack that has as main goal monitoring users paths in order to retrace them.
- repudiation attack is similar to impersonating in the sense that it presumes the identity theft but the motive here is to speak at the same time on behalf of the user's whose identity has been stolen in order to get that user to be repudiated from the system.
- eavesdropping occurs when an attacker listens to the communications from the network in order to get confidential information.

In order to address the problems M. Raya presented different types of specific attacks for VANET, multiple security methods and mechanism have been proposed. See Table 1. A comparison of several approaches has been made in [15] between: ARAN [16], ARIADNE [17], CONFIDANT [18], DCMD [19], SAODV [20], SEAD [21], SLSP [22], SPAAR [23], SOLSR [24] and WATCHDOG-PATHRATER [25]. A comparative summary can be seen in Table 2.

Table 2 Security approaches comparison

Approach	Objectives	Mechanism	Packet overhead	Processing
ARAN	Authentication, integrity and non-repudiation of signaling packets	Certificate authority and timestamps	Average	High
ARIADNE	Authentication and integrity of signaling packets	Symmetric cryptography, hash functions and timestamps	Low	Average
CONFIDANT	Exclude misbehaviour	Reputation system	Average	Low
DCMD	Detect and correct malicious data	Observation and plausibility of events	Average	Low
SAODV	Authentication and integrity of signaling packets	Digital signatures and hash chains	Average	High
SEAD	Authentication and integrity of signaling packets	Hash chains and sequence numbers	High	Average
SLSP	Authentication, integrity and non-repudiation of signaling packets	Certificate authority	Average	High
SPAAR	Authentication, integrity, non-repudiation and confidentiality of signaling packets	Certificate authority and timestamps	Average	High
SOLSR	Authentication and integrity of signaling packets	MACs and timestamps	Average	Low
WATCHDOG-PATHRATER	Excludes misbehaviors	Observation and reputation	Average	Low

The ARAN (Authenticated Routing for Ad hoc Networks) approach is PKI oriented having a trusted certificate authority know to all users. Each user has an IP based certificate granted at the entrance in the network. The advantages of the approach are better protection against malicious nodes by using authentication, message integrity and non-repudiation, elimination of the reply attacks by using timestamps and a good performance for route discovery and maintenance. The disadvantages of the approach are scalability, overhead in the network and latency in the communication.

ARIADNE is a symmetric cryptographic approach in three versions: shared secret between each pair of nodes, between communication nodes combined with broadcast authentication or digital signature. The advantages of the approach are better protection against malicious nodes by using authentication and message integrity, protection against malicious node attacks, routing loops or reply. The main disadvantage of the approach is the increase of package length which in turn decreases the ratio of the packages exchanged between peers.

CONFIDANT (Cooperation of Nodes: Fairness In Dynamic Ad hoc NeTworks) is a trust approach for security formed by: monitor (the entity that observes the traffic), trust manager (the alarm generating entity), reputation system (the ranking of trustworthiness) and path manager (the path decision responsible entity). The advantages of this approach is the fair detection of malicious nodes by constantly monitoring the status of the messages in the network and the reports of the attacks. The disadvantages are scalability and overhead. These are due to the fact that each car introduces a new certificate in the network, that including the tracking for it, the trust list and misbehaviors.

DCMD (Detecting and Correcting Malicious Data in VANET) is rather an application oriented approach than a routing protocol and it is based on the data collected from the sensors in order to determine rightfulness of the received information. The advantages are scalability and mobility by using an event based reputation system. This approach gives more information about the accuracy of an event rather than the users that have reported the incident. The big disadvantage of the network is that the sensors data is not integrity secured. This means that a user can manipulate its sensor data for gathering advantage.

SAODV (Secure Ad hoc on demand Distance Vector) is an asymmetric cryptography oriented approach in which messages are split into signed modifiable fields and fixed fields. The advantages of this approach are integrity and secure efficiency. The main disadvantages are increased overhead of the network and node resource dependability, in the sense that the messages of a node with high computational power will be like a DoS attack for a node with low computational power.

The previous discussed approaches have set a common need for a secure VANET like:

- authentication means that all users have to sign the sent messages and verify the received ones. Signing and verification messages introduce a delay in the network that may be resolved using elliptic curve cryptography.
- availability refers to the network's constant need to be able to provide information for the vehicles. The Denial of Service attacks (DoS) rely on this property.
- non-repudiation practically means that a user cannot state that a message he sent is not his.
- privacy is needed in order to prevent others from tracing one's route or letting out critical information.
- real-time constraints are necessary because of constant and high user movement which makes the time slot for message exchange limited.

- integrity is the property that ensures all the messages that they were received as they were sent without any alterations.
- confidentiality protects the privacy of the drivers.

Besides the properties discussed above about what a network has to provide, there are some extra security challenges characteristic for VANET as follows:

- real time constraint expresses the need of fast processing algorithms.
- data consistency liability refers to the fact that each node may act maliciously, therefor the message exchange need to be correlated with data from other vehicles.
- low tolerance for error is the case when critical information is passed through the network.
- key distribution is a major challenge in VANET because of the high mobility and real time constraints.
- incentives when designing the user application in order to respect the choices each makes.
- high mobility means that the execution time is limited, so in order to implement a security mechanism we have to choose either a low complexity security algorithm or a transport level security protocol.

SEAD (Secure Efficient Ad hoc Distance Vector) is a distance vector routing protocol in which asymmetric cryptography is replaced with one way hash functions of the message in order to grant integrity. The advantages of this approach are good computational performances with limited resources and the fact that it has more recent routing tables thus performing better in high mobility situations. The disadvantage of the approach is the fact that it needs an extra security step in order to prevent message forgeries.

The SLSP (Secure Link State routing Protocol) uses hash functions and public key cryptography in order to secure message routing. The advantage of this method is the fact that it uses division of the vehicle based on areas which makes the approach scalable. Another advantage is the fact that it is easily adaptive to different topologies. The disadvantage is that it is resource consumptive.

The SPAAR (Secure Position Aided Ad hoc Routing) is an asymmetrical cryptography based approach that provides authentication, message integrity, non-repudiation of sent messages and confidentiality. The advantage of this approach is that it is secure efficient but the disadvantage is that it takes twice the time to compute messages and therefor the message exchange ratio is split by half.

The SOLSR (Secure Optimized Link State Routing)uses one way hash MACs in order to grant authentication and prevent reply attacks. The advantage of the approach is that it has low computational requirements whereas the main disadvantages are scalability issues and increased overhead.

The WATCHDOG-PATHRATER is a two step approach based on a watchdog that monitors the network in order to detect nodes that misbehave and on a pathrater that needs to find alternative paths in order to avoid malicious nodes. The main advantage of the system is that it is low resource consumptive while the main disadvantage is that it takes long to exchange messages which decreases the message exchange ratio.

The approaches described above have the same goal of securing VANET even though the methods are different. Another UPB developed approach is presented next. This approach is public key cryptography oriented and implies a PKI. Messages are signed in order to provide accountability and non-repudiation while being encrypted with a shared secret key for confidentiality and message integrity providing. In V2V communication these attributes are obtained using Diffie Hellman key exchange algorithm to obtain a secret shared key, followed by an encryption using the AES algorithm. The testing of proposed approach was made using a traffic simulator developed in the University of Politehnica Bucharest. This allowed observing changes in throughput, bandwidth and message exchange.

3 Security Mechanism and a Theoretical Approach

First of all, an analysis of the characteristics of a vehicular network has been made in order to derive security-related challenges. McGraw [26] enumerates a series of security principles that must be considered, generically, in any network:4efvb

- secure the weakest link principle refers to the fact that adding cryptography in communication does not necessarily mean that the communication is secured. To that respect the author has presented a simple analogy story that reveals the fact that an attacker speculates the easiest way to obtain it's goals. "Imagine you are charged with transporting some gold securely from one homeless guy who lives in a park bench (well call him Linux) to another homeless person who lives across town on a steam grate (well call her Android). You hire an armored truck to transport the gold. The name of the transport company is "Applied Crypto, Inc" Now imagine you are an attacker who is supposed to steal the gold. Would you attack the Applied Crypto truck, Linux the homeless guy or Android, the homeless woman? Pretty easy experiment, huh? The answer is "Anything but the crypto.""
- defend in depth refers to adding redundancy and layered defense in a network in order to make the attacker's work more difficult.
- fail securely refers to what does a system do when an attack takes place, does it simply offer to give all the information or it makes it unavailable? To this respect, in Ref. [26], says that "Any sufficiently complex system will have failure modes. Failure is unavoidable and should be planned for. What is avoidable are security problems related to failure. The problem is that when many systems fail in any way, they exhibit insecure behavior."
- grant least privilege refers to the fact that a system has to start with the guilty presumption and allow to users to improve the reputation and not to grant them root privileges from the start.
- separate privilege is a principle according to which the users are spitted into different categories based on defined roles and that has been proven to be wrong by granting all the users from a role a certain liberty only because one of the users need it.

- economize mechanism expresses the need to keep it simple when talking about the complexity of a network in order to be able to better protect it.
- do not share mechanisms means that each user needs to have its own set of objects in order to avoid shared resources attacks.
- be reluctant to trust talks about the fact that only the authorized users must be able to call the API and have access to information from the network. To this respect, there is something to be remembered and that is that "trust is transitive. Once you dole out some trust, you often implicitly extend it to anyone the trusted entity may trust" [26].
- assume your secrets are not safe expresses the need to have secured devices and to always keep information in a secured form.
- mediate completely refers to the fact that no authentication cache should be kept in order to avoid an attack to explore them.
- make security usable expresses the need to limit security at the point where it can still be used in an efficient way.
- promote privacy is a principle according to which personally identifiable information (PII) should be stored only if necessarily needed and only on secured devices.
- use your resources gives priority to the internal storage one has. For example, one can have multiple mechanism through which it secured a password but if it has put it online in an accessible place to an attacker then all the mechanisms have been overtaken.

For the VANET type of network the characteristics presented above have been adapted as it is stated in [14], leading to the following needs:

- authentication means that each message in the network needs to be authenticated, meaning that each emitter signs the messages and then at the receiver the message is verified. This further steps of signing and verification messages introduce a delay that may be resolved using elliptic curve cryptography.
- availability refers to the network's constant need to be able to provide information for the vehicles. This property makes the system vulnerable to Denial of Service attacks (DoS).
- non-repudiation means that each message can be traced back to the vehicle that had generated him and is needed in order to prevent malicious attackers saying that they did not generate a certain message.
- privacy is needed in order to prevent others from tracing back someone's route or letting out critical information.
- real-time constraints refers to the high speed movement of the vehicles which makes the time slot for message exchange limited.
- integrity is the property that ensures all the messages that they were received as they were sent without any alterations.
- confidentiality protects the privacy of the drivers.
- data consistency liability refers to the fact that each node may act maliciously, therefor the message exchange need to be correlated with data from other vehicles.
- low tolerance for error is the case when critical information is passed through the network.

- key distribution is a major challenge in VANET because of the high mobility and real time constraints.
- incentives when designing the user application in order to respect the choices each makes.
- high mobility means that the execution time is limited, so in order to implement a security mechanism we have to choose either a low complexity security algorithm or a transport level security protocol.

Another objective of the security mechanism is to implement a trust management system, where a trust management system is "an abstract system that processes symbolic representations of social trust, usually to aid automated decision-making process" [27].

Further on, the chapter explains the basic theoretical concepts involved in creating a security approach for a VANET.

3.1 Wireless Communication

The wireless based communication was first recognized and standardized internationally in 1997 and since then has known continuous evolution, developing subsequent amendments in the years that came.

The initial 802.11 standard, obsolete today, had two possible data rates of 1 or 2 megabits per second and forward error correction. As for the how, it could transmit over infrared at 1 MB/s, using frequency hopping spread spectrum or direct sequence spread spectrum at 1 Mbit/s or 2 Mbit/s.

The 802.11a standard uses the 5 GHz frequency, less popular for usage but with serious environmental attenuation problems that degrease significantly the range of usage.

The 802.11b wireless networking standard reaches to 11Mbit/s and represents the first real-connection with the users, being the first widely available solution for users to buy. Its frequency for communication is in the band of 2.4 GHz.

The 802.11g standard s provides a theoretical maximum speed of 54 Mbps at the frequency band of the 802.11b standard of 2.4 GHz frequency.

The 802.11n standard operates at data rates from 54 Mbps to 600 Mbps, and can use the 2.4 and 5 GHz frequencies. The 802.11n standards main novelty is the increased speed of connection. This was able by allowing for bonded channels which doubles the radio spectrum of an 802.11a type of connection, and in turn doubles the data rates. Another speed enhancing technology is MIMO (multiple input multiple output) which uses multiple antennas on the client devices and on the provider's wireless access points in order to achieve diversity gain and reduce fading.

In order to secure the communication over any of the subsequent of the 802.11 standard, one can use a protocol like Wired Equivalent Privacy (WEP), Wi-Fi Protected Access or Wi-Fi Protected Setup (WPS). Each of these has had security

Table 3 Security comparison between wireless protocols

Protocol	Encryption	Authentication	Data integrity	Vulnerabilities	Complexity
WEP	RC4	WEP-Open and WEP-Shared	CRC-32	Chopchop, Bittau fragmentation, FMS, PTW, DoS	Low
WPA	TKIP	WPA-PSK and WPA-Enterprise	Michael	Chopchop, WPA-PSK, Reset, DoS	High for WPA-Enterprise
WPA 2	CCMP and AES	WPA2-Personal and WPA2-Enterprise	CBC-MAC	DoS, MAC spoofing, Offline dictionary in the WPA2-Personal	High for WPA2-Enterprise

breaches, which is why the Wi-Fi Alliance has updated its test plan and certification program. See Table 3.

The WEP protocol is vulnerable because it uses RC4 symmetric stream cipher algorithm which xores a stream of bits: the secret key with the plaintext into obtaining the ciphered text. This is then transmitted to the receiver where it is xored with the same secret key into obtaining the plaintext. The problem is the fact that a zero plaintext xored with a key results in the key itself.

The Wi-Fi Protected Access (WPA) protocol uses two keys: an integrity message check of 64 bits and an encryption key of 128 bits, both derived from a master key. This protocol has resolved the main issues of the WEP protocol but has been proven to be weak against dictionary attacks. These attacks imply the successive testing with words from a predefined list called dictionary and are different from brute force attack because they try only the most probable passwords.

The Wi-Fi Protected Access II (WPA2) protocol introduced the use of counter cipher block chaining message authentication code protocol (CCMP) and uses either the advanced encryption standard (AES) or the temporary key integrity protocol (TKIP). The second implementation was added in order to allow WPA compatibility.

The Wi-Fi Protected Setup (WPS) protocol is used in order to secure the settings of the access point along with those of the devices trying to connect to that access point.

The present proposal has taken into considerations both the advantages and disadvantages of the Wi-Fi protocols mentioned above. Along with the implicit security issues, the power of propagation has also been taken into account, thus leading to the choice of a peer-to-peer wireless 802.11 b/g/n compatible communication.

3.2   Trust Management Models

The concept of trust refers to who are the peers in one's group, which of them can be trusted and based on what can they be trusted.

The first question refers to the users roles in a VANET. The literature has various approaches, in which peers (users) are given roles. These roles can have the same right or are given different levels of trust by a recognized authority.

Therefore, the answer to the second question varies based on the network type implemented. In a subordinated hierarchy there is a root certificate authority that authorizes several descendants and only the certificates given by these entities are recognized as trust-worthy. In a cross-certified mesh, there are a various number of CA that may authorize any other CA, "except if and as naming constraints are applied" [28]. The trustworthiness of this type of network is not uniform, and it varies greatly on the length of the certification chain. The trust lists are a trust design in which the peer is given an initial set of public keys of trusted CA and in order to be validated successfully it must use one of the CA from that set. Various hybrid models can derive from these simple examples, the limit being only the level of creativity and knowledge of the network security designer.

The third question needs to take into consideration whether or not there are prior trusted peers, inactive peers or if the issue of maliciousness is taken into consideration. For example, on eBay, after a transaction, both the buyer and seller send a feedback about the transaction went and rate the each other. The system computes the trustworthiness of someone as an average sum of the rating received over the last 6 months. On e-commerce, generally, the rating of someone is associated with the risks involved in transacting with that person. So, the basic models of managing trust are monitoring or evidence based trust management systems. The first model is based on the idea of constant getting information about the activity of the network from various monitor components in order to spot bad behaviors. If one misbehaves, the reputation system is notified and changes the rating corresponding to that node. If that misbehavior exceeds a threshold, the node is considered to be malicious and added to the blacklist. The other nodes will stop forwarding messages to nodes on the blacklists. If a malicious node is detected, an alarm is sent to the trust system. The second model is based on the idea that trust is the result of relationships where evidence of trust has been given. And by evidence, we mean everything that the policy require in order to establish ones identity. In a public key model, each peer is made responsible for the data/information that it signs and sends forward. The non-repudiation part of the security design has to be strong enough to eliminate any impersonating attacks. An impersonating attack is when a malicious node listens to the network and gets the identity of a legitimate node, and then starts sending bogus information in their behalf.

Generically, trust management models are overviews of networks in terms of the confidence-risks ratio. The most common models for trust management are: public-key systems, resurrecting duckling and distributed trust management. The first model implies the existence of a certificate authority based upon peers can authenticate between themselves. The second model is based on the idea of master-slave, where the master send commands and the slaves have to obey. The third model is based on the idea that trust has to be earned and, moreover, it has to be earned for each peer because it is transitive only if it meets several condition.

Trust management in the context of mobile ad hoc networks has more challenges given the opportunistic communication and the ad hoc nature of the network. Ad hoc networks rely on the active cooperation between all peers for routing and packet forwarding. Communication distance, bandwidth or threshold are various parameters that can influence a peer to act selfishly in the sense that it will only listen to the data from its proximity without forwarding it.

A VANET trust management system has particular properties to comply like:

- dynamicity refers to the aspect of constant mobility which should reflect on the trust system, making it temporary and location dependent.
- subjectivity refers to the fact that mobility affects the way a user may react and such their trust level varies in time.
- incomplete transitivity refers to the fact that receiving a reference of trust from a user does not impose changing the level of trust for the referee.
- context-dependency refers to the fact that there are several types of trust like trust in selflessly, in forwarding packets or computational power and the level of trust-worthiness of one does not impose the trustworthiness of the others.
- asymmetry refers to the fact that trust is not mutual, in the sense that if a user A trusts a user B it does not imply that B trusts as well A.

The proposed security approach implements a hybrid form of trust model between the public key systems and the distributed trust management. In the beginning, the infrastructure gives each node the benefit of a doubt, so they are all considered to be "kind and selflessly". Using the information gathered by the monitoring devices, a list of misbehavior is made. Once a node exceeds a given threshold, like Friedrich Nietzsche said, it will not be trusted again.

3.3   The Cooperative MAC Design

The Cooperative MAC protocol is a multi-rate compatible 802.11 standard protocol based on the idea that users in a VANET have different transfer speeds between sender and receiver. Given low speed communication between a sender and a receiver, if there is a third user between the two that has higher transmission speeds then that user will act as forwarder in order to speed up the overall transmission and decrease the general throughput.

The general usage means that every station needs to find out how close it is to its pair along with the channel and speed at which it is going to communicate. Therefore, Request to Send (RTS) and Clear to Send (CTS) frames need to be introduces in the system for collision avoidance and channel reservation (NAV).

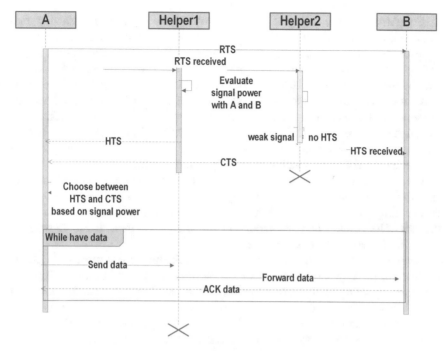

Fig. 2 The cooperative MAC protocol for transferring data

Say there is a peer that wants to send data (Sender), a peer to whom is designated the data (Receiver) and two peers that happen to be in the area (Helper1 and Helper2) like in Fig. 2. The transmission of the data follows the pattern below:

- the Sender sends a RTS packet.
- each peer in the area receives the RTS packet and verifies if the Sender address exists in the proximity table. If it is not, it will be added. Then, each peer computes the data rate at which it could transfer data and stores it along with the MAC address in the proximity table.
- if a peer receives the RTS packet without errors it becomes a possible helper. Each helpers asses if their data rates can improve the transfer. If it can, it will send a helper-ready-to-send (HTS) packet. The helper with the best proximity is chosen as intermediary. Note: HTC is identical to a CTS type of message.
- the Receiver sends a CTS in order to reserve the channel. If there is a helper, it reserves the channel for the time necessary for a transfer with the helper, otherwise it reserves the channel for the time necessary for direct communication.
- the Sender starts sending the data. If received without errors, the Receiver transmits ACK packet. This step repeats until the end of transmission.

The Cooperative Mac Protocol (CoopMAC) presents advantages like higher spatial diversity since any damage in communications between two nodes can be taken over by a third one with better signal power and higher data rates with both nodes.

CoopMAC has the ability to adapt and to mitigate the effects of shadow fading because antenna elements of a cooperative virtual antenna array are separated in space and experience different shadow fading [29]. Another advantage is higher throughput-lower delay since the transmission of a station with low data rates is taken over by stations with high data rates, the overall time for transfer is less, therefor the delay of a third station before finding a free channel is significantly smaller. Another advantage is the fact that it has lower power consumption and lower interference with extended coverage due to the low error rates and high data rates, reflected by high throughput gains. CoopMAC also allows wireless terminals to seamlessly change channel and interference conditions, even opportunistically.

But, all the benefits come with a price. The security issues brought by the protocol start from helper itself that might not want to be so helpful and refuse to forward frames, simply dropping them when they come. In this case, the sender has to notice the lack of helpfulness and find another intermediary or, if there are no, to send the frames itself despite the low data rate.

Another potential issue that might come from the helper if it turns malicious is the attempt to deny service to the source by dropping packets received from the sender but spoofing ACKs on behalf of the destination. But that is not the worst that could happen; the helper turned malicious could end up modifying the payload of the messages and the forward them. This is very hard to detect from the sender's point of view because it does not know anything is wrong with the intermediary in order to change it or to send the frames itself.

The present design intents to take advantage of the benefits of increased communication area at high data rates of the CoopMAC protocol, while fixing the main disadvantage: the selfishness that some peers may present. By adding neutral monitoring devices to scan the network in their proximity at all times, a selfish behavior can be detected and signaled in order to lower the corresponding ranking. If the rank of a peer is reduced under a given threshold then that vehicle will be transferred on to the blacklist upon elimination from the system.

4 Security Approach

The previous work has shown the different perspectives of a VANET with the different advantages and disadvantages. The proposed approach takes into consideration the previous work and uses a public key infrastructure where messages are signed in order to provide accountability and non-repudiation and encrypted for confidentiality and message integrity.

The accountability and non-repudiation properties are obtain by user signing all messages with the private key. The confidentiality and message integrity are gained by AES encryption with a secret shared key obtained with the Diffie Hellman key exchange algorithm.

The Diffie Hellman method for key exchange enables two users, say Alec and Bogdan, with no prior information about each other to generate a common secret key over an unencrypted communication channel. The generated key can then be used as secret key for the following messages between Alec and Bogdan. This is known as "an anonymous (non-authenticated) key agreement protocol" [30] since there is no prior information exchange between the parts involved.

The Diffie Hellman key exchange algorithm is based on the properties of multiplicative group of integers modulo a prime number, generically notated as p. The base is called a primitive root mod p and is notated as g. These values need to be prime and are desired to be very large in order to obtain security. Each party, Alec and Bogdan, then chooses a large random number that remains secret. With that number it computes a key and sends it to the other user. Having the other user's computed key, each computes the shared key.

A simple example is presented in the table below: two participants, Alec and Bogdan, want to set a shared key while a third party is listening to the channel communication.

After achieving a shared secret key, the encryption is done using Advanced Encryption Standard (AES) symmetrical algorithm. A Rijndael based cipher, the AES represent the first cipher used by the National Security Agency (NSA) which has been made available to the world. It has three possible key lengths: 128, 192 and 256 bits and a various number of rounds according to their lengths: 10, 12 or 14.

From a functional point of view, the AES algorithm implies several steps:

- key expansion represents the most resources consuming part of the algorithm. It means that keys for each round are derived from the cipher key, plus one.
- the initial round is different from the other rounds in the sense that it starts with a round key generation.
- based on the key length there are 9, 11 or 13 following rounds each having the following steps:

1. SubBytes represent non-linear byte by byte substitution according to an 8-bit substitution box called S-box.
2. ShiftRows represents a circular shifting phase according to a certain offset given by a permutation box.
3. MixColumns and ShiftRows give the diffusion of the algorithm and it basically means combining the four bytes in each column.
4. AddRoundKey is the step in which a new key is derived from the cipher key.

- the final round is an incomplete round made of only 3 steps:

1. SubBytes represent non-linear byte by byte substitution according to an 8-bit substitution box called S-box.
2. ShiftRows represents a circular shifting phase according to a certain offset given by a permutation box.
3. AddRoundKey is the step in which a new key is derived from the cipher key.

The presentation of the AES algorithm has been made so explicitly because there are several possible implementations of the algorithm (AES Classic, AES Fast, AES Light and AES Wrapper) that differ by the level of key expansion pre-computed and by the key wrap method.

The Diffie Hellman key Exchange algorithm has been chosen over RSA (Rivest Shamir and Adleman) because the production of new DH key pair is extremely fast which is mandatory in mobile networks like VANET. Both "based on supposedly intractable problems, the difficulty of factoring large numbers and exponentiation and modular arithmetic respectively, and with key lengths of 1,024 bits, give comparable levels of security. Both have been subjected to scrutiny by mathematicians and cryptographers, but given correct implementation, neither is significantly less secure than the other" [31].

The V2I communication is secured using X.509 certificates in order to grand accountability, non-repudiation and message integrity. The X.509 represents an International Telecommunication Union for Telecommunication Standardization Sector (ITU-T) standard in cryptography.

In public key cryptography, the standard refers to the structure of certificates and also to a strict hierarchical system for certificate authorities (CA). There are two versions of certificates: a Version1 (v1) certificate used by a root CA and a Version 3 (V3) used for user certification.

The basic difference between them is that a V1 certificate is a self-signed certificate, which means that it has the same id both in the issuer field as in the subject field of the certificate. The V3 certificate has the id of the CA in the issuer field of the certificate and its own id in the subject field of the certificate.

The choice of having a single root certificate authority has the benefit of scalability while the choice of having access points to that certificate authority has the benefit of convenience, making the authority more accessible to the vehicles.

The implementation of the X.509 certificate can be done in several ways, at different network layers: at the transport layer via Secure Socket Layer (SSL), at the message layer via WS-Security Binary Security Token or in the protocol stack between the data link and network layers via IPSec.

In Ref. [32], the implementation of the X.509 certificate at the message layer is stated to have the advantage of high degree of interoperability over the other implementations, although the main disadvantage is the fact that it increases the message processing time. The IPSec implementation has the downside of no fine control of security and the upside of better performances since it is closer to the hardware layer. The SSL implementation has the benefit of performance over the message implementation because it is closer to the OS while the liability of the implementation is the fact that it can not grand message persistence in a secure state and neither verify that all the policy's requirement are respected.

In order to obtain a certificate, one must first generate a certificate request, and then send that certificate request to a certificate authority. Based on the fields of the certificate request, an X509 version 3 certificate is generated. Once the CA is authentic, it awaits for requests from the peers of the network. Peers then start generating version 3 certificate requests which they send to the CA. The CA then responds to the request with a signed certificate which has the credential for the user. From that point on, the user may authenticate himself to other using the given certificate until the expiration date.

Adding X.509 certificates to a network, especially to a VANET, brings several benefits into the network:

- verification of the data when provided with a certificate request form before generating a certificate.
- malicious vehicles manipulation by implementing a revocation list with the certificates of all the users that have acted selfishly and thus blocking their access to the network.
- standard data format which eases the message handler.
- easier management of keys given the customizable fields of the certificate.
- validity constraint ensures that the usage of a certificate is limited in order to prevent identity theft.

Based on the properties stated in subchapter 1 the security mechanism presented respects the following:

- authentication means that each message in the network needs to be authenticated, meaning that each emitter signs the messages and then at the receiver the message is verified. This characteristic has been obtained by using the public key cryptography, each message in the network being signed with the certificate X.509.
- availability refers to the network's constant need to be able to provide information for the vehicles. This property is assured by adding the helper's computational power and data rate advantage.
- non-repudiation means that each message can be traced back to the vehicle that had generated him and is needed in order to prevent malicious attackers saying that they did not generate a certain message. This characteristic is possible by signing all the messages which demonstrates the source of each message.
- privacy is needed in order to prevent others from tracing back someone's route or letting out critical information. This property is obtained by encrypting all the data location with a secret key, always different from the others.
- real-time constraints refers to the high speed movement of the vehicles which makes the time slot for message exchange limited. The characteristic could not be resolved, but a walk around it was found by adding a cooperation model to the network which increases the distance and, therefor, also the time slot.
- integrity is the property that ensures all the messages that they were received as they were sent without any alterations. This property is obtained by signing all the certificates and sending a hash of the message in order to verify that no modifications were made.
- confidentiality protects the privacy of the drivers. This characteristic was not obtained because each vehicle in the network has an unique identifier based on which a unique certificate is generated.

5 Security Mechanism Details

To resume the ideas stated so far, there are two main entities involved in the security mechanism: vehicles and the infrastructure. The vehicles are the basic entity transporting the data while the infrastructure is the CA which is the central processing unit. The infrastructure also contains a new hybrid form represented by the monitoring device which act like a vehicle in the sense that it send data to CA and to the peers, but it also monitor the behavior of the peers in the network.

The messages that are exchanged between the entities of the network can be split into several types:

- alarm for traffic congestion: if a vehicle spend more than a given time in an area, it will save that place as a hot spot.
- alarm for over passing threshold: the monitoring device keeps track of the misbehavior of the vehicles, and if one overpasses a given threshold, the CA will be notified in order to add the vehicle to the blacklist.
- blacklist: a list generated and updated by the CA with the users that have been noticed with bad behavior and therefore must be avoided.
- timestamp: is a message generated only by the CA in order to get correct data; This message also verifies that the GPS satellites timestamp is set correctly (GMT).

The hypothesis is that there is one unique root certification authority that generates certificates, and those certificates are given in a safe environment to users. The expiration date of the CA will be consider undetermined. Another important assumption is the fact that every vehicle in the network will be uniquely identifiable at all times by its MAC address. That address will be take into account in the certificate request sent to the certification authority. Once obtained, in order to lower the possibility of someone else finding out one's private key, the certificate will be placed into a black box that represents a secured hardware device (Fig. 3).

A connection is to be made with the CA, and only if the data rates are not good a peer will be used in order to send the data. That data must be signed by the emitter and encrypted with the public key of the CA. The helper that forwards the data will also sign the message. Only the CA can decide whether a peer is on the blacklist or not, and messages to that respect that are not from it will be rejected.

Communication can be established between two vehicles, no matter if one or both have monitoring devices, when they exchange congestion alarm messages, or it can be established between a vehicle and the CA when transmitting composite messages.

The usual communication has three steps: the first step is represented by the neighbor discovery followed by the actual data transfer and terminated with an end sequence. All messages are entity dependent in the sense that all the users of a VANET signed their messages which grants properties like authentication and accountability.

The discovery stage indicates the need of a vehicle to find out if it has any neighbors (CA or other users). The steps involved in this stage, are being described below:

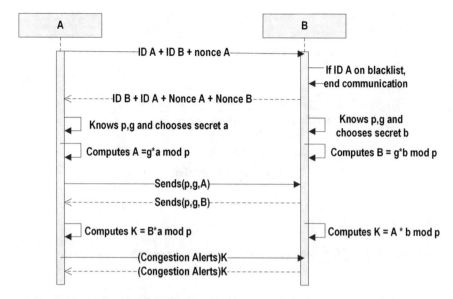

Fig. 3 Security model for vehicle-vehicle exchange of messages

- at regular time frames, a vehicle (Emitter or vehicle 1) generates a "hello" message. This is a broadcast message with whom one discover its neighbor peers. This message is not encrypted, it contains the address of the emitter and also a nonce signed by him.
- each vehicle (vehicles 2, 3, 4 and 5) that receives the "hello" message will respond with the nonce signed with the private key, the clear GPS location and the encrypted self address with the public key of the emitter.
- the initiator (vehicle 1) of this discovery will then gather all the messages and have a relative map of its neighbors. If a CA point is in the proximity, all attention will be diverted to the CA. If not, it can then compute to see which are the best vehicles in order to be used as helpers based on the GPS location and start connecting to each peer in order to exchange congestion alarms messages. The GPS location has been taken into consideration in order to eliminate the selfishness of a peer that may not want to cooperate in forwarding messages.
- after the exchange of messages, a vehicle will not start any conversations with peers with whom it already communicated for a given time frame. This step has been thought of as to limit the effects of a denial of service (DoS) or distributed denial of service (DDoS) attack.

The data exchange is made according to the roles of the entities involves in the transfer (see Figs. 3 and 4).

For every message sent there is an ACK message received, a lack of the ACK message for an idle time meaning that the communication has been terminated and the peer may implicate themselves into other activities. This fact is due to increased

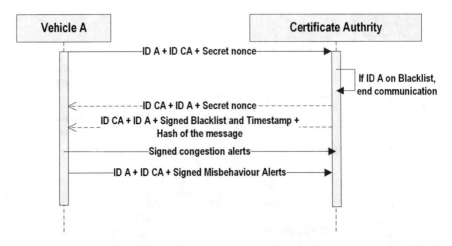

Fig. 4 Security model for vehicle-infrastructure exchange of messages

distance between the peers to the point where the hardware is unable to fulfill the transmission need of the vehicle.

If the data exchange runs without problems, then an end of transmission (EoT) message is sent in order to signal the clearance of the communication channel.

The security of the communication depends on the roles of the parties involved as stated below:

- messages between two vehicle have an initial exchange of message in order to generate a secret shared Diffie Hellman key of 40 bytes length. This exchange means that the first message in the data exchange stage will be a message with a Diffie Hellman request, followed by a Diffie Hellman grant response. Their scope is to share the pair of large prime numbers and the public key obtained from those numbers along with a secret prime random number. As discussed in Sect. 4, the following step is for each to compute the secret shared key. That key will be used to encrypt the following messages using an AES algorithm with 128 key bits and an 128 iv bits obtained by parsing the shared 40 bytes (320 bits) key. See Fig. 4. The message field structure is presented in both cases: direct communication (Fig. 4) and with helper (Fig. 5).
- the messages exchanged between the vehicles and the CA assumes an authentication step in which both parties generate signed messages in order to prove their identity. After this step, the data exchange will be done using signed messages with the private key and encrypted/decrypted with the public key from the x509 certificate. See Fig. 4.

The main change is the integration of a security module where all the messages are encrypted by the sender, while at the reception they are passed through a decryption phase. The integration was made by adding a new package called security that is formed by 6 classes: three of which are different AES engines used in the symmetric

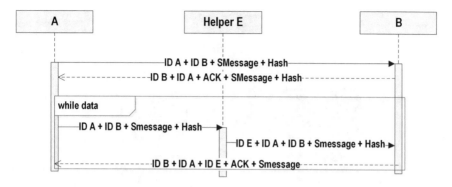

Fig. 5 Message fields

encryption, a Diffie Hellman (DH) parameter class used in DH key exchange, a class designed for the vehicle security and another class for the server security.

By integrating a new package in the simulator, the message exchange state machine has been changed into allowing certificate requests and grants or Diffie Hellman key exchange requests and grants. A request means that the peer has computed his side of the algorithm and that it need the other to do the same while a grant message means that the peer has received a request message according to which it has computed its part of the algorithm and sends his public part of the algorithm.

The process of joining the network implies the step of getting authenticated, thus the need of communication with a certificate authority and the need to obtain a certificate in order to be able to get acknowledgment from the other peers. This step is presented in Fig. 6 where the left side is represented by the vehicles in need of getting certified and the right side represents the actions of the CA called generically Server. As it can be seen, in order to join the network a vehicle must first get in contact with a server or else it will not be recognized and treated like a malicious node. Once in contact with the CA, it will generate a request for entering the network and the processing job then goes to the authentication side. There, the request is assessed and, if it is correct, a certificate is generated and sent back to the requester thus finalizing the stage of joining to the network.

Once a peer is authenticated it can communicate with the other recognized peers of the network and exchange congestion alerts. This kind of exchange implies a Diffie Hellman key exchange algorithm in order to obtain a shared secret key as presented in the state machine from 7. The left side of the figure represent the vehicle that initiates the communication by sending discovery messages while the right side is an abstracting peer in its proximity that receives the discovery message. After reaching a common shared key, the data from the messages is encrypted at the sender and decrypted at the destination (Fig. 7).

From a network point of view, the certificate authorities are considered to be also the servers and their constructor assigns memory for all the class's members and also generates a self signed certificate. Each point of access to the CA is considered

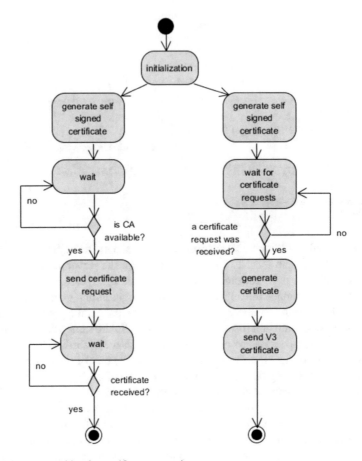

Fig. 6 The state machine for certificate generation

in the program to be a Server entity and it has its own set of credential: a private key–public key pair, a certificate to prove its identity and an index of the certificates that it had generated. Also, the class Server inherits the characteristics of the class it extended so it has information about its identity, a specific location on the map and a set of channels available for communication. Allowing multiple servers and keeping their mobility has been chosen in order for the program to remain compatible with the input parameters.

The vehicles initialization is somewhat simpler since it implies only assignment of memory for all the members. Each vehicle and monitoring devices in the thesis are called generically a car and each has a V3 certificate, a private key-public key pair based on what the certificate is generated, a key agreement needed for the Diffie Hellman key exchange algorithm and a Diffie Hellaman secret shared key–Diffie Hellman public key pair. The class secureCar inherits the characteristics of the class it extended so it has information about the driver's type, the status of the WiFi, the memory size and the location on the map. First of all, all the vehicle in the initial-

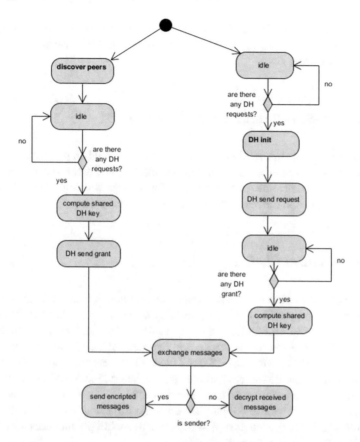

Fig. 7 The state machine for vehicle to vehicle communication

ization phase need to generate a certificate V3 request which is then sent to the CA. The CA will then assess the request and generate a V3 certificate if it is the case. After this step, each vehicle is assigned a certificate and has a public key with which it will be able to verify its identity.

But, initialization does not stop here. Each car has to generate a V3 certificate request that it will send to the CA. For each request it gets, the CA will generates a valid certificate for a period of 90 days and send it to the peer to whom it was issued for.

In order to secure the communication between two vehicles there are a couple of messages to be exchanged. The peer receiving the discovery message generates a shared pair of prime number as a part of the Diffie Hellman initialization along with a Diffie Hellman request.

The vehicle with whom a communication is wanted, then computes its private-public key pair and along with the prime number is able to complete the cycle and find out the shared key. It will then send a Diffie Hellman grant message for the initiator to be able to compute the shared secret key. From that point on, the two vehicle are able to secure their conversation using the AES symmetric algorithm and the shared secret key.

6 Experimental Evaluation and Results

This chapter presents the validation of the security mechanism proposed in the previous chapters with its advantages and disadvantages. A simulator comparison was made in order to determine the better choice for mechanism validation, followed by a study on attacks and the resistance of the mechanism to them.

6.1 Arhitectural Choices

Before developing this thesis we have searched the Internet to see if there is an interest on the topic only to find out that, not only there is a great interest, but people had already made several traffic simulator at microscopic and macroscopic level in order to suit their needs. There are several licensed realistic traffic simulators like VISSIM, PARAMICS, TRANSIMS or CORSIM but their downside is that they do not allow changes on the sources and they do not generate traces of movements. For the past couple of years, open source tools for traffic simulation have become available to people world wide and their characteristics are listed in the following table. The Table 4 resumes the features of several macro-mobility simulators and Table 5 several features of micro-mobility simulators.

In Table 4 GDF is an abbreviation for Geographical Data Files, TIGER represents Topologically Integrated Geographic Encoding and Referencing, AWL means Auto

Table 4 Macro-mobility properties of traffic simulation tools

Simulation tool	Input	Graph	Destination	Acceleration	Velocity
Virtual track	No	user defined (u.d.)	Random on track	No	Uniform
MOVE	TIGER	geographical (geo.)	Random	No	Uniform
IMPORTANT	No	Random	Random	Uniform	Smooth
STRAW	TIGER	geo.	Random in graph	Uniform	Smooth
Canu-MobiSim	GDF, AWL	u.d. or geo.	Random on AP	Uniform	Uniform
VanetMobiSim	TIGER, AWL	Clustered Voronoi	Random on AP	Uniform	Uniform
City	No	Grid	Random	Uniform	Smooth

Table 5 Micro-mobility properties of traffic simulation tools

Simulation tool	Visualization	Human patters	Intersection	Obstacles	Platform
Virtual track	No	No	No	No	QualNet
IMPORTANT	No	CFM	No	Radio	C++
MOVE	Yes	CFM	Stoch turns	Topology graph	C++
VoronoiM	No	No	No	Buildings	C++
Canu-MobiSim	Yes	IDM	No	Graph, building	java
MobiREAL	Yes	CPE	No	Graph, buildings	C++
VanetMobiSim	Yes	AIDM	Traffic signs and lights	Graph, buildings	java
STRAW	No	CFM	Traffic signs and lights	Topology	Swans
;2 GEMM	No	No	No	No	java

White List. These terms are used to describe the input of a network and to determine whether it is similar to real life conditions or if it is a simple theoretical Freeway model or a Manhattan (or Grid) model.

Car Following Model (CFM) is a basic schema for controlling the distance between two cars and their paths. This type of mobility can be seen on simulator tools such as IMPORTANT, MOVE or STRAW.

Attraction Point (AP), Activity and Role reflect the interest of multiple people on a destination, the activities represent the moving process towards the AP and the roles specify mobility tendencies.

In the table below IDM is an abbreviation for Intelligent Driver Model, CPE stands for Condition-Probability-Event and AIDM represents Advanced Intelligent Driver Model. The Street Random Waypoint (STRAW) is based on the open source Scalable Wireless Ad Hoc Network Simulator (SWANS) which takes topology information from TIGER to which it adds micro-mobility support. These models represent a more complex interpretation of the scenario taking into consideration factors like real vehicular movement.

The university Politehnica Bucharest's simulator is SIM2Car and it represents an 802.11b Wireless MAC layer, an UDP transport layer simulator whose routing and addressing schemes have been changed to depend on geographical position.

The input of the simulator consists of vehicle mobility models according to which the vehicles are positioned on the map. Their trace is updated periodically in order to keep a realistic view of the grid. An interesting fact is that it takes into account traffic rules and multiple types of driver behaviors.

The basic architecture of the simulator is shown in Fig. 8 and its basic entities are vehicle and servers and the central unit called an Engine.

Each vehicle has two event handlers for the message exchange: Send Handler and Receive Handler. The engine is the entity that transforms each Send event of a

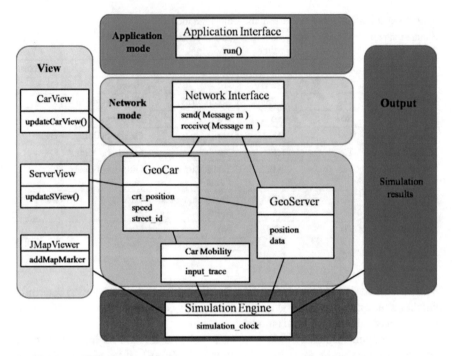

Fig. 8 The UPB VANET simulator structure

vehicle into a Receive event on the destination vehicle. The GPS data of a vehicle is updates regularly according to a scheduler.

The input data of the simulator is given by raw datasets corrections and traces generation and it has Node, Way and Location entity. " A point from an OSM map is represented as a Node object. It stores details about a POI (Point of Interest), such as the GPS coordinates (latitude and longitude), and the node identifier. The Way object is next used to describe a street parsed from an OpenStreetMap file. It contains the nodes (Node objects) forming the street." Based on these files, the simulator then generates the movement of the vehicles. The street graph is built in two phases: parsing and building. At the microscopic level OpenStreetMap (OSM) is used for street graph generation and a correction algorithm for GPS inadvertence.

Bouncy Castle is a cryptographic library collection of the implementations of most common cryptographic algorithms. Its architecture has two basic components in order to support the cryptographic capabilities and these are the low level application programming interface (API) and the java cryptography extension (JCE) provider. Pretty Good Privacy (PGP) support, Secure/Multipurpose Internet Mail Extensions (S/MIME) and similar standards can be built using the JCE provider.

The Bouncy Castle API part is in fact a set of APIs that contain all the cryptographic algorithms, whereas the JCE provider is built upon a set of low-level APIs.

For memory and speed constrained devices like mobile devices or tablets, there is a dedicated version of Bouncy Castle.

The APIs of interest for the thesis are the Diffie Hellman key exchange algorithm, the Advanced Encryption Standard and the X509 Certificate for public key infrastructure.

6.2   Criptographic Experimental Evaluation

The Bouncy Castle collection of libraries contains multiple versions of the AES algorithm: a "classic" version of the algorithm, a "fast" version of the algorithm, a "light" version of the algorithm and a "wrapper" version of it. In order to decide which of the versions is more appropriate for the given thesis several studies have been made.

The "classic" version is an implementation of the AES (Rijndael) algorithm from FIPS-197 that uses one static table of 256 word table for each encryption and decryption for a total of 2K Bytes. It thus adds 12 rotate operations per round the values contained in the other tables from the contents of the first.

The "fast" version is an implementation of the AES (Rijndael) algorithm from FIPS-197 optimized by Dr. Brian Gladman in terms of time consumption by using 8K Bytes of static tables for round precompilation.

The "light" version is an implementation of the AES (Rijndael) algorithm from FIPS-197 optimized by Dr. Brian Gladman in terms of memory usage by using no static tables for round precomputation, which has as result the smallest foot print.

The "wrap" version is an implementation of the AES Key Wrapper from the NIST Key Wrap Specification with the implementation specifications of the "classic" version. Thus, this version was not taken into consideration when performing the AES algorithms comparison.

A simple comparison in terms of time consumption was made on various input parameters with results that can be observed in Table 6. It can be observed that the smallest time resource consumption is obtained with the AES "fast" implementation. But the higher the input size the lower the time difference in encryption between the "classic" and "fast" version due to the fact that the key expansion only occurs once.

Given the fact that the minimum packet size is 9 bytes, the maximum packet is 614 bytes and the average packet size is 177 bytes and the fact that for relative small input size the "fast" version has the best performances, it is recommended to use the "fast" implementation of the AES algorithm.

A variance in output of the thesis was also tested leading to the conclusion that the "fast" implementation of the algorithm had the most amount of messages received. An explanation could be the fact that the time computing the encryption and decryption of the messages is enough for the vehicles to go out of each others range.

A strong security reason in order to avoid or to have caution when using RSA Key Pair Generator are the default setting of the constant $e = 3$. This small value for the

Table 6 A comparison between AES, AES Light ans AES Fast

AES type	Message size (bytes)	Encryption time (ms)	Decryption time (ms)
Classic	1.908	1.525	1.147
Light	1.908	1.328	1.247
Fast	1.908	0.806	1.060
Classic	19.080	5.419	3.468
Light	19.080	6.380	4.317
Fast	19.080	4.578	8.794
Classic	248.040	18.073	10.812
Light	248.040	22.318	15.893
Fast	248.040	10.341	3.751
Classic	2.480.400	35.858	35.521
Light	2.480.400	46.477	63.079
Fast	2.480.400	34.149	35.075

constant makes the key generation easy but increases vulnerability of the security mechanism since any message can be derived from three encrypted values [33].

In order to avoid RSA Key Pair Generator default constant value e = 3 deficiency an alternative implementation has been chosen with generating the key pair for the Diffie Hellman key exchange algorithm with the KeyPairGenerator class that chooses a random different prime, the at each iteration of key generation, making the derivation of the messages harder.

In the case of the advantage in throughput for the CoopMAC protocol a mathematical demonstration does the deal in Ref. [34]. The fundamental assumption for which the demonstration is made is that all stations are uniformly distributed in the coverage area.

The maximum assumed transmission rages $r_{11}, r_{5.5}, r_2$ and Sr_1 are defined for 11, 5.5, 2 and 1 Mbps transition rates.

The transition time of a data packet for a fixed data transition rate of x Mbps is presented in the flowing formulas:

$$T_{11} = T_{count}(n) + T_{overhead} + \frac{8L}{R_{11}} \tag{1}$$

$$T_{5.5} = T_{count}(n) + T_{overhead} + \frac{8L}{R_{5.5}} \tag{2}$$

where $T_{overhead}$ is defined as

$$T_{overhead} = T_{PLCP} + T_{DIFS} + T_{RTS} + T_{CTS} + 3T_{SIFS} + T_{TACK} \tag{3}$$

and $T_{count}(n)$ is defined as the amount of time necessary for a successfully connection

Moreover, for 2 Mbps transition rates the transmitting time if the helper is not available is defined in the flowing formulas:

$$T_2 = T_{count}(n) + (P_{11,11} + P_{5.5,11} + P_{5.5,5.5})T_{CoopOH} + \frac{16P_{11,11}L}{R_{11}} + \frac{8P_{5,11}L}{R_{11}} + \frac{8P_{5,11}L}{R_{5.5}} +$$

$$\frac{16P_{5.5,5.5}L}{R_{5.5}} + (1 - P_{11,11} - P_{5.5,11} - P_{5.5,5.5})(T_{overhead} + \frac{8L}{R_2}) \tag{4}$$

where

$$R_x = x \, Mbps \text{ and } T_{CoopOH} = 2T_{PLCP} + T_{DIFS} + 5T_{SIFS} + T_{RTS} + 2T_{CTS} + T_{ACK}. \tag{5}$$

Based on the formulas presented above it can be written in similar mode the equations for average transmission time T_1.

The CSMA/CA protocol guarantees the fact that each station in the network have the same number of packet for a long period of time. Therefore the average transmission time per packet is calculated as:

$$T = f_{11}T_{11} + f_{5.5}T_{5.5} + f_2T_2 + f_1T_1 \tag{6}$$

where

$$f_{11} = \frac{r_{11}^2}{r_1^2} \tag{7}$$

$$f_{5.5} = \frac{(r_{5.5}^2 - r_{11}^2)}{r_1^2} \tag{8}$$

$$f_2 = \frac{(r_2^2 - r_{5.5}^2)}{r_1^2} \tag{9}$$

$$f_1 = \frac{(r_1^2 - r_2^2)}{r_1^2} \tag{10}$$

Taking into account equation (6) and the fact that the maximum range is about $r_{11} \simeq 36m < r_{5.5} \simeq 45m < r_2 \simeq 48m < r_1 \simeq 51m$ we can say that the CoopMac average time is better than T_1 and T_2 but worse than T_{11} and near to T_{55}. Given the fact that the CoopMAC is used for bad communication speed transfers, thus interfering in the T_1, T_2 and more rarely to T_5 we can state that the method improves both timing and throughput.

6.3   Results

In order to demonstrate the validity of the proposed security mechanism, several simulation have been made using 802.11 p technology with two different map scenarios: San Francisco (121 km^2) and Beijing ($16.807.8 \text{ km}^2$).

The test have taken into account a number of 500 vehicles and 10 infrastructure point on the map. A number of 50 vehicles has been randomly chosen and their results plotted in the simulation.

The decreased number in the tiles exchanged both between the vehicles and between vehicles and the infrastructure is motivated by the increased overhead that the security mechanism has introduced to the network. An average length of the message without the security mechanism was computed as being around 100 bytes, whereas with the security mechanism has increased the number of bytes of the message to 177 bytes.

Based on the attacks presented in Sect. 4 with an adaptation on the security mechanism presented in the paper, we can state that the following attacks have been resolved:

- alteration attacks occur when the messages exchanged are different from the receiver to the destination, thus creating a false image of the traffic. This type of attack has been resolved by introducing a hash of the message along with the message which is verified at the destination. If the computed hash of the message is different from the sent hash, then the message is dropped. This solution has considered that communication error are of at most 2 bits and can be resolved automatically by the wireless transceiver.
- replay attacks are the ones that collect data over a time slot only to reuse that data at later times in order obtain certain privileges. This attack has been resolved by the introduction of pair of GPS position into the encrypted data filed and the message packet ID, therefor, if a helper wants to resend information the GPS encrypted data will show that the location is different from the receiver's location and that the difference between the timestamp of the message and the current time is more that one minute. Thus the messages will be dropped.
- sybil attacks are part of the impersonating attacks and happen when an attacker uses a different set of identification at the same time. This attack was limited by imposing that all messages should be signed and the message data has information about the GPS location of the emitter. In order to apply this attack, one should get the certificates of more than one vehicle from the network which means attacking the infrastructure. This type of attack has not been treated because of the unlimited resource capability of the infrastructure.
- ID disclosure attack adds prejudice to users privacy, revealing their secret data. This type of attack is possible only if the attacker is in the proximity of the victim at all times and it corresponds with it at the discovery stage.
- eavesdropping occurs when an attacker listens to the communications from the network in order to get confidential information. This attack is useless in the net-

work in the message exchange phase because of the shared key algorithm imposed which determine the use of a different key at all times.

Attacks like the ones stated below have been taken into consideration and the solution found was to introduce monitoring devices in the network which should detect the misbehavior and signal to the infrastructure. This, in turn, will add the vehicle to a revocation list or a blacklist and announce it to all the vehicle in the network in order to prevent further malicious acts. The disadvantage of the proposed solution is that the elimination of malicious vehicles is made in time and therefor their actions can affect the network for a longer time.

- denial of service or distributed denial of service are attacks meant to make the network unavailable and thus users uninformed about the traffic status. This attack is very hard to mitigate because it does not affect the security of the system, but the availability of it. It does not rely on the system's cryptography weaknesses but on the wireless communication environment.
- fabrication attacks happen when a user creates false information in order to obtain certain privileges. These bogus information inserted into the system are mitigated at the infrastructure level when correlating the information from various sources. The vehicles that uses this approach will be found by the signature of the message. Messages that have been stored from other users lose their viability if the sender to the infrastructure is not the same as the message generator vehicle.

7 Conclusions and Future Work

In the context of constant growth of the number of vehicles on the roads, the super-saturation of the roads and the traffic congestion are more and more obvious thus leading to the need of intelligent traffic systems (ITS). ITS are advanced applications that imply multiple technologies of electronics, control and sensing, computer and others along with innovative services in order to allow users to be better informed about the status of the roads, to improve the safety and efficiency on the streets, to relieve traffic congestion and reduce air pollution.

Imagine an intersection where all the vehicles can be uniquely identifiable and each car has the ability to give information about the traffic it experienced until that point through the GPS position, traffic alerts, pollution or economic driving. Then, all the vehicles could have an image of the traffic status and could decide further paths in order to avoid traffic jams and even accidents caused by the correlation of fast speed driving with little space between the vehicles. But if a car says that a certain path was crowded then all the cars that wanted to pass on that street will reconsider and try using a different path, which leads to traffic jams on the alternative route. This motivates a vehicle to listen to the information about traffic jams but not to pass it forward so that he will be able to circulate faster than the others. In order to prevent such behaviors, we need to be able to distinguish the messages each car sends and protect their content. This preventive behavior is part of a bigger mechanism called the security mechanism.

The security mechanism described in the thesis is based on wireless communicating devices in which vehicles act like clients that send information about their GPS location or from their sensors and clients that need to receive information from a higher instance about the bigger picture of the network, that bigger picture being the infrastructure.

The novelty of the project is given by the fact that it tries to protect nodes from one another, in the sense that if one node is malicious, its action need to be limited. Translated to the real life events, it means that all vehicles will provide secured information about their GPS position and about the traffic jams that they experienced to the Infrastructure. In this way, all the vehicles have a trusted communication with the Infrastructure and each car can be held responsible for the information that it brings to the network.

Another advantage introduced by the security mechanism presented in this thesis is that all the processing data is made on the infrastructure side, which eases the complexity of the application existing on vehicles. All the vehicles generate raw data about their location and traffic alerts when they remain in a range for more then a quantum of time, and then send that data to the infrastructure. The infrastructure then correlates the data to the data existing in the network to that point in order to determine the rightfulness of the data and to estimate the risks that it involves. If traffic alerts have been spotted for a certain path, the infrastructure has the duty of finding alternative paths and to split the vehicles so that those alternative routes do not become also crowded.

Another idea brought by the thesis is the security of the communication when using third parties to forward messages without allowing them to alter the messages. This idea was based on the Cooperative MAC protocol that allows helpers to pass along messages from a sender to a receiver in order to help the communication between the two communication parties but the novelty was the fact that all the senders sign the messages they send and then add a hash to the messages, the helper takes the message from the sender and also signs the message. Thus, in case of a difference between the received hash of the message and the computed hash of the message, the receiver can determine where things went wrong and signal it to the infrastructure.

The security mechanism presented in the thesis takes into account the high mobility and opportunistic aspects of VANET in order to grant several characteristics for the vehicles involved in the network, such as:

- authentication means that each message in the network needs to be authenticated, meaning that each emitter signs the messages and then at the receiver the message is verified. This characteristic has been obtained by using the public key cryptography, each message in the network being signed with the certificate X.509.
- availability refers to the network's constant need to be able to provide information for the vehicles. This property is assured by adding the helper's computational power and data rate advantage.

- non-repudiation means that each message can be traced back to the vehicle that had generated him and is needed in order to prevent malicious attackers saying that they did not generate a certain message. This characteristic is possible by signing all the messages which demonstrates the source of each message.
- privacy is needed in order to prevent others from tracing back someone's route or letting out critical information. This property is obtained by encrypting all the data location with a secret key, always different from the others.
- real-time constraints refers to the high speed movement of the vehicles which makes the time slot for message exchange limited. The characteristic could not be resolved, but a walk around it was found by adding a cooperation model to the network which increases the distance and, therefor, also the time slot.
- integrity is the property that ensures all the messages that they were received as they were sent without any alterations. This property is obtained by signing all the certificates and sending a hash of the message in order to verify that no modifications were made.
- confidentiality protects the privacy of the drivers. This characteristic was not obtained because each vehicle in the network has an unique identifier based on which a unique certificate is generated.

Based on the attacks presented in Sect. 4 we can state that the security mechanism presented in the thesis is capable to mitigate the following attacks:

- alteration attacks occur when the messages exchanged are different from the receiver to the destination, thus creating a false image of the traffic. This type of attack has been resolved by introducing a hash of the message along with the message which is verified at the destination. If the computed hash of the message is different from the sent hash, then the message is dropped. This solution has considered that communication error are of at most 2 bits and can be resolved automatically by the wireless transceiver.
- replay attacks are the ones that collect data over a time slot only to reuse that data at later times in order obtain certain privileges. This attack has been resolved by the introduction of pair of GPS position into the encrypted data filed and the message packet ID, therefor, if a helper wants to resend information the GPS encrypted data will show that the location is different from the receiver's location and that the difference between the timestamp of the message and the current time is more that one minute. Thus the messages will be dropped.
- sybil attacks are part of the impersonating attacks and happen when an attacker uses a different set of identification at the same time. This attack was limited by imposing that all messages should be signed and the message data has information about the GPS location of the emitter. In order to apply this attack, one should get the certificates of more than one vehicle from the network which means attacking the infrastructure. This type of attack has not been treated because of the unlimited resource capability of the infrastructure.
- ID disclosure attack adds prejudice to users privacy, revealing their secret data. This type of attack is possible only if the attacker is in the proximity of the victim at all times and it corresponds with it at the discovery stage.

- eavesdropping occurs when an attacker listens to the communications from the network in order to get confidential information. This attack is useless in the network in the message exchange phase because of the shared key algorithm imposed which determine the use of a different key at all times.

Attacks like the DoS or Fabrication have been taken into consideration and the solution found was to introduce monitoring devices in the network which should detect the misbehavior and signal to the infrastructure. This, in turn, will add the vehicle to a revocation list or a blacklist and announce it to all the vehicle in the network in order to prevent further malicious acts. The disadvantage of the proposed solution is that the elimination of malicious vehicles is made in time and therefor their actions can affect the network for a longer time.

Denial of Service or Distributed Denial of Service are attacks meant to make the network unavailable and thus users uninformed about the traffic status. This attack is very hard to mitigate because it does not affect the security of the system, but the availability of it. It does not rely on the system's cryptography weaknesses but on the wireless communication characteristics.

Fabrication attacks happen when a user creates false information in order to obtain certain privileges. These bogus information inserted into the system are mitigated at the infrastructure level when correlating the information from various sources. The vehicles that use this approach will be found by the signature of the message. The infrastructure takes into account when computing the status of the network only the messages received directly from vehicles.

The main disadvantage to the network is the fact that it cannot grant anonymity, and that the infrastructure can restore any paths of the participating vehicles. If an attack takes place on the server side, and the infrastructure cannot resist, it will reveal data about the participants, along with the identity of those who have monitoring devices.

Further development, implies making a Wi-Fi application compatible for different operating systems in order to ensure better device compatibility and an infrastructure approach more feasible.

Acknowledgements The research in this paper is supported by national project MobiWay—Mobility beyond Individualism (PN-II-PT-PCCA-2013-4-0321), and by national project DataWay—Real-time Data Processing Platform for Smart Cities: Making sense of Big Data (PN-II-RU-TE-2014-4-2731). The authors would like to thank the reviewers for their time and expertise, constructive comments and valuable insight.

References

1. Schagrin, M.: Vehicle-to-vehicle (V2V) communications for safety. http://www.its.dot.gov/research/v2v.htm (2013). Accessed 01 Aug 2015
2. Wireless networking speed: Ideals and experiences. https://www.cites.illinois.edu/wireless/speed.html. Accessed 01 Aug 2015

3. Kamvar, S.D., Schlosser, M.T., Garcia-Molina, H.L.: The eigentrust algorithm for reputation management in P2P networks. In: Proceedings of the 12th International Conference on World Wide Web, pp. 640–651. ACM (2003)
4. Perrig, A., Newsome, J., Shi, E., Song, D.: The sybil attack in sensor networks: analysis and defenses. In: Third International Symposium on Information Processing in Sensor Networks (2004)
5. Raya, M., Hubaux, J.-P.: The security of vehicular ad hoc networks. In: Proceedings of the 3rd ACM Workshop on Security of Ad hoc and Sensor Networks, pp. 11–21. ACM (2005)
6. Sampigethaya, K., Huang, L., Li, M., Poovendran, R., Matsuura, K., Sezaki, K.: CARAVAN: providing location privacy for VANET. Technical Report, DTIC Document (2005)
7. Ma, S., Wolfson, O., Lin, J.: A survey on trust management for intelligent transportation system. In: Proceedings of the 4th ACM SIGSPATIAL International Workshop on Computational Transportation Science, pp. 18–23. ACM (2011)
8. Hubaux, J.-P., Capkun, S., Luo, J.: The security and privacy of smart vehicles. IEEE Secur. Priv. 3, 49–55 (2004)
9. Gollan, L., Gollan, I.L., Meinel, C.: Digital signatures for automobiles?! In: Systemics, Cybernetics and Informatics. SCI, Citeseer (2002)
10. Furgel, I., Lemke, K.: A review of the digital tachograph system. In: Embedded Security in Cars, pp. 69–94. Springer (2006)
11. Raya, M., Papadimitratos, P., Hubaux, J.-P.: Securing vehicular communications. IEEE Wirel. Commun. Mag. Spec. Issue Inter-Veh. Commun. 13(LCA-ARTICLE-2006-015), 8–15 (2006)
12. Samara, G., Al-Salihy, W.A., Sures, R.: Security analysis of vehicular ad hoc nerworks (VANET). In: 2010 Second International Conference on Network Applications Protocols and Services (NETAPPS), pp. 55–60. IEEE (2010)
13. Dak, A.Y., Yahya, S., Kassim, M.: A literature survey on security challenges in VANETs. Int. J. Comput. Theory Eng. 4(6), 1007 (2012)
14. Raw, R.S., Kumar, M., Singh, N.: Security challenges, issues and their Solutions for VANET. Int. J. Netw. Secur. Appl. (IJNSA) 5(5) (2013)
15. Fonseca, E., Festag, A.: A survey of existing approaches for secure ad hoc routing and their applicability to VANETS. NEC Netw. Lab. 28, 1–28 (2006)
16. Sanzgiri, K., Dahill, B., Levine, B.N., Shields, C., Royer, E.M.B.: A secure routing protocol for ad hoc networks. In: 10th IEEE International Conference on Network Protocols, 2002. Proceedings, pp. 78–87. IEEE (2002)
17. Hu, Y.-C., Perrig, A., Johnson, D.B.: Ariadne: a secure on-demand routing protocol for ad hoc networks. Wirel. Netw. 11(1–2), 21–38 (2005)
18. Buchegger, S., Le Boudec, J.-Y.: Performance analysis of the CONFIDANT protocol. In: Proceedings of the 3rd ACM International Symposium on Mobile Ad hoc Networking and Computing, pp. 226–236. ACM (2002)
19. Golle, P., Greene, D., Staddon, J.: Detecting and correcting malicious data in VANETs. In: Proceedings of the 1st ACM International Workshop on Vehicular Ad hoc Networks, pp. 29–37. ACM (2004)
20. Zapata, M.G.: Secure ad hoc on-demand distance vector routing. ACM SIGMOBILE Mobile Comput. Commun. Rev. 6(3), 106–107 (2002)
21. Hu, Y.-C., Johnson, D.B., Perrig, A.: SEAD: secure efficient distance vector routing for mobile wireless ad hoc networks. Ad hoc Netw. 1(1), 175–192 (2003)
22. Papadimitratos, P., Haas, Z.J.: Secure link state routing for mobile ad hoc networks. In: 2003 Symposium on Applications and the Internet Workshops, Proceedings, vol. 2003, pp. 379–383. IEEE (2003)
23. Carter, S., Yasinsac, A.: Secure position aided ad hoc routing (2003)
24. Adjih, C., Clausen, T., Jacquet, P., Laouiti, A., Muhlethaler, P., Raffo, D.: Securing the OLSR protocol. In: Proceedings of Med-Hoc-Net, pp. 25–27 (2003)
25. Marti, S., Giuli, T.J., Lai, K., Baker, M.: Mitigating routing misbehavior in mobile ad hoc networks. In: Proceedings of the 6th Annual International Conference on Mobile Computing and Networking, pp. 255–265. ACM (2000)

26. McGraw, G.: Software Security: Building Security, vol. 1. Addison-Wesley Professional (2006)
27. Ruohomaa, S., Kutvonen, L.: Trust management survey. In: Trust Management, pp. 77–92. Springer (2005)
28. Linn, J.: Trust models and management in public-key infrastructures. RSA Lab. **12** (2000)
29. Boncella, R.J.: Wireless security: an overview. Commun. Assoc. Inf. Syst. **9**(1), 15 (2002)
30. X.509 public key certificate and certification request generation. http://www.bouncycastle.org/wiki/display/JA1/X.509+Public+Key+Certificate+and+Certification+Request+Generation. Accessed 01 Aug 2015
31. Kocher, P.C.: Cryptanalysis of Diffie-Hellman, RSA, DSS, and other systems using timing attacks. In: Advances in Cryptology, CRYPTO95: 15th Annual International Cryptology Conference. Citeseer (1995)
32. Solo, D., Housley, R., Ford, W.: Internet x. 509 public key infrastructure certificate and CRL profile (1999)
33. Kaufman, C., Perlman, R., Speciner, M.: Network Security: Private Communication in a Public World. Prentice Hall Press (2002)
34. Liu, P., Tao, Z., Panwar, S.: A cooperative MAC protocol for wireless local area networks. In: 2005 IEEE International Conference on Communications, 2005. ICC 2005, vol. 5, pp. 2962–2968. IEEE (2005)

Concurrency Control for Mobile Collaborative Applications in Cloud Environments

Moulay Driss Mechaoui and Abdessamad Imine

Abstract As the world is progressing quickly towards more connected mobile devices, the use of mobile collaborative applications is gaining an increasing popularity. For instance, real-time data streams and web applications (such as social networking and ad-hoc collaboration) are seamlessly incorporated in mobile applications. Despite this powerful evolution, the resource limitation (energy consumption and unstable connectivity) remains a serious problem against a safe concurrency control for an efficient and continuous use of mobile collaboration. In this chapter, we describe the data consistency issues when mobile applications support collaboration through the cloud. Based on human factors (such as high interactivity and data consistency), we present two concurrency control techniques for offloading and ensuring data synchronization among mobile devices and the cloud. The first technique relies on a client-server style to ensure safe coordination, while the second one supports a peer-to-peer mechanism to achieve a decentralized data synchronization.

1 Introduction

The spectacular development of mobile devices (smartphones, tablets, PDA) and the rapid progression of mobile communications in these last few years have offered a new environment of development for mobile applications. These mobile devices have changed the way we interact with our social environment and become the devices of choice to collaborate with family members, friends and business colleagues and/or customers. However, deploying ad-hoc collaboration around mobile applications requires increasing amounts of computation, data storage and network communica-

M.D. Mechaoui (✉)
University of Sciences and Technology Oran 'Mohamed Boudiaf' USTO-MB,
Oran, Algeria
e-mail: moulaydriss.mechaoui@univ-usto.dz

A. Imine
Université de Lorraine and INRIA-LORIA Grand Est, Nancy, France
e-mail: abdessamad.imine@loria.fr

© Springer International Publishing Switzerland 2017
C.X. Mavromoustakis et al. (eds.), *Advances in Mobile Cloud Computing
and Big Data in the 5G Era*, Studies in Big Data 22,
DOI 10.1007/978-3-319-45145-9_11

tions. Moreover, preserving the consistency of the manipulated shared data (such as the shared document in mobile collaborative editor) under constraints of the mobile applications, namely the freshness and the energy consumption, remains still a serious problem.

In this case, resorting to cloud computing becomes a necessity. Cloud computing is a multi-purpose paradigm that aggregates several technologies such as virtualization, peer-to-peer networks and autonomic computing. It is an emerged model based on virtualization for efficient and flexible use of hardware assets and software services over a network. Virtualization extends the mobile device resources by offloading execution from the mobile to the cloud where a clone (or virtual machine) of the mobile is running. It provides a seamless and rich functionality to mobile applications regardless of the resource limitations of mobile devices. Cloud computing allows users to build virtual networks "à la peer-to-peer" where a mobile device may be continuously connected to other mobiles to achieve a common task.

In this chapter, we provide a global view of mobile collaborative applications in the cloud, while highlighting the specific issues of data consistency in mobile cloud computing. We present the principle and drawbacks of two concurrency control techniques (centralized and decentralized) for offloading and preserving data consistency between mobile devices and the cloud. More precisely, we describe the components of two existing collaborative editing protocols, CloneDoc [1] for the centralized control concurrency and OptiCloud [2, 3] for the distributed one.

The remainder of this chapter is organized as follows: Data consistency issues related to mobile collaboration through the cloud are given in Sect. 2. Section 3 presents a concurrency control scheme supporting the client-server style for consistency maintenance of the shared data. In Sect. 4, we describe another concurrency control scheme based on a pure peer-to-peer model. We discuss the related work in Sect. 5 and conclude in Sect. 6.

2 Data Consistency Issues

In this section, we present the collaborative model for manipulating shared data regardless of spatial and temporal constraints using mobile devices in the cloud environment, and we illustrate this model by two use cases. Finally, we highlight data inconsistency problems that mobile users are likely to face due to mobile-to-clone and clone-to-clone interactions.

2.1 Collaboration Model

The aim of the collaboration in cloud environments is to allow many geographically dispersed users to manipulate the shared data at anytime and anywhere. This collaboration model involves a set of mobile devices and a set of clones (or virtual

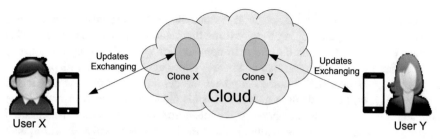

Fig. 1 Collaboration model

machines) in such a way each mobile user owns her/his clone in the cloud as illustrated in Fig. 1. The clone is characterized by machine features (e.g. CPU frequency, memory size) and a virtual image to be set up (e.g. softwares like the operating system and synchronization protocols).

To achieve continuous collaboration with high data availability, each user owns two copies of the shared data where the first one is stored in the mobile device whereas the second one is on its clone (at the cloud level). The collaboration between users is performed as follows: each user modifies the mobile copy and then sends local modifications to her/his clone in order to update the second copy and propagate these local modifications to other mobile users by means of their clones.

Based on human factors, a mobile collaborative application is characterized by the following requirements [2, 3]:

1. *High local responsiveness*: the application has to be as responsive as if it is based on single-user;
2. *High concurrency*: users must be able to concurrently and freely modify any part of the shared data at anytime and anywhere;
3. *Consistency*: all concurrent updates must be synchronized in such a way users must eventually be able to see an identical view of all copies;
4. *Scalability*: a group must be dynamic in the sense that users may join or leave the group at any time;
5. *Failure recovery*: users have to recover easily all shared documents when technical hitch (e.g. crash, theft or loss of mobile device) happens, and continue seamlessly the collaboration.

In this chapter, we focus only on features 2 and 3 to present how existing systems achieve these features.

In the following, we present two use cases that illustrate how this collaboration model can be deployed over mobile cloud network to overcome some problems:

Assisting tourists. Suppose that a group of tourists want to make a tour in London city. They are already provided with a mobile application based on the proposed collaboration model helping them to visit the city using a map. One member of the group creates a collaborative group in the cloud and downloads the map with relevant information about the city. The other members join the created group and share the

London map. When visiting museums, restaurants or historic places, tourists can share, in real-time, their opinions by writing their comments in the local map. This allows enriching and updating the map content. For instance, during the journey one of them realizes that, instead of the art gallery appearing on his map, there is a pharmacy. Hence, she/he corrects/updates the description of this localization in her/his map, and then she/he sends it to her/his clone in the cloud in order to be broadcast to other group members. At the meantime, one tourist was disconnected when sending the update information. In this case, her/his clone will notify the update information after her/his re-connection to the group as if she/he did not quit it.

Social networking. With the widespread use of online social networks (e.g. Facebook, MySpace, and Twitter), which have become a lifestyle in our society, allowing users to keep in touch with families and friends and also extending for business purposes such as searching for new career opportunities. Recently, these social networks are increasingly going mobile, and propose a new trend of social networks called Mobile Social Networks (MSN) [4–6], that has emerged and attracted considerable attention from the academic and industrial communities. A user can share and synchronize social data (such as a list of friends, comments, a set of song lyrics, etc.) across a group of friends. However, MSN should address the constraints of mobile devices, i.e., limited energy, low memory capabilities, limited processing power, scalability, and heterogeneity [7]. To satisfy the constraints of mobile devices in MSN, a cloud network can be used where each mobile device creates its own clone. The clone stays connected and reachable for the other clones in the cloud whether its mobile device is connected or not. The processing, storage and dissemination of data are delegated to the clone. The clone can also send possible updates/notifications (e.g. recommendations, nearby friends, prizes, etc.) back to the other clones or to its mobile device either when requested or as a response to the events created by other mobile devices. This cloning-based model does not require the mobile device to be online all the time. Therefore, it retains the power consumption of the mobile device.

However, this collaborative model is not free from problems. The main challenge is: how to maintain consistency and properly resolve conflicts throughout the shared data, while several users are simultaneously updating the same data? Data inconsistency may appear in situations related to clone-to-clone and mobile-to-clone interactions.

2.2   Interaction Between Clones

The inconsistency problem occurs when two or several clones produce simultaneous updates. To illustrate this problem, consider the following example of collaborative editor where two clones, CLONE 1 and CLONE 2, contain the same document "ABC". CLONE 1 executes editing operation $o_1 = Ins(2, X)$ to insert the character 'X' at position 2 and ends up with "AXBC". Concurrently, CLONE 2 performs editing operation $o_2 = Del(2)$ to remove the character 'B' at position 2 and obtains the

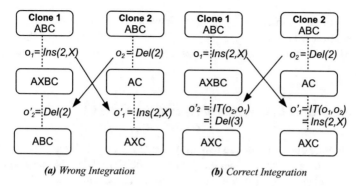

(a) *Wrong Integration*   **(b)** *Correct Integration*

Fig. 2 Clone to Clone synchronization scenario

state "AC". After exchanging operations among clones, the CLONE 1's document becomes "ABC" but the CLONE 2's document has a different state "AXC" as shown in Fig. 2a.

Several cloud-based collaborative editors such as Google Docs,[1] Cloud9,[2] Zoho Suite[3] use Operational Transformation (OT) approach [8, 9] that provides a general model for synchronizing the shared data, while allowing each user to apply local updates optimistically. OT is considered as the efficient and safe method for consistency maintenance in the literature of collaborative editors. It consists to transform the parameters of an operation to include the effects of previously concurrent operations so that the transformed operation can lead to consistent document. In the OT approach, each site is equipped by two main components [8, 10]: the integration component and the transformation component. The integration component determines how an operation is transformed against a given operation sequence (e.g., the log buffer). It is also responsible for receiving, broadcasting and executing operations. It is rather independent of the type of the shared data. The transformation component is a set of *transformation functions* which is responsible for merging two concurrent operations defined on the same state. Every transformation function is specific to the semantics of a given shared data. The most known OT-based theoretical framework is established by Ressel et al. [10]. They define two consistency criteria:

- *Causality*: If one operation O_1 causally precedes another operation O_2, then O_1 must be executed before O_2 at all sites.
- *Convergence*: When all sites have performed the same set of operations, the copies of the shared data must be identical.

---

[1]https://www.google.fr/intl/fr/docs/about/.
[2]https://c9.io/.
[3]https://www.zoho.com/.

Thus, at CLONE 1, operation o_2 needs to include the effect of o_1 using a transformation function IT: $o'_2 = IT(o_2, o_1) = Del(3)$. As for CLONE 2, operation o_1 is left unchanged. Accordingly, both of them get the same state "AXC" as shown in Fig. 2b.

2.3   Interaction Between Mobile and Its Clone

Since the mobile and its clone are two entities that are physically and geographically separated. Each entity has its own local copy of the shared data to be synchronized. Therefore, a delay of applying the same operations on both sides with the same copy is possible. This may lead to document inconsistency.

Consider the example of collaborative editing illustrated in Fig. 3. Given a mobile M and its clone C that have the same initial document "ABC". Mobile M performs a local update operation $o_1 = Ins(2, X)$ to insert the character 'X' at position 2 and results in state "AXBC". Simultaneously, clone C executes two operations o_2 and o_3 coming from other clones: operation $o_2 = Ins(2, X)$ adds the character 'X' at position 2 and gives the state "AXBC"; operation $o_3 = Del(3)$ removes the character "B" at position 3 and obtains the state "AXC".

When Mobile M and its Clone C decide to synchronize, they commit their operations that have been applied locally. After exchanging operations, clone C performs operation o_1 and its document becomes "AXXC". At the meanwhile, Mobile M executes operations o_2 and o_3 to result in the state "AXBC". As illustrated in Fig. 3, the mobile and its clone have different states.

It is clear that preserving data consistency between the mobile and its clone requires some additional treatments that have to be performed by either the mobile

Fig. 3 Mobile to clone synchronization scenario

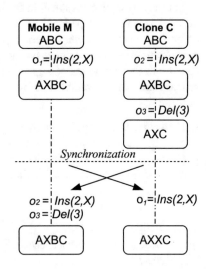

device, its clone or both. Note that the more computing tasks are executed on mobile device the more battery life is reduced.

To deal with data consistency problems caused by clone-to-clone and mobile-to-clone interactions, two kinds (centralized and decentralized) of OT-based concurrency control protocols will be presented in the following sections.

3   Centralized Concurrency Control

In this section, we give a general presentation of concurrency control protocols designed in the client-server style and using OT approach to coordinate all concurrent updates.

3.1   Principle

To maintain consistency, many centralized concurrency control protocols (such as CloneDoc [1], SPORC [11] and Google Docs[4]) are based on a single server (or super clone), that is the backbone of the collaboration with the following features:

- It relies on OT technique to enforce continuous and global order on all updates to avoid the divergence of user's document view from the server.
- It enables users to join or/and leave any collaborative group;
- It ensures the availability of shared documents for all mobile users.
- It manages the synchronization and propagation of updates between clones.

Two layers are used to maintain the data consistency: the first layer ensures synchronization between clones and the second one consists in synchronizing the mobile with its clone.

Clone-to-Clone synchronization. The clone is used to (i) submit the operation coming from its mobile to the super clone (or central server), (ii) transform the operations of the other clones received from the super clone and (iii) handle its queues so that its state of the shared document is coherent to that of other clones. Note that the super clone serializes all operations according to a total order.

For instance, the clone in CloneDoc [1] maintains two states: *The pending queue* contains update operations received from its mobile that have already been applied to its local state, but not sent yet to the super clone (i.e. not yet serialized). *The committed queue* contains operations already ordered by the super clone.

When a clone receives an operation from its mobile, it applies it immediately over its local state, saves it locally and then sends it to the central server to be ordered. In the case of CloneDoc, the operation is saved in *the pending queue*.

---

[4]https://www.google.fr/intl/fr/docs/about/.

To serialize operations, two total ordering schemes may be enforced by the super clone [12]:

- *The implicit total ordering scheme*: it is based on a central server for broadcasting operations among collaborating clones where each clone sends local operations to the super clone via a FIFO (First In First Out) communication channel. Then, the super clone serializes operations and broadcasts them among all clones. The operation serialization order at the super clone *implicitly* forms a total order among all operations. This total ordering scheme is used in Google Docs.
- *The explicit total ordering scheme*: it is based on a special sequencer which does not broadcast operations but only generates continuous sequence numbers (tickets) for ordering operations. After generating a local operation, a collaborating clone requests the central sequencer for a sequence number to timestamp this operation. In this way, all operations are totally ordered by sequence numbers. This total ordering scheme is used in CloneDoc [1].

When the clone receives an ordered operation form the central server, two cases are possible:

- If the received operation is generated by another clone then, the clone transforms it over its precedent ordered operations (w.r.t the total order enforced by the super clone), applies it on its local state and sends it to its mobile. In the case of Clone-Doc, the operation is transformed over *the committed queue*, then the resulted operation is applied on the local state of the clone and sent to the mobile device.
- If the clone received its own operation from the central server, the clone stores it locally and sends it to its mobile without executing it because it is already executed. In the case of CloneDoc, the clone extracts the received operation from *the pending queue*, adds it in *the committed queue* and sends it to its mobile.

Mobile-to-Clone synchronization. The mobile device and its clone are not physically the same. This leads to an inevitable delay in their communications which may introduce data inconsistency (as shown in Sect. 2.3). Therefore, an additional consistency protocol based on OT approach is deployed on mobile device to solve this problem. For example, a user applies several operations on her/his mobile to edit the local copy of the shared document in disconnected mode; these operations are logged in a local queue of the mobile device (a *pending queue* in the case of Clone-Doc). At meanwhile, the clone can receive operations from other clones. When the mobile joins the cloud, its clone sends it the operations received and integrated from other clones. Then, the mobile transforms these operations over its local operations stored in its local queue.

Consider the scenario illustrated in Fig. 4 where User 1 exchanges operation O with User 2. First, User 1 executes immediately the operation O on the local copy and sends it to his clone namely Clone 1. Then, Clone 1 performs O, saves it in the pending queue and sends it to *SuperClone* in order to serialize it according to the used total ordering scheme. Next, Super Clone broadcasts the ordered operation O' to all the clones of the cloud, including Clone 1.

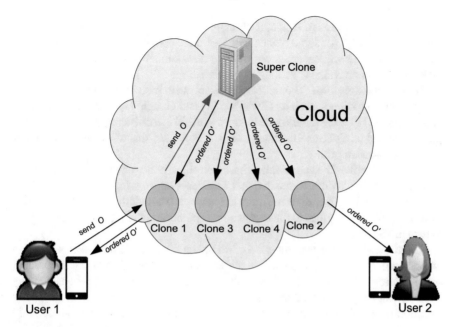

Fig. 4 Centralized synchronization

When Clone 2 receives the ordered operation O', it transforms it over its committed queue. Then, the clone executes the transformed operation over its local state and sends it to the mobile of User 2.

On the other side, when Clone 1 receives its own ordered operation O' from *SuperClone*, it extracts the operation from the pending queue, adds the ordered operation in the committed queue and sends it to the mobile of User 1.

The clones send back the ordered operation O' to their real devices (smartphones) to be transformed (if there are operations in pending queue) and executed (except in User 1) such that the copy of User 1 is coherent with the copies of the other users in the system.

3.2 Drawbacks

The strong dependence of different users (clones) to the coordination server (which plays a central and important role in the collaboration) can cause serious problems that may negatively affect the collaboration. Hereafter, we list some disadvantages of centralized concurrency control.

Failure. Based on central coordination server, mobile collaborative applications deployed on cloud environments are expected to suffer from bottlenecks and are more prone to faults. For example, if all clones running in the cloud keep making lot of updates to the super clone, the performance will be drastically downgraded at the expense of responsiveness. Again, if the super clone fails (due to hardware failure or denial of service [13]), some clones will lose their updates. Moreover, a malicious user can flood the sequencer server with a great number of operations in order to make the server unavailable as long as possible.

Energy consumption. The collaborative software based on centralized concurrency control protocols, such as CloneDoc [1] and Google Docs, use an additional processing of OT on the mobile side in order to avoid inconsistency problem between the mobile and its clone as illustrated in Fig. 3. However, this will cause supplementary energy consumption.

Network traffic. The coordination based on super node (or central server) incurs more network bandwidth to receive and broadcast updates from/to clones respectively. Larger the number of users, more the network bandwidth is consumed. Therefore, this problem may lead to the loss of some operations of clones and consequently to data divergence.

Intention violation. The intention of an operation o means the effect which can be obtained by executing o on the document state from which o was generated. If one operation o is generated at clone i, then its intention should be preserved at any clone j (with $i \neq j$) regardless the transformations that o undergoes. Preserving users intention using OT approach is a hard task in collaborative editing applications. In the following, we present a scenario of violation of intention in Google Docs.

Google Docs allows users to modify and update the same document in the same time by using a central coordination server. Google Docs is based on Jupiter [14] which is a client-server collaborative system. The Jupiter server maintains multiple 2D state-spaces, one for every client. A state-space consists of a local dimension for operations generated by the corresponding client, and a global dimension for operations from all other clients. To avoid using 2D state-spaces, Google adapted Jupiter system [14] by adding a *Stop-and-Wait* protocol between the client and the server. As a consequence, a single 1D buffer at the server is sufficient to maintain all transformation states.

However, Google Docs inherits the main flaw of Jupiter system, namely intention violation [15]. As illustrated in Fig. 5, three users concurrently execute different operations on the same document that contains initially the state "A". User 1 performs operation $Ins(1, X)$ to add 'X' at the position 1. Simultaneously, User 2 executes operation $Ins(2, Y)$ to insert 'Y' at position 2 and User 3 performs $Del(1)$ to delete 'A'.

Then, the users send their operations to the central server in order to be transformed. The Google Docs server uses the reception order of operations to determine the priority among operations. Consequently, we might get different results depending on the reception order of operations.

In Fig. 5, we have three different reception orders which result in two divergent states "XY" and "YX" for the same document.

For example, If the server executes the delete operation o_3 first, the two insert operations o_1 and o_2 will be transformed to insert different characters at the same position in the document.

Fig. 5 Intention preservation scenario in Google Docs [15, 16]

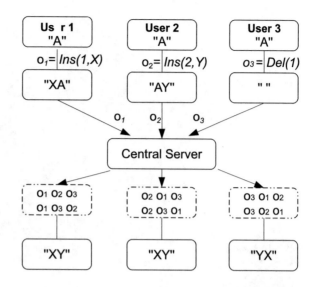

Moreover, if we assume that the Google Docs server is potentially malicious (i.e. in the sense that it might diverge the states of different users at a given point of time), then it is very hard (or even impossible) for users to be aware about this problem and to recover back to the point of time when their views were consistent.

Security risks. The central server may be vulnerable to malicious side-channel attacks [17]. This may affect the synchronization process. For instance, BYOD (Bring Your Own Devices)[5] practice is increasingly becoming a global phenomenon. Indeed, it allows employees/students to bring personally owned mobile devices (laptops, tablets, and smartphones) to their workplace, and to use those devices to access privileged company information/education platforms. What happens if BYOD practice is used for accessing a collaborative application based on central coordination? It is clear that BYOD brings significant security risks. The users could be susceptible from attacks originating from compromised web sites that may contain harmful malware and compromise the proper functioning of the application.

4 Decentralized Concurrency Control

In this section, we give a general presentation of decentralized concurrency control scheme designed "à la Peer-to-Peer" and supporting an unconstrained collaborative work (without the necessity of central coordination). Using OT approach, synchronization of divergent copies is fulfilled automatically at each clone. To better present

---

[5]http://www.ibm.com/mobilefirst/us/en/bring-your-own-device/byod.html.

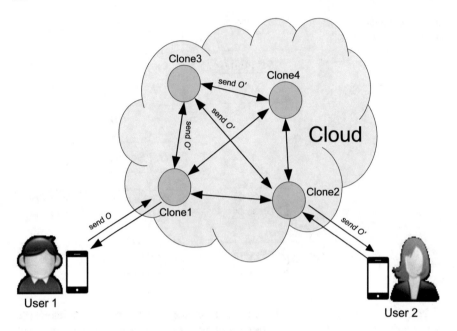

Fig. 6 Distributed synchronization

this decentralization, we describe the main components of OptiCloud [2, 3] as it is the best representative of decentralized concurrency control protocols combining mobile and cloud environments.

4.1 Principle

As illustrated in Fig. 6, the mobile collaborative application, based on decentralized coordination, provides a pure *peer-to-peer virtual private network* (without any server role assigned to some clone) platform where users can form ad-hoc groups based on their clones to achieve a common objective. It allows users to cooperate as follows (see Fig. 6): each user has a local copy of the shared document in her/his mobile and another copy of the same document in the clone; the user's operation O is locally executed in the mobile device and then is sent to its clone in order to be transformed (O' is the transformed form of O) and executed on other mobile devices (via their clones).

In [2, 3], a collaborative editing service is presented for manipulating the shared data, regardless of spatial and temporal constraints, where mobile users can edit collaboratively shared documents in peer-to-peer mode. The advantages of this model are (i) the availability of data anytime and anywhere, and (ii) the optimal use of mobile devices resources. In fact, the collaboration and communication tasks are seamlessly turned on the cloud. Moreover, it is equipped with mechanisms to transparently manage the user departure, the arrival of new users joining the collaboration group.

Each clone has a local copy of the shared document S_C and its log L_C storing all updates received from other clones and updates generated from its mobile. The log L_R contains remote updates transformed against L_C, executed on S_C and in synchronization pending with the mobile. Updates sent by the mobile and generated locally in the clone are stored in log L_M. An index S_P of log L_C is used to indicate the last synchronization point between the clone and its mobile.

To preserve data consistency in peer-to-peer mode between mobile users, Opti-Cloud [2, 3] is constituted from two synchronization layers: The first layer ensures synchronization between clones (as back-end) and the second one consists in synchronizing the mobile with its clone (as front-end).

4.1.1 Clone-Clone Synchronization

Two events can be triggered in the clone: (i) receiving operations from mobile to be generated and integrated in the clone; (ii) receiving and integrating remote operations coming from other clones.

Generation of local operations. Once the clone receives operations from its mobile, it performs the following steps:

1. Computes the minimal execution context of each operation O received from mobile. An operation may depend on previous operations according to the execution order. Tracking this dependency inside a log enables to identify operations that must be executed on all clones according to the same order. The clone synchronization protocol uses a minimal dependency relation which is independent of the number of users, and accordingly, it is well suited for dynamic groups. In other words, instead of considering O as being dependent of all L_C operations, this step reduces this context by excluding as much as possible some operations of L_C to give the transformed operation O'. For more details, we can refer to [15, 18].
2. Sends O' to other clones and adds it in L_M which contains all operations (coming from the mobile) executed on L_C.
3. Determines the operations that are concurrent to O' in log L_R and calls the transformation component in order to get operation O'' that is the transformed form of O' according to the concurrent operations;
4. Applies operation O'' over its local state S_C and adds it to log L_R.

Integration of remote operations. When a clone receives a remote operation O from another clone, the integration of this operation proceeds by the following steps:

Fig. 7 Scenario of collaboration

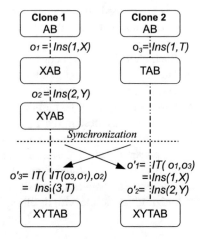

1. From the log L_C, it determines the sequence *seq* of concurrent operations to O in order to transform O against *seq* and obtain O';
2. After added O' to the local log L_C, it determines from the log L_R the sequence *seq'* of operations that are concurrent to O';
3. It calls the transformation component in order to get operation O'' that is the transformation of O' according to *seq'*;
4. It executes O'' on the current state and adds it to the local log L_R.

Illustrative example. Given two clones, Clone 1 and Clone 2 editing a shared document described in Fig. 7. Initially, each Clone has a copy that contains "AB". Two local insertion operations O_1 and O_2 have been executed by Clone 1. Concurrently, Clone 2 has executed another insertion operation O_3. The added characters are 'X', 'Y' and 'T' respectively.

There is a dependency relation between operations O_1 and O_2 in such a way O_1 must be executed before O_2 in all clones. This is due to the fact that their added characters are adjacent (positions 1 and 2) and created by the same clone (for more details see [18]). This dependency relation is minimal in the sense that when O_2 is broadcast to all clones, it holds only the identity of O_1 as it depends on directly. As illustrated in Fig. 7, the execution order is as follows.

At Clone 1, O_3 is considered as concurrent. It is then transformed against O_1 and O_2. The sequence $[O_1 = \text{Ins}(1, X), O_2 = \text{Ins}(2, Y), O'_3 = \text{Ins}(3, T)]$ is executed and logged in Clone 1, where O'_3 results from transforming O_3 to include the effect of operations O_1 and O_2 (i.e. $O'_3 = \text{IT}(\text{IT}(O_3, O_1), O_2) = \text{Ins}(3, T)$ using transformation function IT given in [18]).

At Clone 2, O_1 and O_2 are concurrent with respect to O_3. They must be transformed before being executed after O_3 according to their dependency relation. Thus, the following sequence is executed and logged in Clone 2: $O_3 = \text{Ins}(1, T)$ and $O'_1 = \text{IT}(O_1, O_3) = O_1$ and $O'_2 = O_2$.

4.1.2   Mobile-to-Clone Synchronization

In OptiCloud [2, 3], the shared document is considered as a critical section between the mobile and its clone. Indeed, when the mobile tries to commit/synchronize w.r.t its clone, only one of them will have the exclusive right to access in synchronizing mode to its document copy. This distributed mutual exclusion protocol is achieved by the exchange of messages (i.e., token). Initially, the mobile device has the right to be the first to commit/synchronize with its clone. To ensure a safe synchronization between mobile and its clone, we use two tokens T_M and T_C:

- Token T_M gives the mobile state: (i) $T_M = 1$ means that mobile is editing its local copy; (ii) $T_M = 0$ indicates that mobile is synchronizing with its clone.
- Token T_C indicates the state of the clone: (i) $T_C = 1$ states clone is integrating remote operations received from other clones; (ii) $T_C = 0$ means that clone is synchronizing with its mobile.

Whatever where the exclusive access right is, the mobile device and its clone can edit independently their local copies. The clone continues to receive remote operations from other clones to integrate them later on the local state. At the meantime, the mobile user can work on her/his copy in unconstrained way. But, once she/he decides to synchronize with her/his clone, all local editing operations are sent to her/his clone and the exclusive access right is released to enable the clone to start the synchronization with the mobile device as illustrated in Fig. 8. Thus, the clone performs the operations issued by the mobile device on its local state, includes their effects by transformation in its local (and not yet seen by the mobile) operations, and sends the resulting (or transformed) operations to the mobile device in order to integrate them.

In the mobile side, OptiCloud uses just the local state and the *Log* that contains performed operations. Note that the mobile device does not perform any specific

Fig. 8 Sequence diagram of synchronization process among mobile and clone

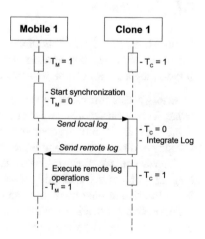

treatment; it updates only its state. The steps of synchronization between the mobile and its clone can be represented as follows:

Generation of local operations. If the mobile is in editing state (i.e. $T_M = 1$) then it creates operation o (e.g. $o = ins(1, X)$), executes this operation directly on its local state and adds o to its log. When the user wants to synchronize with her/his clone (i.e. $T_M = 0$ indicating that the mobile is in synchronizing state), the log is sent to the clone. The mobile cannot generate local operations up to the end of synchronization.

Reception of mobile operations by the clone. When the clone starts the synchronization process with its mobile, after receiving log L_M from the mobile, the clone generates locally each received operation (as previously explained in step ***Generation of local operations*** of Sect. 4.1.1). After integrating all operations coming from the mobile, the clone prepares the operations (not yet seen by the mobile) to be sent to the mobile. These operations are included in the interval from S_P to $|L_C|$ in log L_C, where S_p is the last synchronization point between the mobile and its clone and $|L_C|$ the length of log L_C. Each operation O in the interval $[S_P, |L_C|]$ inside log L_C is integrated over L_M as follows:

1. Defines the operations that are concurrent to O in the log L_M;
2. Transforms O according to the defined operations to result O';
3. Adds O' to the local log L_M;

Next, operation O' is added to a *Log*. Then, the *Log* is sent to the mobile in order to be applied on its local state.

The reception of operations from the clone. When the mobile receives operations from its clone, it applies them over its local state and makes $T_M = 1$ to indicate that it is available to generate local operations.

4.2 Drawbacks

Although the cloud-based mobile collaborative applications using distributed concurrency control avoid several flaws of the ones based on the centralized coordination server, they also have some weaknesses:

Energy consumption. Since the heavy computing tasks are delegated to the clone, it consumes more energy in the cloud in order to ensure communication among clones and also its mobile, compute transformation procedures to maintain data consistency and manage join and leave events of its mobile.

Access control. Ensuring a distributed access control to a shared data is a challenging problem in the cloud-based mobile collaborative application. The availability of the shared data in anytime and anywhere is one of the main requirements of collaborative applications, whereas access control looks to guarantee this availability only to users with proper authorization. Moreover, high responsiveness of local updates is required. However, when adding an access control layer, high responsiveness is lost because every update must be guaranteed by some authorization coming from a distant user (such as central server) [19].

5 Related Work

The massive development of cloud technology encourages users to delegate their heavy computing treatments to cloud platforms. The mobile applications are the most concerned candidates to benefit from the huge computing power of Cloud to satisfy their constraints in terms of resources (battery life, date storage and application speed up).

Several approaches [20–22] have been proposed to offload parts of their heavy tasks to the cloud since execution in the cloud is considerably faster than the one on mobile devices and mobile-cloud offloading mechanisms delegate heavy mobile computation to the cloud.

SPORC [11] is a collaborative system where several users can edit shared documents using smartphones. To maintain consistency, it relies on operational transformation technique and a single server to give global order to concurrent user updates. However, SPORC exchanges a large number of messages between users and the server. Therefore, it is not well adequate to mobile devices due to short battery life.

Inspired from SPORC, CloneDoc [1] is a secure real-time which enables collaboration for mobile devices that are cloned in the cloud in order to lessen the heavy computing tasks of collaborative editing works on mobile devices. CloneDoc is implemented upon C2C platform [23] which is a distributed peer-to-peer platform for cloud clones of smartphones. It is based on Operational Transformation (OT) approach and a single server to enforce a continuous and global order to avoid the divergence of client's document view from the server. Unfortunately, a server failure could stop the collaboration between mobile devices. Moreover, CloneDoc needs additional treatment of OT on the mobile side to ensure convergence between the mobile and its clone.

rbTree-Doc [24] is a collaborative editor framework for cloud environments. It allows multiple users to share and edit online documents. The document is represented as Red-Black tree [25]. The user can download a part of a document to be updated from the cloud service, and this enables rbTree-Doc to reduce the amount of data that needs to be encrypted by focusing on the analysis of content that has been updated by the collaborative services. However, rbTree-Doc is based on a client-server model and uses a complex structure to represent the shared document, and this from one hand, cannot support a server failure situation and from the other hand, the complex structures are hardly treated in mobile devices.

Hermes [26] is a transparent approach to interoperate between heterogeneous collaborative editing services in the cloud. Users are enabled to use their familiar services to participate in the cross-cloud document collaboration. Hermes uses an OT driven approach to resolve conflicts and maintain data consistency for cross-cloud document synchronization. Like Hermes, the collaborative editing service proposed in [2, 3] can be extended to manage the mobile collaboration on several clouds.

The platform presented in [2, 3, 27] is cloud service-based approach, and it is composed of two level systems. The first level, cloning engine, provides an automatic chain for preparing a dynamic platform that can contain multiple virtual private networks. Each network corresponds to a group of users and contains clones of their mobiles. It provides web services to manage users groups and creates clones of mobiles in the cloud. The second layer, OptiCloud, provides group collaboration mechanisms for editing shared documents in fully decentralized way. Unlike SPORC [11], CloneDoc [1] and rbTree-Doc [24], the platform of [2, 3] furnishes group collaboration mechanisms in real-time without any role assigned to the server. Procedures for maintaining consistency of shared documents are executed on the clone side.

6   Conclusion

Designing concurrency control for mobile collaborative applications is considered as a challenge, since mobile devices are constrained by insufficient resources which must be regarded when combining mobile and cloud environments.

In this chapter, we have presented the data consistency issues when mobile users are cloned in the cloud and they update simultaneously the shared data replicated in the mobile device and its clone. Two kinds of concurrency control scheme are described with their drawbacks. The first control uses central coordination server to maintain data consistency. As for the second one, it provides a synchronization mechanism in fully decentralized way.

As future research direction, it is interesting to add a new layer for security within mobile collaborative applications in the cloud. It consists in developing a protocol for managing distributed access rights and adding cryptographic mechanisms to ensure maximum security of shared resources.

Appendix

- *Replica*: is a copy of the shared data that can be modified at will by the user.
- *Data Consistency*: means that data values must be the same for all replicas when there is no updates in transit. This term is used to indicate that the system is able to reflect correctly the updates performed on a copy to all other copies of the shared data.
- *Data Concurrency*: means that many users can access simultaneously shared data to perform read and write operations.

- *Data Dependency*: an update operation applied on replica may depend on previously performed operations. In other words, the effect of such operation may be influenced by previous operations.
- *Clone*: in our case, the clone is a virtual machine Android X86[6] running in the cloud and has the same features as a physical mobile device.

References

1. Kosta, S., Perta, V.-C., Stefa, J., Hui, P., Mei, A.: Clonedoc: exploiting the cloud to leverage secure group collaboration mechanisms for smartphones. In: IEEE INFOCOM, Italy (2013)
2. Guetmi, N., Mechaoui, M.D., Imine, A., Bellatreche, L.: Mobile collaboration: a collaborative editing service in the cloud. In: SAC 2015, pp. 509–512, Spain (2015)
3. Mechaoui, M.D., Guetmi, N., Imine, A.: Lightweight and mobile collaboration across a collaborative editing service in the cloud. Peer-to-Peer Networking Appl. (2016)
4. Buchegger, S., Datta, A.: A case for p2p infrastructure for social networks—opportunities and challenges. In: WONS'09, pp. 149–156, Piscataway, NJ, USA (2009)
5. Vastardis, N., Yang, K.: Mobile social networks: architectures, social properties, and key research challenges. IEEE Commun. Surv. Tutorials **15**(3), 1355–1371 (2013)
6. Karam, A., Mohamed, N.: Middleware for mobile social networks: a survey. In: HICSS 2012, pp. 1482–1490, Maui, Hawaii (2012)
7. Borcea, C., Gupta, A., Kalra, A., Jones, Q., Iftode, L.: The mobisoc middleware for mobile social computing: challenges, design, and early experiences. In: MOBILWARE '08, pp. 1–27, ICST, Brussels, Belgium (2007)
8. Ellis, C.A., Gibbs, S.J.: Concurrency control in groupware systems. In: SIGMOD Conference, pp. 399–407, Portland, USA (1989)
9. Sun, C., Jia, X., Zhang, Y., Yang, Y., Chen, D.: Achieving convergence, causality-preservation and intention-preservation in real-time cooperative editing systems. ACM Trans. Comput. Hum. Interact. **5**(1), 63–108 (1998)
10. Ressel, M., Nitsche-Ruhland, D., Gunzenhauser, R.: An integrating, transformation-oriented approach to concurrency control and undo in group editors. In: ACM CSCW'96, pp. 288–297, Boston, USA, November 1996
11. Feldman, A.J., Zeller, W.P., Freedman, M.J., Felten, E.W.: Sporc: group collaboration using untrusted cloud resources. In: OSDI, pp. 337–350, Vancouver, BC, Canada (2010)
12. Yi, X., Sun, C.: Conditions and patterns for achieving convergence in OT based collaborative editing systems. IEEE Trans. Parallel Distrib. Syst. **27**(3), 695–709 (2015)
13. Subashini, S., Kavitha, V.: Review: a survey on security issues in service delivery models of cloud computing. J. Netw. Comput. Appl. **34**(1), 1–11 (2011)
14. Nichols, D.A., Curtis, P., Dixon, M., Lamping, J.: High-latency, low-bandwidth windowing in the jupiter collaboration system. In: Proceedings of the 8th Annual ACM Symposium on User Interface and Software Technology (UIST), pp. 111–120, New York, USA (1995)
15. Abdessamad, I.: Conception Formelle d'Algorithmes de Réplication Optimiste. Vers l'Edition Collaborative dans les Réseaux Pair-à-Pair. Phd thesis, University of Henri Poincaré, Nancy, France, December 2006
16. Paull, D.: Google wave: intention preservation, branching, merging and TP2 (2010). http://www.thinkbottomup.com.au/site/blog/Google_Wave_Intention_Preservation_Branching_Merging_and_TP2
17. Baig, M.B., Fitzsimons, C., Balasubramanian, S., Sion, R., Porter, D.E.: Cloudflow: cloud-wide policy enforcement using fast vm introspection. In: IC2E '14, pp. 159–164, Washington, DC, USA (2014)

---

[6]http://www.android-x86.org/.

18. Imine, A.: Coordination model for real-time collaborative editors. In: Coordination Models and Languages, 11th International Conference, COORDINATION 2009, Lisboa, Portugal, June 9–12, 2009. Proceedings, pp. 225–246 (2009)
19. Cherif, A., Imine, A., Rusinowitch, M.: Practical access control management for distributed collaborative editors. Pervasive Mob. Comput. **15**, 62–86 (2014)
20. Fangming, L., Peng, S., Hai, J., Linjie, D., Yu., J., Di, N., Bo, L.: Gearing resource-poor mobile devices with powerful clouds: architectures, challenges, and applications. IEEE Wirel. Commun. **20**(3), 1–10 (2013)
21. Cuervo, E., Balasubramanian, A., Cho, D.K., Wolman, A., Saroiu, S., Chandra, R., Bahl, P.: Maui: making smartphones last longer with code offload. In: MobiSys, pp. 49–62, San Francisco, CA, USA (2010)
22. Chun, B.G., Ihm, S., Maniatis, P., Naik, M., Patti, A.: Clonecloud: elastic execution between mobile device and cloud. In: EuroSys, pp. 301–314, Paris, France (2011)
23. Kosta, S., Perta, V.C., Stefa, J., Hui, P., Mei, A.: Clone2clone (c2c): peer-to-peer networking of smartphones on the cloud. In: HotCloud, San Jose, USA (2013)
24. Yeh, S.-C., Ming-Yang, S., Chen, H.-H., Lin, C.-Y.: An efficient and secure approach for a cloud collaborative editing. J. Netw. Comput. Appl. **36**(6), 1632–1641 (2013)
25. Bayer, R.: Symmetric binary b-trees: data structure and maintenance algorithms. Acta Inform. **1**(4), 290–306 (1972)
26. Xia, H., Lu, T., Shao, B., Ding, X., Gu, N.: Hermes: on collaboration across heterogeneous collaborative editing services in the cloud. In: IEEE CSCWD, pp. 655–660, Hsinchu, Taiwan (2014)
27. Mechaoui, M.D., Guetmi, N., Imine, A.: Mobile co-authoring of linked data in the cloud. In: New Trends in Databases and Information Systems—ADBIS, pp. 371–381, Poitiers, France (2015)

Resource Management Supporting Big Data for Real-Time Applications in the 5G Era

Konstantinos Katzis and Christodoulos Efstathiades

Abstract The main storage schemes used in mobile cloud computing require appropriate resources and infrastructure to operate. It is forecasted that with the deployment of IoT devices, large volumes of mobile data will be generated constantly at a very high rate. Current and future 5G wireless mobile networks are called to support the operation of such databases. This chapter presents the most widely used database models in the context of mobile cloud, followed by the state of the art technologies in streaming data access and processing methods. It then demonstrates the importance of Cognitive Radios forming the basis for the operation of 5G wireless networks. Finally, a number of IEEE 802 standards are presented as possible candidates for delivering 5G wireless services.

1 Introduction

With the proliferation of the Internet of Things (IoT) and the recent advances in computer systems and communication technologies, and particularly mobile network technologies such as 5G, network-enabled devices constitute equipment for data acquisition. Large volumes of mobile data are being generated constantly at a very high rate. The high volume of the data produced by mobile devices as well as their variety and heterogeneity [1] render their management challenging. The mobile devices are expected to be dispersed in a wide geographical area exchanging information anytime of the day. It is expected that by 2020, more than 40 trillion gigabytes will be generated, replicated and consumed [2]. The cloud provides storage and access to these data, but because of their high volume and real-time requirements, both current hardware and storage infrastructure are experiencing problems in coping up. An additional characteristic of such data is that they are usually accompanied with a set of coordinates in space generated by the mobile device. Location-based services use these data for numerous applications. This

K. Katzis (✉) · C. Efstathiades
European University Cyprus, Egkomi, Cyprus
e-mail: k.katzis@euc.ac.cy

© Springer International Publishing Switzerland 2017
C.X. Mavromoustakis et al. (eds.), *Advances in Mobile Cloud Computing and Big Data in the 5G Era*, Studies in Big Data 22,
DOI 10.1007/978-3-319-45145-9_12

enables cloud storage infrastructures to cope with such multi-dimensional data and provide efficient access methods and querying capabilities on such data. To support all these cloud services and associated applications, it is necessary to develop an unimpeded network infrastructure that will gather the data and move it to data centers for effective knowledge discovery.

Data aggregation and processing occurs mostly in data centers. This process heavily depends on inter-data-center networks, access networks, and the Internet backbone [3], as depicted in Fig. 1. The network architecture required here must be carefully planned and deployed in order to become the digital highway that bridges data sources, data processing platforms, and data destinations. There is a strong demand for a large part of this network infrastructure to perform the data collection and distribution wirelessly by providing connectivity to the devices listed in Fig. 1. Such devices can be geographically dispersed while generating large amounts of data on a daily basis [3]. In order to achieve this wirelessly, a wireless access network must be deployed that will be able to handle the high traffic, signaling and the large amount of data generated, before they are tunneled to the data center via the internet backbone.

The 5th generation of wireless systems, also known as 5G, is expected to fill a significant gap in wireless network technologies as it will be called to handle the overwhelming amount of data generated in our lives. The 5G wireless mobile network environment is expected to have polymorphic characteristics since it might not necessarily require to deliver 'gigabit experience' across its coverage area but users might operate at lower data-rates depending on the application/device in reference. In any case, future 5G wireless mobile networks as well as current wireless networks will have to provide connectivity to billions of devices by 2020 [4].

In order to support the polymorphic characteristics of future 5G wireless mobile networks, it is imperative that the spectrum requirements are first addressed. It is expected that for 5G wireless mobile networks, more than 1 GHz of new frequency bands should be identified by 2020 [5]. New authorized shared access policies should be explored for better spectrum utilization in order to enable 5G networks to address the challenging user requirements.

Nokia [6] and Ericsson [7] introduced a number of use-cases that will drive the technology such as mobile broadband, mobile media, connected and self-driving cars, heavy machinery controlled over distances, IoT and finally massive machine

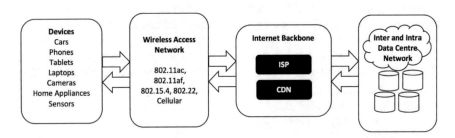

Fig. 1 Wireless mobile network architecture

type communications (very large number of meters/sensors embedded in the field). All these use-cases define the system requirements of future 5G wireless networks. As illustrated in Fig. 2, these parameters are: throughput, capacity, number of devices, cost, latency and reliability. Like any other wireless network, performance is subject to spatial and temporal variations [8].

Depending on the application, in our case supporting Cloud Storage Schemes, optimization is required focusing on multiple parameters or just a single parameter with one key performance indicator (KPI). 5G networks are requested to support such diversity in performance optimization in a flexible and reliable way. More specifically, 5G is expected to fulfill the following key performance indicators (KPI's) [9]:

- Provide 1000 × higher wireless area capacity
- Enhance service capabilities
- Save up to 90 % of energy per service provided
- Reduce the average service creation time cycle from 90 h to 90 min
- Create a secure, reliable and dependable Internet with a "zero perceived" downtime for services provision
- Facilitate highly dense deployments of wireless communication links to connect over 7 trillion wireless devices serving over 7 billion people.
- Enabling advanced user-controlled privacy.

To support the KPI listed above, a new, revolutionary communication architecture must be deployed that will address the following requirements [10]:

- 1–10 Gbps connections to end points in the field
- 1 ms end-to-end round trip delay (latency)

Fig. 2 5G main operation parameters

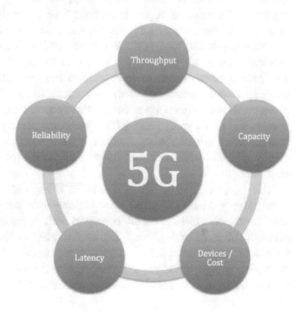

- $1000\times$ bandwidth per unit area
- $10–100\times$ number of connected devices
- (Perception of) 99.999 % availability
- (Perception of) 100 % coverage
- 90 % reduction in network energy usage
- Up to ten-year battery life for low power, machine-type devices

Technically it is difficult for a single platform to address all 8 requirements simultaneously. As discussed in [10], it is not necessary for 5G wireless networks to address all 8 requirements since no use-case, service or application has been identified that requires all eight performance attributes across an entire network. To support some of the most critical requirements above such as high data-rates, latency, bandwidth and number of devices, it is necessary to improve the radio spectrum utilization.

Radio spectrum is a natural resource that is currently being used and reused for delivering mobile wireless communications. The radio technology employed requires that the radio bands allocated (by regulators) are fixed and used only by authorized providers in order to maintain the desired level of quality of service. Across the countries, spectrum is strictly regulated so that different bands are allocated exclusively to particular services. It has been common practice to allocate a band to a single system in any given location and it is generally illegal to transmit without an explicit license within those bands/areas. The reason for this is to eliminate harmful interference. This is effectively the driving force behind the drafting of new policies for spectrum licensing. Current spectrum allocation map indicates that most of the usable radio spectrum has already being allocated leaving almost no space for introducing new mobile (radio spectrum demanding) technologies [11]. This licensing scheme, also known as the exclusive use model, has experienced various changes throughout time since the interest in spectrum has changed. At the early stages, the regulator decided on the radio services to be offered and selected who gets the licenses, and which technology is used on each spectrum band. This technique is known as the command and control model [12, 13]. Recent studies however, have showed that this is not the case and most of the radio spectrum allocated is somehow underutilized concluding that the current resource allocation techniques are limiting and not efficient when managing radio spectrum [11]. Cognitive radio (CR) networks are a promising radio technology that addresses in many ways the three basic system parameters which are spectral, energy and cost efficiency and makes them ideal for delivering the 5G network requirements.

This chapter aims to outline the main storage schemes used in mobile cloud computing and its basic requirements along with the current and future wireless mobile networks that are called to support their operation. More specifically, it first presents the most widely used database models in the context of mobile cloud, followed by the state of the art technologies in streaming data access and processing methods. It then demonstrates the importance of radio resource management techniques in mobile communications outlining the operation of future Cognitive

Radio (CR) systems as a possible enabler of 5G wireless mobile networks. Finally, a number of IEEE 802 standards are presented as candidates for delivering 5G wireless services in the context of mobile cloud computing.

2 Storage and Processing of Data on Mobile Clouds

As suggested in [14], a high level architecture for a big data ecosystem consists of three layers: (a) the data ingestion layer, where data is gathered from various sources and data centers, (b) the data analytics layer that consists of scalable systems for streaming or batch processing of big data and (c) the data storage layer that consists of scalable database systems with advanced indexing and querying capabilities for big data. Figure 3 shows this high level architecture.

Commercial mobile storage clouds exist that enable the users to store the data generated by their devices. Examples of such mobile storage clouds include Apple's iCloud, Microsoft's OneDrive, Google Drive and Dropbox. These services enable the users to synchronize the data they have stored on their mobile device

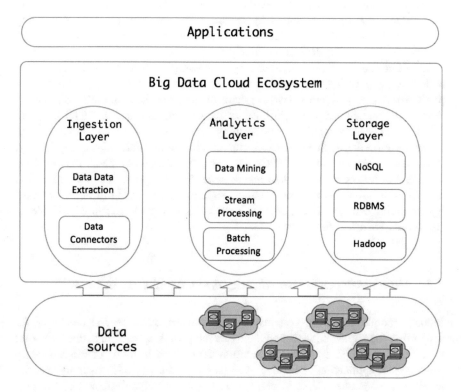

Fig. 3 The high level architecture for a big data cloud-based ecosystem

with the storage cloud, through their storage as a service (STaaS) infrastructure. This type of infrastructure enables the portability of the user's data across heterogeneous devices. In spite of their popularity, such services don't come without limitations. Considering the pay-as-you-go model of cloud computing and the huge amount of data produced by mobile devices, it becomes infeasible for users to store all their data using such services.

In addition, mobile cloud computing as an infrastructure for big data, with its distributed storage capabilities, enables the efficient storage and processing of such data. Disk resident approaches are being used to tame big data, such as the Hadoop distributed file system [15] that leverages the use of the MapReduce distributed data processing model, which operates on a large cluster of machines. Part of the Hadoop project are a set of subprojects that together form a stacked architecture constituting a big data infrastructure. Hadoop uses the Hadoop Distributed File System (HDFS), which is a highly fault-tolerant file system that operates on commodity hardware. It is designed mostly for batch processing of data with an emphasis on high throughput of data accessing. HDFS achieves data replication across machines in a large cluster of computers and it is therefore very robust in case of failures. Hive [16] is a data warehousing solution built on top of Hadoop and is an open-source project created by Facebook. It includes HiveQL, a SQL-like declarative language that supports high level queries which it translates into map-reduce jobs to be executed by Hadoop. The majority of cloud computing infrastructures as well as companies like Yahoo and Facebook use Hadoop for the storing and processing of their extremely large datasets.

In addition to Hadoop, main-memory processing approaches exist that support both batch processing and processing of real-time streaming data, such as data produced by mobile devices. These systems include Apache Spark [17, 18], Apache Storm [19] and Stratosphere [20]. Spark uses a data abstraction called Resilient Distributed Datasets (RDDs) for performing in-memory fast computations on large clusters, supporting fault tolerance. It supports interactive data mining applications that leverage large amounts of streaming data. Spark performs better than Hadoop even in batch processing of data and Apache Spark reports that more than 500 organizations are using it in production, rendering it one of the most widely used big streaming data processing engine. It can be used in combination with HDFS or any other networked file system.

2.1 Databases and Data Management in Mobile Clouds

Modern mobile operating systems support the use of embedded databases for supporting the applications that are being developed. In spite of this, there are a lot of limitations imposed by the operating systems, both in terms of the secondary storage each application can use as well as the amount of main memory that the application consumes. This, in combination with the limited amount of the processing power by mobile devices, necessitates the use of mobile cloud databases, where the storage as well as the computation needed for any applications is off-loaded to the mobile cloud. This way, complex computations on both historic and streaming data needed by the applications can be achieved without burdening the mobile device, either considering storage or computation power and battery consumption.

Since the scale of data that is being generated as well as consumed by mobile devices is tremendously increasing, scalable database management systems (DBMSs)—both for update intensive application workloads, as well as decision support systems—are a critical part of the cloud infrastructure [21]. Cloud platforms serve a diverse set of applications resulting therefore in a variety of databases (*tenants*) having very different schemas, workload types, data access patterns, and resource requirements [22]. The efficient integration of these tenants and the resource sharing between them in order to achieve an effective utilization of resources is called *multitenancy*. In such a model, a multitenant database management system ensures good performance for the tenants and at the same time tries to balance between performance and resource sharing. Further, in order to optimize the operating cost, another goal is added, the one of *elasticity*. Elasticity refers to one of cloud computing's main characteristics and it is about managing the limited resources offered by the cloud by scaling up and down as needed by the providers.

Database management systems have traditionally been used for serving enterprise applications for short, atomic and isolated transactions. They are aimed at low latency and on maximizing the transaction processing throughput. Traditional relational database management systems (RDBMSs) are designed to reflect the operational characteristics of applications and therefore do not take into consideration mobile cloud computing characteristics such as elasticity and scale-out. The design of RDBMSs lacks scalability and expandability, and therefore they cannot meet the performance requirement of such data. On the other hand, NoSQL systems (i.e. non-relational databases) have proven their advantages when it comes to the management, processing and analysis of large volumes of unstructured data. NoSQL databases, as suggested in [23], have the following features: (a) they are able to horizontally scale the throughput over many servers, (b) they can replicate and partition data over many servers, (c) they provide a simple interfaces, (d) they provide a weaker concurrency model than the ACID (Atomicity, Consistency, Isolation, Durability) transactions supported by RDBMSs, (e) they employ efficient distributed indexes for data storage, (f) they provide the ability to dynamically add new attributes to data records. It thus appears advisable that they should be used in a mobile cloud computing storage infrastructure. What follows is a description of the most widely used database management systems on the mobile cloud both in academia as well as in the industry.

RDBMSs for the mobile cloud
Since relational database management systems are not built for the characteristics of data generated by mobile devices, several architectures were proposed in order to enhance their functionality. In order to achieve scale-out, these systems (such as Cloud SQL Server proposed by [24]), use a partitioned database where transactions are constrained to execute on one partition, avoiding the use of locking protocols.

Additionally, they replicate the database in order to support fault tolerance. However, these systems do not consider elasticity in order to reduce the operation cost of the cloud database.

Key-value storage systems

Key-value stores are database management systems aiming at high scalability and high availability. In order to support these goals, they are confined in supporting simple, key-based access to data, and guarantees for consistency are supported on single rows. One of the first key-value storage system that was proposed was Dynamo [25] by Amazon. Dynamo only supports operations on single data items, does not provide any isolation guarantees and permits only single key updates. There are no security requirements such as authentication and authorization, since it was targeted and used only by Amazon's internal services. Subsequent research efforts have proposed designs that support transactions on multiple rows, such as the works of [26–29]. The ElasTraS system proposed by [22] uses the Hadoop distributed file system (HDFS) in its storage layer, and manages to effectively deal with large numbers of small tenants, while providing scalability for large tenants.

Document stores

Document stores are systems that support more complex data than key-value stores. They support secondary indexes and multiple types of documents per database [23], however they do not provide the ACID transactional properties. An example of such a database is Couchbase [30], which is used by a variety of research works such as in [31], where they maintain a big data repository on a mobile cloud for data that are being generated by mobile devices. Couchbase can store and serve the most frequent queries by leveraging its Memcache system that fetches the respective data immediately from the main memory.

Relational database management systems are being used by commercial systems, however when it comes to streaming data of high velocity such as mobile data, they are restricted by the relational model, therefore they cannot cope with data with very complex relationships. Both key-value stores and document stores are being used by cloud storage systems, depending on the type of data they support, but this is limited by the fact that in the future mobile cloud computing paradigm mobile data are generated by heterogeneous sources, and have different forms depending on the source. With the emergence of mobile cloud computing, commercial mobile cloud databases have appeared to incorporate high volumes of streaming mobile data. Reference [32] provides a comparison of four such databases, two databases that conform to the relational model (database.com, CloudAnt) and two document stores (nuoDB, Couchbase). In addition, they propose MCloudDB, a mobile database framework that is able to provide mobile data services by integrating mobile applications with mobile data clouds. Depending on the data that the cloud service is using, MCloudDB's database abstraction layer chooses the appropriate database and allocates resources for the data.

2.2 Spatial Data Access Methods in Mobile-Aware Cloud Databases

Mobile cloud computing poses challenges when compared to traditional cloud computing. These challenges stem from the inherent challenges of mobile computing such as low bandwidth, mobility and limited storage [33]. The number of cloud services as well as the demand for accessing content on the cloud is increasing. The challenges therefore for storing, managing and accessing data on the cloud increase proportionally. The limitations, both in the storage and main memory, for mobile devices are opposed to the fact that these devices generate large amounts of data that need to be processed and stored. In the majority of the cases, these data have spatial characteristics (i.e. locations), since mobile devices that people use, contain a spatial location generated by various sensors on the device (e.g. GPS, WiFi, 4G and 5G sensors). Additionally, indoor-based localization systems such as [34] are being used to find the location of the device and in consequence the position of the generated data. Therefore, one inherent challenge for a cloud database is the efficient storage and retrieval of streaming mobile data with spatial characteristics, where the location expires very frequently (i.e. devices change location). Current NoSQL database systems provide some geospatial data support, however their data partitioning techniques to the nodes of the cluster are very simple and do not cope with very frequent updates on the data.

In order for cloud databases to incorporate such data, appropriate access methods and indexing mechanisms should be used. Work in indexing spatial data has been extensively studied by the database community, and efficient access methods have been proposed. A spatial access method organizes spatial objects in a particular space in a way so that, given a query, only specific subsets of all of the objects stored in the spatial database will be considered as a possible answer [35]. Typical spatial index structures include the R-tree [36] and its variants (such as the R*-tree [37]) as well as the Quadtrees [38], which are also employed by modern DBMSs such as Oracle, MS SQL Server and PostgreSQL. The problem with these data structures is that they are designed assuming a centralized architecture and also they suffer from the search/update tradeoff [39]. On the one hand, the R-tree and its variants are data-oriented and don't handle updates efficiently and on the other hand Quadtrees are space-oriented. Assuming that in the majority of the cases the space distribution of real datasets is skewed, the index performs very poorly for searching. Therefore, datasets that change dynamically in time cannot be handled efficiently by such data access methods. There have been proposed various indexing structures to cope with dynamic updates, but they do not perform well in a mobile cloud-based context where the volume as well as the speed of the data stream is tremendous.

Recent research works propose techniques and access methods for the efficient indexing of multi-dimensional data in cloud systems. The work in [40] proposes RT-CAN, an indexing scheme that incorporates different types of indexes to serve various application scenarios. In this work, each node in the cloud uses an R-tree variant structure for the storing of the data in its local repository. Using Content Addressable Network (CAN) as an overlay peer-to-peer network, the data for each storage node are mapped to a point in the coordinate space. This global index supports different types of spatial queries such as point, range and nearest neighbor queries. The use of R-trees in RT-CAN renders the index not being able to cope with many frequent updates as this is the case with mobile data. A variety of other approaches have been proposed, such as QT-Chord [41] and SD-Rtree [42], which are also based on R-tree and quadtree variants, but they all suffer since they do not scale well in the cases of frequent updates.

Data structures for indexing methods in the mobile cloud should therefore be able to cope with updates on high velocity streaming data. Therefore, one dimensional structures such as space-filling curves and two-dimensional or multi-dimensional grids have been used widely in the literature as spatial indexes. However, these structures either provide approximate answers to the queries because of the dimensionality reduction (in the case of space filling curves), or inefficiency in query execution in the case of 2d-grids (due to the skewness of the data distributions in space). One work towards managing dynamic location data in cloud databases is ToSS-it [43], where a scalable distributed cloud-based index is being proposed. This index can handle frequent updates as well as efficient searching over this big mobile location data that change with massive speed in real time. The data structure is a Voronoi diagram, that decomposes space into cells and is able to cope with very fast updates. The framework distributes the objects in the cloud servers, and for each server it maintains a Voronoi index. Then a global Voronoi index is generated for all cloud servers. ToSS-it supports range and nearest neighbor queries and manages to scale linearly with the number of nodes in the cluster due to the exploitation of multicore CPUs in each node. Reference [44] points out that one disadvantage of RT-CAN is that the distribution of the data in the nodes by the CAN system leaves approximately half of the cluster idle if applications are used in different time zones. They propose the CEPS method for a cost-efficient partitioning of spatial data, aiming in a better utilization of the servers in order to save cost. Therefore, they consider not only the spatial proximity of data points but also access patterns that are associated with the objects, thus achieving to reduce the cost by up to 40 %.

SpatialHadoop [45] is an effort to integrate the support for spatial data with the Hadoop framework. It adopts several index structures such as grids, R-trees and its variants and supports several spatial operations such as range searching, nearest neighbor queries and spatial joins. Spatial awareness is injected in all of the Hadoop layers, and the support of several spatial partitioning techniques [46] and index structures in the storage layer, as well as various spatial queries render it a scalable and complete solution for mobile data in the cloud that have spatial characteristics.

3 Cognitive Radio Networks Supporting Mobile Cloud Computing in 5G

3.1 Mobile Cloud Computing QoS Requirements

In recent days there has been a dramatic increase in the number of mobile devices allowing users to run powerful applications making use of their built-in sensing and better data exchange. As a result, mobile applications are expected to seamlessly integrate with real-time data streams and Web 2.0 applications. Current applications are used for gaming, video capturing, editing and uploading, managing personal health, micro payments etc. and these applications are expected to grow in variations and resource requirements [47]. Although, mobile device hardware continues to evolve and to improve, it will always be behind on resources and security but most of all energy since the devices are powered by batteries. Resource poverty can be restrictive for the seamless and optimal operation of many applications [48]. In this context, 5G is expected to support Mobile Cloud Computing, in order to guarantee a minimum Quality of Service (QoS) requirements. According to NIST, cloud computing is expected to provide ubiquitous, convenient and on-demand network access to a shared pool of configurable computing resources. This must be achieved within a short time period and with minimal management effort or service provider interaction [49]. The five essential characteristics of NIST proposed cloud model are: on-demand self-service, broad network access, resource pooling, rapid elasticity and measured service. The first characteristic, is on-demand self-service where the user is expected to be able to unilaterally provision computing capabilities (server time and network storage), automatically and without interacting with the service provider. The second characteristic is broad network access allowing access by heterogeneous thin or thick platforms such as mobile phones, tablets, laptops etc. while network capabilities must support the platform requirements. The third characteristic is resource pooling, where the provider's computing resources are pooled. The main priority here is to serve multiple users employing different physical and virtual resources, which are dynamically assigned and reassigned according to user's demand. Such resources are storage, processing, memory and network bandwidth. The fourth characteristic is rapid elasticity, referring to the capabilities available to the consumer that can be elastically provisioned and released. Finally, the fifth characteristic is measured service, referring to the monitoring, controlling and reporting of resources in a transparent way for both the provider and the consumer of the utilized service. This allows resource re-allocation and optimization based on the consumer's type of required service (storage, processing, bandwidth, etc.) on a pay-per-use or charge-per-use basis.

3.2   Cognitive Radio Networks in 5G

There are many ways of allocating resources in a mobile communication system. Depending on its architecture and the application is designed for, this could vary from a loosely distributed controlled scheme to a highly centralized one. The channel allocation techniques matured through time in one shape or another but it was truly developed through the genesis of cellular mobile networks. Coordination of the resources was mandatory to maintain a healthy system that satisfied the minimum quality of service (QoS) requirements. From the first generation of cellular mobile networks, channel allocation techniques managed resources for new call arrivals or handoff attempts. The main purpose of implementing a channel allocation scheme is to achieve high spectral efficiency while satisfying the minimum grade of service. In some cases, it allows maintaining the planned cell boundaries intact, speeds up the allocation process and rearranges resources to highly populated areas to relief the system from network congestion. In mobile cellular systems, three basic types of channel allocation techniques can be identified. These are fixed, flexible and dynamic [50].

One of the main forces pushing towards the deployment of cognitive radios (CRs) is what appears to be a spectrum shortage. Current radio technologies employ portions of the radio spectrum through long-term licenses and it is impossible for someone else to use them. Although the radio spectrum seems to be highly occupied with hundreds of bands allocated to various companies, organisations etc., spectrum scarcity largely depends not on how many frequencies are available but on the technologies that can be deployed and how these frequencies will be used.

Currently there is a high demand for high-speed heterogeneous broadband technologies. These technologies require a substantial radio spectrum for their operation as well as they must be able to communicate using various currently available wireless standards. Furthermore, operating frequencies must be able to support a mobile, busy, urban propagation environment such as we find in 3G and LTE technologies. Alternatively, technologies such as IEEE802.11 (Wi-Fi) and IEEE802.16 (WiMax) are examples of modern broadband wireless networks. Both of them operate in the industrial, scientific and medical (ISM) bands. These bands are internationally reserved for purposes other than telecommunications which can in some occasions cause electromagnetic interference with communication systems that are using them. Since these frequencies have been free to use (unlicensed), Wi-Fi has been a cheap alternative to the wired network for allowing connectivity to the internet or even locally for devices such as laptops, smartphones, TVs, cameras etc. It is therefore clear that the technological market is currently driving the research and development of communications towards an evermore wireless, high speed, high capacity type of network. The bad news for the community is that the spectrum map is highly congested (in terms of allocation) and it is almost impossible to increase the bandwidth of the existing operating wireless networks without affecting other wireless systems operating in adjacent radio bands. The good news is that recent studies showed that the spectrum map is also underutilized

[11]. This has lead to the revolutionary idea of using a new type of radios that will be capable of using the underutilized spectrum in order to deliver the requirements of future wireless mobile communications.

CR has been defined in [51] as an intelligent wireless communication system that is aware of its surrounding environment and uses the methodology of understanding-by-building to learn from the environment and adapt its internal states based on new statistical variations. CR has become a reality thanks to the available reconfigurable platforms that are currently available in the market. The platforms have been based on digital radio and computer software. In fact, software CR platforms can be defined as the evolution of Software Defined Radios (SDRs). SDRs have been around for more than 20 years starting in the analogue modem industry where manufacturers implemented the modulating and de-modulating algorithm in software. This allowed them to upgrade/change the communication standards without having to change any of the hardware components. Based on this concept, SDRs have been further evolved to better utilize the radio spectrum by allowing real-time reconfigurability and improve compatibility and coexistence with different wireless standards. This has been achieved by employing software to perform the modulation and demodulation of the radio signals. Currently there is a considerable number of available software and hardware CR platforms mainly used for experimental purposes. Some of the main software platforms are: GNU Radio, IRIS, ASGARD. Combining these with the appropriate hardware RF front such as USRP, BEE2, VESNA, WARP it is possible to create what is known as a CR test-bed [52]. Using these test-beds it is possible to experimentally evaluate radio resource management algorithms for CR networks.

3.3 Resource Management Techniques for Cognitive Radio Enabled Networks

A CR type of wireless network is expected to be employing a loose type of channel assignment algorithm that enables the nodes to freely choose the best possible channel for their communication. After all, CR networks philosophy is to embrace the freedom of frequency allocation. Nevertheless, taking advantage of this kind of freedom does not suggest anarchy for the radio spectrum usage. Not at all; as each CR node, is expected to follow a list of rules. Avoiding any of these rules can end up in denying services to the node in reference or to causing severe interference to other nodes.

Simply put, the primary purpose of managing radio spectrum is to develop an adaptive strategy for the efficient and effective use and reuse of radio spectrum. In the case of CR, this will lead to highly reliable communications whenever and wherever needed. Inspired on existing wireless communication systems, whether these are cellular or not, the channel assignment algorithm for CR networks must be able of building on the spectrum holes detected by the radio-scene analyzer and the output of transmit-power controller [53]. Then, select the modulation strategy that adapts to the time-varying conditions of the radio environment. The radio scene analyzer proposed in [53] involves the estimation of interference temperature and detection of spectrum holes. Information gathered based on these two techniques are sent back to the transmitter through the feedback channel. It also involves the deployment of an adaptive beamforming mechanism that saves power by not radiating in all directions thus minimizing interference due to the action of other transmitter. What exactly are spectrum holes? Spectrum holes are presented in [53] as potential opportunities for non-interfering use of spectrum and can be considered as multidimensional regions within frequency, time, and space. This is provided that the CR systems, also known as the secondary system, is able to sense these holes within a given range of frequencies. Spectrum holes are classified into three categories [53]. The black spaces, the grey space and he white spaces. Black spaces represent the spectra that is occupied by high power local interferers some of the time. Furthermore, grey spaces refer to partially occupied spectra by low power interferers. Finally, white spaces are free of interferers except from any ambient noise in the area such as thermal noise, transient reflections, impulse noise and broadband thermal noise.

Detecting spectrum holes can be tricky and requires capable hardware and software to carry out this task. Some of the main issues regarding spectrum hole detection are listed in [53] as the environmental factors, exclusive zones and pre-diction algorithms. Environmental factors such as path-loss can reduce significantly the received signal power whereas shadowing can cause fluctuations about the path loss by a multiplication factor. In [54], authors propose quintile models for uncertain probability distributions (e.g. for shadowing) while secondary radio positions have been considered unconstrained. From the results, assuming multiuser settings, the degree of shadowing correlation has proven highly uncertain. Authors suggest that it might be easier to achieve a firm consensus regarding the correlation of shadowing across different frequencies for a single radio than it is to achieve a consensus regarding the shadowing correlation across users. "Weighted Probability of Area Recovered" (WPAR) is the proposed metric that employs a discounting-function to weigh the probability of recovering area at a given distance away from a single primary transmitter. While exponential discounting has been used by authors for convenience, it remains an interesting open question to deter-mine what the right discounting functions are for different application scenarios.

Some issues that are addressed in [55] disclose areas of spectrum hole detection that must be addressed in the future. These are the cooperative sensing strategies, the tradeoffs between the time-overheads and space-overheads. In addition, how the signal to noise ratio (SNR) walls must be understood in the context of the proposed WPAR algorithm. In [55] authors present issues regarding the critical design problem of CR systems having to process multi-gigahertz wide bandwidth and reliably detect spectrum occupancy. This places severe requirements on the future CR platforms that will have to feature high sensitivity, linearity, and dynamic range of the circuitry in the RF front-end.

3.4 IEEE Standards Supporting 5G Wireless Networks

Some of the recently developed/under development standards, feature architectural characteristics of CR engines. Such wireless standards could become integral part of 5G wireless networks that support cloud storage schemes. In this section, such standards are presented.

802.11ac

802.11ac evolved from 802.11n to become a much faster and more scalable wireless standard developed by IEEE [56]. Wireless Local Area Network (WLAN) sites are expected to experience significant improvements in the number of clients supported by an access point (AP). Individual users will also realize better experience since there is higher bandwidth availability for parallel video streams. Device battery life is extended, since the device's Wi-Fi can wake up and exchange data with its AP [56]. Compared to 802.11n, 802.11ac has a significantly improved performance thanks to the increased channel bonding offered which can reach up to 160 MHz. Furthermore, it employs 256 quadrature amplitude modulation (QAM) for a 33 percent speed burst at shorter ranges. Finally, a more complex multiple-input multiple-output (MIMO) architecture supports up to 8 spatial streams, twice as many than 802.11n. 802.11ac is a 5 GHz-only technology. It is therefore expected that APs will be dual-band supporting 802.11ac at 5 GHz and 802.11n at 2.4 GHz.

802.11af

IEEE 802.11af is the standard defined for spectrum sharing among unlicensed white space devices and licensed services in TV white space [57, 58]. This standard which is also known as Super Wi-Fi or WhiteFi protects the licensed users by applying a geolocation database mechanism. This is carried out using a white space map (WSM) which is defined in [57] as the information on identified available frequencies that is obtained from a geolocation database (GDB) and that is used by IEEE 802.11af stations (STAs). IEEE 802.11af envisions a geolocation database that stores the frequencies and operating parameters of white space devices by their geographic location to fulfill the regulatory requirements. IEEE 802.11af channels are 6, 7 and 8 MHz wide, thus allowing backward compatibility with existing international TV band allocations. Operation may vary from one to four channels, either contiguously or in two non-contiguous blocks. This way, devices are able to aggregate sufficient spectrum in a fragmented TV band spectrum to achieve high data rates. For Cloud Storage applications, although IEEE 802.11af has lower coverage range compared to cellular solutions, its lower costs due to the spectrum price are making it a promising candidate.

802.15.4

802.15.4 provides for ultra low complexity, ultra low cost, ultra low power consumption, and low data rate wireless connectivity among inexpensive devices. The raw data rate is high enough (250 kb/s) to satisfy a set of applications but is also scalable down to the needs of sensor and automation needs (20 kb/s or below) for wireless communications [59]. While most ongoing work in IEEE 802 wireless working groups is geared to increase data rates, throughput and QoS, the 802.15.4 LR-WPAN task group is aiming for other goals [60]. The advantages of this type of network is that it is easy to install, it has reliable data transfer, short-range operation, low cost and reasonable battery life [61]. Consequently, this standard is used to effectively deliver solutions for a variety of areas including consumer electronic device control, energy management and efficiency, home and commercial building automation as well as industrial plant management [62]. The application of 802.15.4 consists an example of a wireless network architecture that will be used in for supporting IoT type of applications.

802.22

IEEE 802.22, is a standard for wireless regional area networks (WRAN) that enables broadband wireless access using CR technology and spectrum sharing in white spaces [63]. It utilizes unused TV broadcast bands without interfering with other devices in frequency bands already licensed for specific uses. This is considered as a cost effective method of providing broadband access to sparsely populated areas where a wired connection would prove to be too costly. The 802.22 network is consisted of a number of base stations (BS) and client stations also called customer premises equipment (CPE) operating in a point to multi-point arrangement. This way, a BS manages its own cell and all associated CPE stations. Each BS in 802.22 network controls the medium access in its cell and maintains communication with its clients. Furthermore, it manages a unique feature of distributed sensing. This means that it is capable of observing the radio frequency spectrum and processing the observation to determine if a channel is occupied by a licensed transmission. The BS controls when sensing is performed by instructing various CPE to perform distributed measurement of different TV channels. Based on the reading collected, the BS decides the next action to be taken regarding the allocation of a channels.

4   Conclusions

Mobile cloud computing is an emerging paradigm in an era where communication technologies evolve to support connectivity between a vast numbers of heterogeneous devices. The exponentially increasing generation of data requires that the cloud should support scalable database management systems with efficient storage and processing capabilities. To deliver mobile clouds it is imperative that 5G wireless mobile networks are prepared to address the various types of wireless devices generating the data. Currently, there is a significant number of wireless standards that aim to fulfill the requirements set by the mobile cloud. These standards could be considered as part of the future 5G wireless mobile networks.

References

1. Chen, M., Mao, S., Liu, Y.: Big data: a survey. Mob. Netw. Appl. **19**(2), 171–209 (2014)
2. Gantz, J., Reinsel, D.: The Digital Universe In 2020: Big data, bigger digital shadows, and biggest growth in the far east, IDC analyze the future (2012). https://www.emc.com/collateral/analyst-reports/idc-the-digital-universe-in-2020.pdf. Accessed 24 Feb 2016
3. Yi, X., Liu, F., Liu, J., Jin, H.: Building a network highway for big data: architecture and challenges. IEEE Netw. **28**(4), 5–13. doi:10.1109/MNET.2014.6863125
4. Advanced 5G network infrastructure for the future internet public private partnership in horizon 2020, creating a smart ubiquitous network for the future internet. https://5g-ppp.eu/wp-content/uploads/2014/02/Advanced-5G-Network-Infrastructure-PPP-in-H2020_Final_November-2013.pdf. Accessed 13 March 2015
5. Hattachi, R., Erfanian, J.: NGMN 5G White Paper, NGMN 5G Initiative, 17 Feb 2015. https://www.ngmn.org/uploads/media/NGMN_5G_White_Paper_V1_0.pdf. Accessed 10 Feb 2016
6. Nokia Networks: Future works 5G use cases and requirements. White paper—5G Use Cases and Requirements. http://networks.nokia.com/sites/default/files/document/5g_requirements_white_paper.pdf. Accessed 13 Dec 2015
7. Ekudden, E.: Head of technology strategies, "Ericsson, talks about 5G use cases", 5G use cases. http://www.ericsson.com/news/150708-5g-user-cases_244069645_c?query=5g. Accessed 23 Feb 2016
8. Mavromoustakis, C.X., Mastorakis, G., Batalla, G.: Internet of Things (IoT) in 5G Mobile Technologies. Hardcover. Springer (2016). ISBN:978-3-319-30911-8
9. Advanced 5G network infrastructure for the future internet public private partnership in horizon 2020. Creating a smart ubiquitous network for the future internet. https://5g-ppp.eu/wp-content/uploads/2014/02/Advanced-5G-Network-Infrastructure-PPP-in-H2020_Final_November-2013.pdf. Accessed 13 Feb 2016
10. GSMA Intelligence, ANALYSIS Understanding 5G: Perspectives on future technological advancements in mobile. https://gsmaintelligence.com/research/?file=141208-5g.pdf&download. Accessed 13 Dec 2016
11. Shukla, A., Alptekin, A., Bradford, J., Burbidge, E., Chandler, D., Kennet, M., Levine, P., Weiss, S.: Cognitive Radio Technology: A Study for Ofcom—Summary Report. Ofcom, London (2007)
12. FCC: Connecting America: The National Broadband Plan. http://download.broadband.gov/plan/national-broadband-plan.pdf. Accessed 13 Dec 2016
13. Lehr, W., Crowcroft, J.: Managing shared access to a spectrum commons. In: IEEE DYSPAN '05, Nov 2005, pp. 420–44
14. Ranjan, R.: Streaming big data processing in datacenter clouds. IEEE Cloud Comput. **1**, 78–83 (2014)
15. Apache Hadoop website. http://hadoop.apache.org/. Accessed 23 Feb 2016
16. Thusoo, A., Sarma, J.S., Jain, N., Shao, Z., Chakka, P., Zhang, N., Antony, S., Liu, H., Murthy, R.: Hive-a petabyte scale data warehouse using hadoop. In: 2010 IEEE 26th International Conference on Data Engineering (ICDE), pp. 996–1005. IEEE (2010)
17. Zaharia, M., Chowdhury, M., Das, T., Dave, A., Ma, J., McCauley, M., Franklin, M.J., Shenker, S., Stoica, I.: Resilient distributed datasets: A fault-tolerant abstraction for in-memory cluster computing. In: Proceedings of the 9th USENIX Conference on Networked Systems Design and Implementation, pp. 2–2. USENIX Association (2012)
18. Apache Spark website http://spark.apache.org/. Accessed 23 Feb 2016
19. Apache Storm website. http://storm.apache.org/. Accessed 23 Feb 2016
20. Alexandrov, A., Bergmann, R., Ewen, S., Freytag, J.C., Hueske, F., Heise, A., Kao, O., Leich, M., Leser, U., Markl, V., Naumann, F.: The stratosphere platform for big data analytics. VLDB J. **23**(6), 939–964 (2014)

21. Agrawal, D., Das, S., El Abbadi, A.: Big data and cloud computing: current state and future opportunities. In: Proceedings of the 14th International Conference on Extending Database Technology. ACM (2011)
22. Das, S., Agrawal, D., El Abbadi, A.: ElasTraS: An elastic, scalable, and self-managing transactional database for the cloud. ACM Trans. Database Syst. (TODS) 38.1, 5 (2013)
23. Cattell, R.: Scalable SQL and NoSQL data stores. ACM SIGMOD Record **39**(4), 12–27 (2011)
24. Bernstein, P.A., Cseri, I., Dani, N., Ellis, N., Kalhan, A., Kakivaya, G., Lomet, D.B., Manne, R., Novik, L., Talius, T.: Adapting microsoft SQL server for cloud computing. In: 2011 IEEE 27th International Conference on Data Engineering (ICDE), pp. 1255–1263. IEEE (2011)
25. DeCandia, G., Hastorun, D., Jampani, M., Kakulapati, G., Lakshman, A., Pilchin, A., Sivasubramanian, S., Vosshall, P., Vogels, W.: Dynamo: amazon's highly available key-value store. ACM SIGOPS Oper. Syst. Rev. ACM **41**(6), 205–220 (2007)
26. Baker, J., Bond, C., et al.: Megastore: Providing scalable, highly available storage for interactive services. In: Proceedings of the 5th Biennial Conference on Innovative Data Systems Research, pp. 223–234 (2011)
27. Peng, D., Dabek, F.: Large-scale incremental processing using distributed transactions and notifications. In: Proceedings of the 9th USENIX Conference on Operating Systems Design and Implementation. USENIX Association, Berkeley, CA (2010)
28. Vo, H.T., Chen, C., Ooi, B.C.: Towards elastic transactional cloud storage with range query support. Proc. VLDB Endow. **3**(1), 506–514 (2010)
29. Das, S., Agrawal, D., El Abbadi, A.: G-Store: a scalable data store for transactional multi key access in the cloud. Proceedings of the 1st ACM Symposium on Cloud Computing, pp. 163–174. ACM, New York, NY (2010)
30. Couchbase: http://www.couchbase.com/. Accessed 23 Feb 2016
31. Larkou, G., Mintzis, M., Andreou, P.G., Konstantinidis, A., Zeinalipour-Yazti, D.: Managing big data experiments on smartphones. Distrib. Parallel Databases **34**(1), 33–64 (2016)
32. Lei, L., Sengupta, S., Pattanaik, T., Gao, J.: MCloudDB: A Mobile Cloud Database Service Framework. In: 2015 3rd IEEE International Conference on Mobile Cloud Computing, Services, and Engineering (MobileCloud), pp. 6–15. IEEE (2015)
33. Fernando, N., Loke, S.W., Rahayu, W.: Mobile cloud computing: a survey. Fut. Gen. Comput. Syst. **29**(1), 84–106 (2013)
34. Konstantinidis, A., Chatzimilioudis, G., Zeinalipour-Yazti, D., Mpeis, P., Pelekis, N., Theodoridis, Y.: Privacy-preserving indoor localization on smartphones. IEEE Trans. Knowl. Data Eng. **27**(11), 3042–3055 (2015)
35. Güting, R.H.: An introduction to spatial database systems. VLDB J: Int. J. Very Large Data Bases **3**(4), 357–399 (1994)
36. Guttman, A.: R-trees: a Dynamic Index Structure for Spatial Searching, vol. 14, no. 2, pp. 47–57. ACM (1984)
37. Beckmann, N., Kriegel, H.P., Schneider, R., Seeger, B.: The R*-tree: an Efficient and Robust Access Method for Points and Rectangles, vol. 19, no. 2, pp. 322–331. ACM (1990)
38. Samet, H.: The quadtree and related hierarchical data structures. ACM Comput. Surv. (CSUR) **16**(2), 187–260 (1984)
39. Akdogan, A., Shahabi, C., Demiryurek, U.: ToSS-it: a cloud-based throwaway spatial index structure for dynamic location data. In: 2014 IEEE 15th International Conference on Mobile Data Management (MDM), vol. 1, pp. 249–258. IEEE (2014)
40. Wang, J., Wu, S., Gao, H., Li, J., Ooi, B.C.: Indexing multi-dimensional data in a cloud system. In: Proceedings of the 2010 ACM SIGMOD International Conference on Management of Data, pp. 591–602. ACM (2010)
41. Ding, L., Qiao, B., Wang, G., Chen, C.: An efficient quad-tree based index structure for cloud data management. In: WAIM (2011)
42. Du Mouza, C., Litwin, W., Rigaux, P.: Sd-rtree: a scalable distributed rtree. In: IEEE 23rd International Conference on Data Engineering. ICDE 2007, pp. 296–305. IEEE 2007

43. Akdogan, A., Shahabi, C., Demiryurek, U.: ToSS-it: a cloud-based throwaway spatial index structure for dynamic location data. In: 2014 IEEE 15th International Conference on Mobile Data Management (MDM), vol. 1, pp. 249–258. IEEE (2014)
44. Akdogan, A., Indrakanti, S., Demiryurek, U., Shahabi, C.: Cost-efficient partitioning of spatial data on cloud. In: Big data (Big Data), pp. 501–506. IEEE (2015)
45. Eldawy, A., Mokbel, M.F.: SpatialHadoop: a MapReduce framework for spatial data. In: 2015 IEEE 31st International Conference on Data Engineering (ICDE), pp. 1352–1363. IEEE (2015)
46. Eldawy, A., Alarabi, L., Mokbel, M.F.: Spatial partitioning techniques in Spatial Hadoop. In: Proceedings of the VLDB Endowment, vol. 8, issue no. 12, pp. 1602–1605 (2015)
47. Kovachev, D., Cao, Y., Klamma, R.: Mobile Cloud Computing: A Comparison of Application Models (2011). arXiv:1107.4940
48. Satyanarayanan, M., Bahl, P., Cáceres, R., Davies, N.: The case for VM-based cloudlets in mobile computing. IEEE Pervasive Comput. **8**(4), 14–23 (2009)
49. Mell, P., Grance, T.: The NIST definition of cloud computing, pp. 800–145. http://nvlpubs. nist.gov/nistpubs/Legacy/SP/nistspecialpublication800-145.pdf. Accessed 23 Feb 2016
50. Stüber, G.L.: Principles of Mobile Communication. Kluwer Academic Publishers (2002). ISBN:978-0-7923-7998-0 (Print) 978-0-306-47315-9
51. Mitola, J., III; Maguire, G.Q. Jr: Cognitive radio: making software radios more personal. IEEE Personal Commun. **6**(4), 13–18 (1999)
52. Katzis, K., Perotti, A., De Nardis, L.: Testbeds and implementation issues. In: Di Benedetto, M.-G., Bader, F. (eds.) Cognitive Communication and Cooperative HetNet Coexistence. ISBN:978-3-319-01401-2 (Print) 978-3-319-01402-9. Accessed 17 Jan 2014
53. Haykin, S.: Cognitive radio: brain-empowered wireless communications. IEEE J. Sel. Areas Commun. **23**(2), 201–220 (2005). doi:10.1109/JSAC.2004.839380
54. Tandra, R., Mishra, M., Sahai, A.: What is a spectrum hole and what does it take to recognize one? In: Proceedings of the IEEE, pp. 824–848 (2009)
55. Cabric, D., Mishra, S.M., Brodersen, R.W.: Implementation issues in spectrum sensing for cognitive radios. In: Conference Record of the Thirty-Eighth Asilomar Conference on Signals, Systems and Computers, 2004, vol. 1, pp. 772, 776. Accessed 7–10 Nov 2004
56. 802.11ac: The Fifth Generation of Wi-Fi Technical White Paper, 2014 Cisco. http://www. cisco.com/c/en/us/products/collateral/wireless/aironet-3600-series/white_paper_c11-713103. html. Accessed 20 Feb 2016
57. IEEE Std 802.11afTM-2013. https://standards.ieee.org/getieee802/download/802.11af-2013. pdf. Accessed 23 Feb 2013
58. Flores, A.B., Guerra, R.E., Knightly, E.W., Ecclesine, P., Pandey, S.: IEEE 802.11 af: a standard for TV white space spectrum sharing. IEEE Commun. Mag. **51**(10), 92–100 (2013)
59. IEEE Std 802.15.4TM-2011 (Revision of IEEE Std 802.15.4-2006. https://standards.ieee.org/ getieee802/download/802.15.4-2011.pdf. Accessed 23 Feb 2016
60. Gutierrez, J., Naeve, M., Callaway, E., Bourgeois, M., Mitter, V., Heile, B.: Ieee 802.15.4: a developing standard for low-power low-cost wireless personal area networks. IEEE Netw. **15**(5), 12–19 (2001)
61. Baronti, P., Pillai, P., Chook, V.W., Chessa, S., Gotta, A., Hu, Y.F.: Wireless sensor networks: A survey on the state of the art and the 802.15.4 and zigbee standards. Comput. Commun. **30**(7), 1655–1695 (2007)
62. Han, D.-M., Lim, J.-H.: Smart home energy management system using ieee 802.15.4 and zigbee. IEEE Trans. Consum. Electron. **56**(3), 1403–1410 (2010)
63. IEEE802.org, IEEE 802.15.4, 2016. http://www.ieee802.org/15/pub/TG4.html. Accessed 23 Feb 2016

Smartphone-Based Telematics for Usage Based Insurance

Prokopis Vavouranakis, Spyros Panagiotakis, George Mastorakis
and Constandinos X. Mavromoustakis

Abstract In this chapter we study and introduce a smartphone-based telematics system for usage-based insurance (UBI). The smartphone has been identified as an enabler for future UBI, replacing the in-vehicle telecommunication hardware devices with a ubiquitous device with a plurality of sensors, means for data processing and wireless communication. We implemented and developed an end-to-end system including a telematics android-based application for client's smartphones and a portal to collect, analyze and record driving patterns and score drivers. Also monitoring driver behavior, recording their driving events (safe and aggressive) and giving feedback of recorded events can enhance driver safety. So we developed an android-based application, which can estimate driving behavior, using data only from the accelerometer sensor or using orientation data of a sensor fusion method. Complementary, we developed a portal, where we can have access to an overall dashboard of all registered drivers. In the portal are presented scores, behaviors, trips reports and routes in maps of all recorded trips. With this way, we help the insurance carriers to assess better the risk of the drivers.

Keywords UBI · Usage based insurance · Driving behavior · Smartphone · Android · Pervasive computing · IoT · Big data

P. Vavouranakis · S. Panagiotakis (✉) · G. Mastorakis
Department of Informatics Engineering, Technological Educational
Institute of Crete, 71004 Heraklion, Crete, Greece
e-mail: spanag@teicrete.gr

P. Vavouranakis
e-mail: akisvavour@gmail.com

G. Mastorakis
e-mail: gmastorakis@staff.teicrete.gr

C.X. Mavromoustakis
Department of Computer Science, University of Nicosia, Nicosia, Cyprus
e-mail: mavromoustakis.c@unic.ac.cy

© Springer International Publishing Switzerland 2017
C.X. Mavromoustakis et al. (eds.), *Advances in Mobile Cloud Computing and Big Data in the 5G Era*, Studies in Big Data 22,
DOI 10.1007/978-3-319-45145-9_13

1 Introduction

Usage-based insurance (UBI) [1] is a type of vehicle insurance whereby the costs are dependent upon type of vehicle used, measured against time, distance, behavior and place.

This differs from traditional insurance, which attempts to differentiate and reward "safe" drivers, giving them lower premiums and/or a no-claims bonus. However, this rewarding is a reflection of history that may do not correspond to the present pattern of driving behavior. This means that it may take a long time before safer (or more reckless) patterns of driving and changes in lifestyle feed the insurance premiums.

The simplest form of usage-based insurance bases the insurance costs simply on the number of miles driven. However, the general concept of pay as you drive includes any scheme where the insurance costs may depend not just on how much you drive but how, where, and when one drives [2].

Pay as you drive (PAYD) means that the insurance premium is calculated dynamically, typically according to the amount driven. There are three types of usage-based insurance:

- Coverage is based on the odometer reading of the vehicle.
- Coverage is based on mileage aggregated from GPS data, or the number of minutes the vehicle is being used as recorded by a vehicle-independent module transmitting data via cellphone or RF technology [3].
- Coverage is based on other data collected from the vehicle, including speed and time-of-day information, historic riskiness of the road, driving actions in addition to distance or time travelled.

The formula can be a simple function of the number of miles driven, or can vary according to the type of driving or the identity of the driver. Once the basic scheme is in place, it is possible to add further details, such as an extra risk premium if someone drives too long without a break, uses their mobile phone while driving, or travels at an excessive speed. Telematics usage-based insurance provides a much more immediate feedback loop to the driver [1] by changing the cost of insurance dynamically with a change of risk. This means drivers have a stronger incentive to adopt safer practices. For example, if a commuter switches to public transport or to working at home, this immediately reduces the risk of rush hour accidents. With usage-based insurance, this reduction would be immediately reflected in the cost of car insurance for that month.

So the proposal for the user is simple. The user has to download an application in his smartphone, open it when he is driving and let his insurance company to monitor his driving behavior: where his vehicle is driven, how fast he drives, how hard he breaks, how hard he corners and so on. In return the insurance will give him a discount on insurance premiums.

The Progressive Insurance [2], the largest insurance company using such methods in the USA, found after analysis of billion of miles and relevant data that key points in driving behavior such as actual miles traveled, braking and time of the day give more than the twice predictabilities compared to tradition variables such as age of the driver, gender, the manufacturer and the model of the insured vehicle. The average discount on insurance premiums for a driver who agrees to record his driving behavior amounts to 10–15 %.

In the future, anyone who does not agree to record his driving behavior may not be required to pay higher insurance premiums, but most companies will not even accept to insure his car. Insurance based on telematics can find great appeal, as no longer requires special devices installed in the car and a simple download of an application in the driver's smartphone is enough. In addition, smartphones are made especially for communication. On the other hand special devices need some kind of transmitter.

Insurance programs based on telematics [2] would mean big changes for road safety. The insurance application on the mobile phone will notify you when you brake too abruptly or run too much, and tame the way you drive. The fact is that people drive more carefully, simply because they know that their driving behavior is recorded. The more expensive insurance premiums act as penalty for recklessness driving behavior.

In this chapter we study and introduce a smartphone-based telematics system for usage-based insurance. The smartphone has been identified as an enabler for future UBI, replacing the vehicle mounted dedicated hardware with a ubiquitous device with a plurality of sensors, means for data processing and wireless communication. In this context we have implemented and developed an end-to-end system including a telematics android-based application for client's smartphones and driving behavior recognition, as well as a portal to collect, analyze and record driving patterns and score the drivers. The rest of the paper is organized as follows: Sect. 2 details in UBI concepts, Sect. 3 presents the methodology for detecting drivers' behavior using smartphone sensors, Sects. 4 and 5 introduce the Native Android Application we have developed for Usage Based Auto Insurance and the e-Platform of a potential Auto UBI System for drivers management, respectively, and, finally, Sect. 6 concludes the chapter.

2 Usage Based Insurance (UBI)

2.1 Definition of UBI

Usage-based insurance (UBI) [2] also known as pay as you drive (PAYD) and pay how you drive (PHYD) and mile-based auto insurance is a type of vehicle insurance whereby the costs are dependent upon type of vehicle used, measured against time, distance, behavior and place.

Usage-Based Insurance is a recent innovation by auto insurers that more closely aligns driving behaviors with premium rates for auto insurance. Mileage and driving behaviors are tracked using odometer readings or in-vehicle telecommunication devices (telematics) that are usually self-installed into a special vehicle port or already integrated in original equipment installed by car manufactures. The basic idea of telematics auto insurance is that a driver's behavior is monitored directly while the person drives. These telematics devices measure a number of elements of interest to underwriters: miles driven; time of day; where the vehicle is driven (GPS); rapid acceleration; hard breaking; hard cornering; and air bag deployment. The level of data collected generally reflects the telematics technology employed and the policyholders' willingness to share personal data.

The insurance company then assesses the data and charges insurance premiums accordingly. For example, a driver who drives long distance at high speed will be charged a higher rate than a driver who drives short distances at slower speeds. With UBI, premiums are collected using a variety of methods, including utilizing the gas pump, debit accounts, direct billing and smart card systems.

The first UBI programs began to surface in the U.S. about a decade ago, when Progressive Insurance Company and General Motors Assurance Company (GMAC) began to offer mileage-linked discounts through combined GPS technology and cellular systems that tracked miles driven. These discounts were (and still are) often combined with ancillary benefits like roadside assistance and vehicle theft recovery. Recent accelerations in technology have increased the effectiveness and cost of using telematics, enabling insurers to capture not just how many miles people drive, but how and when they drive too. The result has been the growth of several UBI variations, including Pay-As-You-Drive (PAYD), Pay-How-You-Drive (PHYD), Pay-As-You-Go, and Distance-Based Insurance.

2.2 Pricing of UBI

The pricing scheme for UBI deviates [2] greatly from that of traditional auto insurance. Traditional auto insurance relies on actuarial studies of aggregated historical data to produce rating factors that include driving record, credit-based insurance score, personal characteristics (age, gender, and marital status), vehicle type, living location, vehicle use, previous claims, liability limits, and deductibles. Premium discounts on traditional auto insurance is usually limited to the bundling of insurance on multiple vehicles or types of insurance, insurance with the same carrier, protection devices (like airbags), driving courses and home to work mileage.

Policyholders tend to think of traditional auto insurance as a fixed cost, assessed annually and usually paid for in lump sums on an annual, semi-annual, or quarterly basis. However, studies show that there is a strong correlation between claim and loss costs and mileage driven, particularly within existing price rating factors (such as class and territory). For this reason, many UBI programs seek to convert the fixed costs associated with mileage driven into variable costs that can be used in conjunction with other rating factors in the premium calculation. UBI has the advantage of utilizing individual and current driving behaviors, rather than relying on aggregated statistics and driving records that are based on past trends and events, making premium pricing more individualized and precise.

2.3 Advantages of UBI

UBI programs offer many advantages [2] to insurers, consumers and society. Linking insurance premiums more closely to actual individual vehicle or fleet performance allows insurers to more accurately price premiums. This increases affordability for lower-risk drivers, many of whom are also lower-income drivers. It also gives consumers the ability to control their premium costs by incenting them to reduce miles driven and adopt safer driving habits. Fewer miles and safer driving also aid in reducing accidents, congestion and vehicle emissions, which benefits society.

The use of telematics helps insurers more accurately estimate accident damages and reduce fraud by enabling them to analyze the driving data (such as hard breaking, speed, and time) during an accident. This additional data can also be used by insurers to refine or differentiate UBI products. Additionally, the ancillary safety benefits offered in conjunction with many telematics-based UBI programs also help to lower accident and vehicle theft related costs by improving accident response time, allowing for stolen vehicles to be tracked and recovered, and monitoring driver safety. Telematics also allow fleets to determine the most efficient routes, saving them costs related to personnel, gas and maintenance.

2.4 Challenges

The practice of tracking mileage and behavior information in UBI programs has raised privacy concerns. As a result, some states have enacted legislation requiring disclosure of tracking practices and devices. Additionally, some insurers limit the data they collect. Although not for everyone, acceptance of information sharing is growing as more mainstream technology devices (such as smartphones, tablets, and GPS devices) and social media networks (such as Facebook and MySpace) enter the market.

Implementing a UBI program, particularly one that utilizes telematics, can be costly and resource intensive to the insurer. UBI programs rely heavily on costly technology to capture and sensitize driving data. Additionally, UBI is an emerging area and thus there is still much uncertainty surrounding the selection and interpretation of driving data and how that data should be integrated into existing or new price structures to maintain profitability. This is particularly important, as the transitioning of lower-risk drivers into UBI programs that offer lower premium could put pressure on overall insurer profitability.

Insurers must also manage regulatory requirements within the states that they do business. Many states require insures to obtain approval for the use of new rating plans. Rate filings usually must include statistical data that supports the proposed new rating structure. Although there are general studies demonstrating the link between mileage and risk, individual driving data and UBI plan specifics are considered proprietary information of the insurer. This can make it difficult for an insurer who does not have past UBI experience. Other requirements that could prevent certain UBI programs include the need for continuous insurance coverage, upfront statement of premium charge, set expiration date, and guaranteed renewability. However, it should be noted that a Georgia Institute of Technology survey of state insurance regulations (2002) found that the majority of states had no regulatory restrictions that would prevent PAYD programs from being implemented.

2.5   Implementations in USA

Metromile. Metromile is a California-based insurance startup funded by New Enterprise Associates, Index Ventures, National General Insurance/Amtrust Financial, and other investors. It offers a driving app and a pay-per-mile insurance product using a device that connects to the OBD-II port of all automobiles built after 1996. Metromile does not use behavioral statistics like type of driving or time of day to price their insurance. They offer consumers a fixed base rate per month plus a per-mile-rate ranging from 2 to 11 cents per mile, taking into account all traditional insurance risk factors. Drivers who drive less than the average (10,000 miles a year) will tend to save. Metromile allows users to opt out of GPS tracking, never sells consumer data to 3rd parties, and does not penalize consumers for behavioral driving habits. Metromile is currently licensed to sell auto insurance in California, Oregon, Washington, Virginia and Illinois (as of July 2015) [4]. More states are expected to roll out shortly.

Progressive. Snapshot is a car insurance program developed by Progressive Insurance in the United States [5]. It is a voluntary, behavior-based insurance program that gives drivers a customized insurance rate based on how, how much, and when their car is driven. Snapshot is currently available in 46 states plus the District of Columbia. Because insurance is regulated at the state level, Snapshot is currently not available in Alaska, California, Hawaii, and North Carolina.

Driving data is transmitted to the company using an on-board telematic device. The device connects to a car's Onboard Diagnostic (OBD-II) port (all automobiles built after 1996 have an OBD-II) and transmits speed, time of day and number of miles the car is driven. Cars that are driven less often, in less risky ways and at less risky times of day can receive large discounts. Progressive has received patents on its methods and systems of implementing usage-based insurance and has licensed these methods and systems to other companies. Progressive has service marks pending on the terms Pay As You Drive and Pay How You Drive.

Allstate. Allstate announced on October 8, 2012 that it has expanded its usage-based auto insurance product, Drive Wise, to four additional states including New York and New Jersey [5]. As of October 2012 Drive Wise is currently available in: Colorado, Michigan, New Jersey, New York, Arizona, Illinois, and Ohio. Allstate's usage-based insurance product, Drive Wise, gets installed into a car's onboard diagnostic port, near the steering column in most cars. Allstate said its usage-based insurance measures things such as mileage, braking, speed, and time of day when a customer is driving. Using that data, Allstate calculates a driving discount for each customer using its telematics technology.

One of the big advantages with Drive Wise is that it can constantly provide feedback to the consumers for as long as they keep the device in the car. Allstate's Drive Wise utilizes data from a monitoring device plugged into a car's onboard diagnostic port. Of the drivers earning a discount, the average savings is nearly 14 % per vehicle. More than 10 % of all new Allstate customers are opting to participate in this coverage.

Liberty Mutual Insurance. Onboard Advisor is a commercial lines pay-how-you-drive, PHYD, or "safety-driven" insurance product by Liberty Mutual Agency Corporation. It offers up to 40 % discount to commercial and private fleets based on how safely they actually drive [1].

National General Insurance. National General Insurance is one of the first and largest auto insurance companies to institute a Pay-As-You-Drive (PAYD) program in the United States back in 2004 [6]. The National General Insurance Low-Mileage Discount is an innovative program offered to OnStar subscribers in 34 states, where those who drive less pay less on their auto insurance. This opt-in program is the first of its kind [7] leveraging state-of-the-art technology using OnStar to allow customers who drive fewer miles to benefit from substantial savings. Eligible active OnStar subscribers sign up to save on their premiums if they drive less than 15,000 miles annually. Subscribers who drive even less than that can save even more (up to 54 %).

Under the program, new National General Insurance customers receive an automatic insurance discount of approximately 26 % upon enrollment [8] (existing OnStar customers receive a discount based on historical mileage). With the subscriber's permission, the odometer reading from his or her monthly OnStar Vehicle Diagnostics report is forwarded to National General Insurance. Based on those readings, the company will decrease the premium using discount tiers corresponding to miles driven.

Information sent from OnStar to National General Insurance pertains solely to mileage, and no additional data is gathered or used for any purpose other than to help manage transportation costs. Customers who drive more than 15,000 miles per year are not penalized and all OnStar customers receive an insurance discount simply for having an active OnStar subscription.

2.6 Future of UBI

UBI is poised for rapid growth in the U.S. According to SMA Research [2], approximately 36 % of all auto insurance carriers are expected to use telematics UBI by 2020. Based on a May 2014 CIPR survey of 47 U.S. state and territory insurance departments, in all but five jurisdictions—California, New Mexico, Puerto Rico, Virgin Islands, and Guam—insurers currently offer telematics UBI policies. In twenty-three states, there are more than five insurance companies active in the telematics UBI market. The CIPR survey is part of a recently released *CIPR study: "Usage-Based Insurance and Vehicle Telematics: Insurance Market and Regulatory Implications"* [2], on how technological advances in telematics are driving changes in the insurance market and its impact on insurers.

Telematics-based UBI growth is being propelled by technology advances, which continue to substantially improve the cost, convenience, and effectiveness of using telematics devices. It is through the use of telematics that insurers are able to collect driving data that better enable them to more closely link a driver's individual risk with premium. Through UBI programs, insurers are able to differentiate products, gain competitive advantage, and attract low-risk policyholders. Recognition of the societal benefits and growing consumer acceptance of personal data collection will only serve to further increase demand for telematics-based UBI products in the future.

3 Detecting Driver Behavior Using Smartphone Sensors

There are a lot of applications, which try to detect and analyze the driving behavior of the driver. Many of them have target to stimulate the drive in order to improve his driving style and by this way to achieve lower fuel consumption and to decrease the risk of road accidents [9]. These systems or applications use expensive car-dependent information such as engine power, pedal pressure, wheel position etc. Nowadays the smartphones are already well integrated in our life. Because of that and the various embedded sensors (accelerometer, orientation sensor etc.) in them, the smartphones represent a suitable platform to compute the driving behavior of the driver [10].

We implement a usage based auto-insurance information system, which consists of two elements. The first element is an android-based application for smartphones and tablets, which detects the driver's behavior by analyzing the collected data from device's sensors. For this reason we study various driving detection methods using smartphone's sensor in order to find which method or methods are the best for our implementation. The second element is an e-platform, where someone can have access to all data (trip's information, routes of trips, graphs of sensor's data) of all drivers, of the insurance company.

3.1 Calibration of Device

In order someone to get accurate measurements from a smartphone to evaluate the driving quality, it is required at first to calibrate the position of the device relatively to that of the vehicle [11]. This means virtually rotating the three-axis of the accelerometer sensor of the smartphone to meet the vehicle's orientation (Fig. 1). To do this we need to know all the angles of rotation. To calculate the angles of rotation, we use the atan2 function [12] to measure the angle between two given coordinates.

The calibration is always initiated before monitoring, and is carried out in three steps. The first step of the calibration is to calculate the rotation angles according to the vehicle's level. The second step is to calculate the angle of the driving direction of the vehicle. When these two steps are completed, the final step of the calibration is executed, updating the rotation angles. Figure 1 illustrates the calibration angles.

The first step, explained above, is done while the vehicle is standing still. The angles calculated in the first step are both the roll- and pitch angles. The XY magnitude offset is calculated next. Calculating the average magnitude between the

Fig. 1 Illustration of the calibration angles

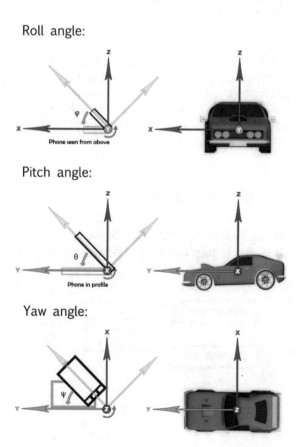

X and Y-axis does this. This offset is used to check if the vehicle is in motion. If the offset is varying widely, this means that the vehicle is in motion. If the vehicle is in motion at this stage, the first step is restarted and the offset is reset.

When the XY magnitude offset is calculated, we can proceed to the second step; calculating the yaw angle. First we check if the vehicle is in motion by checking if the magnitude of the XY axis (subtracted by the XY offset) is above a certain value. If this is true, we calculate the yaw angle based on this magnitude, and add the angle in an average buffer. This process is repeated until we have enough angles to calculate the average yaw angle. We use the average yaw angle to be certain that it is accurate and valid. If we just set the first calculated angle as the yaw angle, this has got the potential of being invalid.

3.2 Sensor Fusion Orientation Data

Sensor Fusion [13, 14] is the more appropriate method that can be used to determine the orientation of a device. With this method, we combine data from 3 embedded smartphone sensors: the accelerometer, the magnetometer and the gyroscope, in order to get the three orientation angles. The low noise gyroscope data is used only for orientation changes in short time intervals. The accelerometer and magnetometer data is used as support information over long time intervals. With this way we filter out the gyro drift (a slow rotation of the calculated orientation), with the accelerometer/magnetometer data, which do not drift over long time intervals. This process is similar to high pass filtering of the gyroscope data and to low pass filtering of the accelerometer/magnetometer data. Figure 2 illustrates this process.

3.3 Detection Methods

Our algorithm characterizes the behavior of the driver as *Excellent, Very Good, Good, Bad* or *Very Bad* and computes the average speed of the vehicle at the end of every trip.

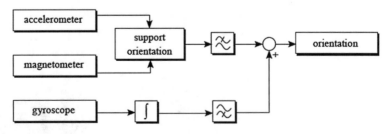

Fig. 2 Flow of the sensor fusion with complementary filter

Our algorithm for the detection of driver behavior uses two types of data. The acceleration output data of the accelerometer sensor and the orientation output data of the sensor fusion method. In the first method we use the three-axis accelerometer sensor in order to record and analyze driver's behavior. In particular, we utilize the x-axis accelerometer data to detect the left/right direction of the vehicle and therefore driving events like safe or sharp turns. For the detection of front/rear direction of vehicle and therefore to measure how the driver accelerates and applies the brakes we utilize the y-axis data. In the second method we use the three-axis orientation output data of sensor fusion. In particular, we use roll data in order to detect the left/right direction of the vehicle. For the detection of front/rear direction of vehicle and therefore to measure how the driver accelerates/decelerates, we utilize pitch data.

Our experience with these two methods shows that the best practice for recognizing driving patterns is the use of accelerometer data. The reason is not that sensor fusion methodology is not accurate or reliable but rather that sensor fusion is not the best option for the detection of sharp turns or sharp lane changes. When we use only the accelerometer, the algorithm detects sharp turns or sharp lane changes faster than when we use the orientation data of sensor fusion method. On the other hand, with sensor fusion the system detects faster the safe/hard accelerations or decelerations of the vehicle than when we use only the acceleration data.

After having chosen the preferred detection method, the algorithm analyzes the behavior of the driver applying specific thresholds over the data collected as he drives [10]. These thresholds have been acquired by testing the data of the detection methods under various driving events and maneuvers. This is the methodology followed in most similar works [10, 15, 16]. With these thresholds (Table 1) we can distinguish and detect 12 driving events: Safe Acceleration, Safe Deceleration, Safe Left Turn, Safe Right Turn, Safe Left Lane Change, Safe Right Lane Change, Hard Acceleration, Hard Deceleration, Sharp Left Turn, Sharp Right Turn, Sharp Left Lane Change and Sharp Right Lane Change.

Every driving event or maneuver, that our system can detect, has a counter. When the algorithm detects that the driver makes one of the already talked maneuvers or driving events, the counter for this type of event is incremented. If the algorithm detects a safe driving event, a counter for this driving event is incremented. When detects a dangerous driving event, then a counter for this event is incremented as well.

When the driver has finished his trip, our algorithm computes the percentage of the penalty for every dangerous driving event. The computed penalties are: hard acceleration penalty, hard deceleration penalty, sharp left turn penalty, sharp right turn penalty, sharp left lane change penalty and sharp right lane change penalty.

The equations for the Hard Acceleration and Hard Deceleration penalties are:

$$HardAccPenalty = \frac{HardAccCounter}{SafeAccCounter + HardAccCounter}$$

Table 1 Thresholds and data Used for the detection of various driving events using accelerometer's data or sensor fusion orientation data

Driving event	Data used (Accelerometer)	Threshold (m/s^2)	Data used (Sensor fusion)	Threshold (rad/s)
Safe Acceleration	Y-axis data	1.3–2.5	Pitch angle	−0.08–0.12
Safe Deceleration	Y-axis data	−1.3–2.5	Pitch angle	0.08–0.12
Safe Left Turn	X-axis data	−1.8–3.0	Roll angle	0.10–0.30
Safe Right Turn	X-axis data	1.8–3.0	Roll angle	−0.10–0.30
Hard Acceleration	Y-axis data	>2.5	Pitch angle	<−0.12
Hard Deceleration	Y-axis data	<−2.5	Pitch angle	>0.12
Sharp Left Turn	X-axis data	<−3.0	Roll angle	>0.30
Sharp Right Turn	X-axis data	>3.0	Roll angle	<−0.30

$$HardDecPenalty = \frac{HardDecCounter}{SafeDecCounter + HardDecCounter}$$

The equations for the Sharp Left Turn and Sharp Right Turn are:

$$SharpLeftTurnPenalty = \frac{SharpLeftTurnCounter}{SafeLeftTurnCounter + SharpLeftTurnCounter}$$

$$SharpRightTurnPenalty = \frac{SharpRightTurnCounter}{SafeRightTurnCounter + SharpRightTurnCounter}$$

The equations for the Sharp Left Lane Change and Sharp Right Lane Change are:

$$SharpLeftLCPenalty = \frac{SharpLeftLCCounter}{SafeLeftLCCounter + SharpLeftLCCounter}$$

$$SharpRightLCPenalty = \frac{SharpRightLCCounter}{SafeRightLCCounter + SharpRightLCCounter}$$

Using the above penalties we can compute the Total Sharp Turn Penalty and the Total Sharp Lane Change Penalty with the following equations:

$$SharpTurnPenalty = SharpLeftTurnPenalty + SharpRightTurnPenalty$$
$$SharpLCPenalty = SharpLeftLCPenalty + SharpRightLCPenalty$$

The equation for the total penalty is:

$$TotalPenalty = HardAccPenalty + HardDecPenalty + HardTurnPenalty + HardLCPenalty$$

Except the penalties at the end of the trip, the algorithm computes the driving score of the driver. The total score depends of the penalties of the dangerous driving events (Total Penalty). The maximum score a user can achieve is ten (10) and the minimum is zero (0).

The equation for the computation of the total score, achieved from the user at the end of the trip is:

$$TotalScore = 10 - TotalPenalty$$

According to the total score (at the end of the trip) of the user, the algorithm characterizes the user for his driving behavior for the current trip. There are various behaviors depending the total score of the user (Table 2). The behavior of the driver at the end of the trip can be *Excellent, Very Good, Good, Bad* or *Very Bad*.

4 Native Android Application for Usage Based Auto Insurance

Putting the above steps in practice, we have developed a native android-based application that can be used by a usage-based insurance company. The application via smartphone's sensors can detect and evaluate the driving behavior of the user for all his trips. All trip data that contains statistics, routes and graphs is saved in smartphone's local memory. Also these data is sent to a company's server, where can be accessed by the employees of the company. The company evaluates the data of all the trips of the user, whose payment to the company is depending to his

Table 2 Driver behavior categories based on the total score of the driver

Driving behavior	Total score
Excellent	Score > 9.75
Very Good	9 < Score ≤ 9.75
Good	7.5 < Score ≤ 9
Bad	5 < Score ≤ 7.5
Very Bad	Score ≤ 5

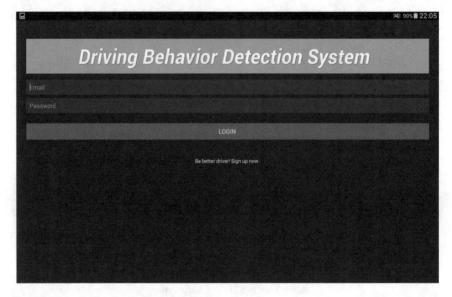

Fig. 3 The login screen of the application

average behavior and to total distance of his trips. In the subsections below we can see how the application works, what data are recorded and how they presented to the user.

4.1 Driver's Login and Registration System

After the user installs and enters the application in his device, the first thing that he sees is a login system (Fig. 3). If the user has already registered in the system, by entering his e-mail and password in the respective fields, he can be entered into the system. If he is not registered in the system, he has to sign up by pressing the "Be better driver! Sign up now" button. In the registration form (Fig. 4) he has to enter the following data: Full name, e-mail, password, address, phone number and the vehicle license plate. When the user completes his registration process he can login successfully in the system.

4.2 Main Menu

When the user is successfully connected, the main menu is presented (Fig. 5) to him. In the main menu there are 4 options. The first option is the "New Trip". We choose this option when we want to start a new trip. The second option is the "My

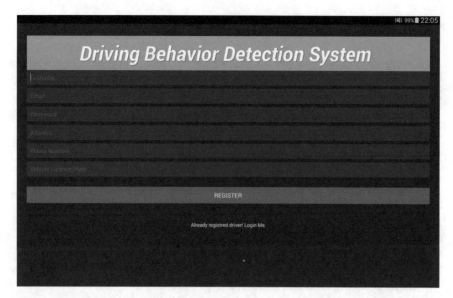

Fig. 4 The registration screen of our application

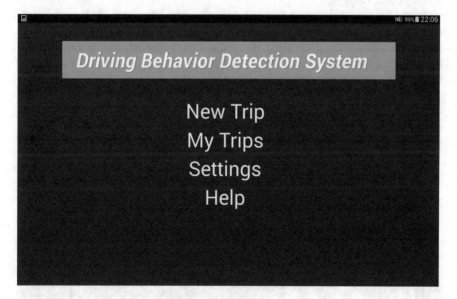

Fig. 5 The main menu of our application

trips" and we choose it when we want to review our trips (routes and statistics). In the third option there are our "Settings" and the last option is "Help" where we can find tutorials and information about how we use the application.

4.3 New Trip

As already motioned above, we choose the "New Trip" option when we want to start a new trip. Before we start driving, the application will tell us to follow some instructions for the calibration of the device (Fig. 6). We set the calibration procedure for the device in order to get accurate readings from the sensors in any fixed orientation of the device inside the vehicle.

During the calibration process the device must be attached in a fixed position and the vehicle must be steel. The whole process takes about 5 s to complete. When the appropriate message is given we can start driving to our destination (Fig. 7).

In the beginning of the process the signal to keep vehicle steel is displayed and in a few seconds "Drive vehicle forward" signal is displayed. When we start driving forward the calibration is completed and the application starts monitoring and detecting our driving behavior until the end of our trip.

The user can select between two options for the presentation of his driving behavior. The first option is the basic option where in the main screen is displayed a "Monitoring Driving Behavior" message (Fig. 8).

When the system detects a safe driving event (maneuver) the main screen color changes to green and the message also changes to the name of the current safe driving event (Fig. 9).

When the system detects a dangerous driving event the color of the main screen changes to yellow color and the message changes to "Attention!" followed by the name of the dangerous event (Fig. 10). For all the dangerous driving events except the attention and the name of the event there is displayed a hint. The hints help drivers to improve their driving behavior and achieve better scores in their trips.

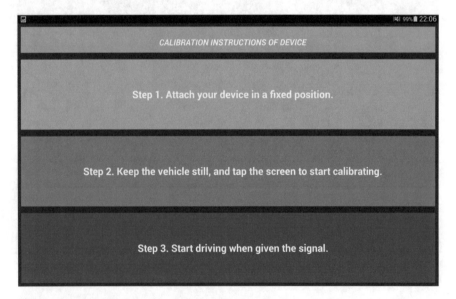

Fig. 6 The calibration instructions screen

Fig. 7 The given calibration messages as they are presented in our application

Fig. 8 The basic monitoring option. The system is not detecting anything safe or dangerous driving events

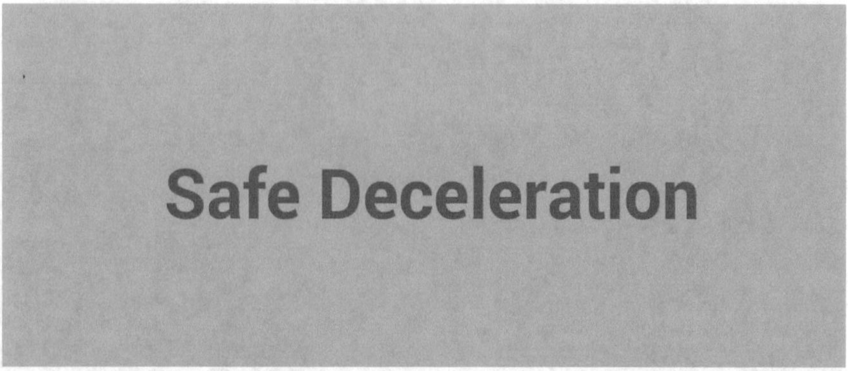

Fig. 9 The system detects a safe deceleration

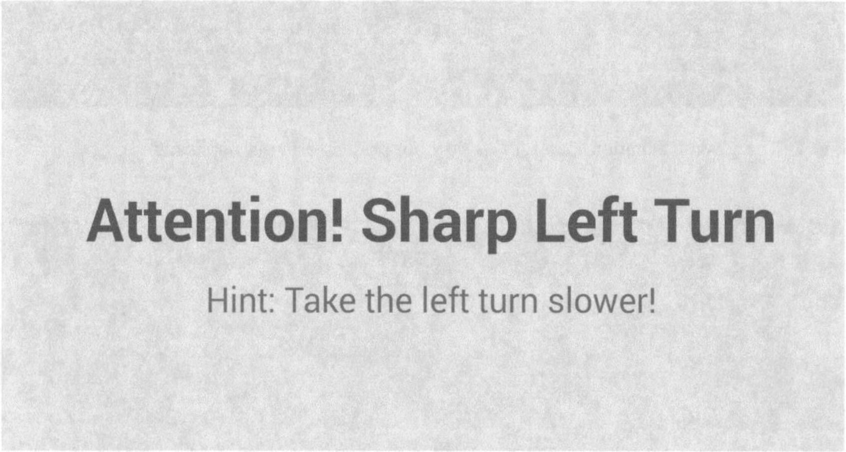

Fig. 10 The system detects a dangerous Sharp Left Turn

The hints that are displayed are: "try to maintain uniform acceleration" (or deceleration), "try to take left (or right) turn slower" and "try to change the left (or right) lane slower". The attention message and hint is followed by a notification sound.

The second option for displaying the monitoring of the user's driving detection is the Map option (Fig. 10). In this screen option a map is displayed with the car's spot into the map (Fig. 11). When we start driving, our route is outlined in the map. Also, it is displayed with pointers, the starting point of our trip and any bad-driving event (maneuver) that happens.

When the system detects a safe driving event, then a pop up window opens and displays the current safe driving event. In the same way when a dangerous driving event is detected a pop-up window with the name and the hint of the event is displayed for some seconds (Fig. 12). Also, a notification sound is being listened and a pointer in the map with the name of the dangerous driving event is created.

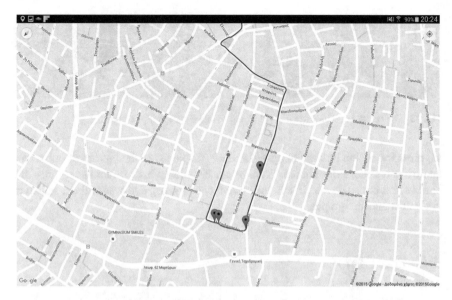

Fig. 11 The map monitoring option of our application

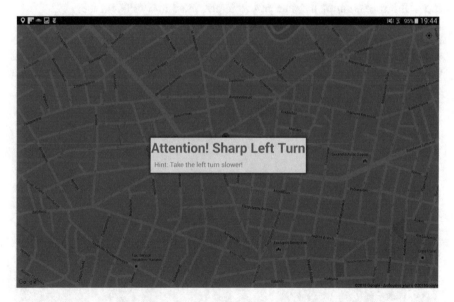

Fig. 12 The system detects a Sharp Left Turn during the map monitoring option

We can alternate between the 2 options for displaying the monitoring of our driving behavior (Basic Monitor to Map Monitor and the opposite) any time we want during the trip by pressing the option button from our device (Fig. 13). When

Fig. 13 Option menu, where we can select the monitoring option we prefer

we press the option button, an option menu is displayed in the bottom of our screen and we can choose our choice. When the user finishes his trip, he has to press the back button on his device. Then, the system ends the trip and loads all the statistics, the routes and the graphs.

4.4 Trip's Info

As we mentioned before, by pressing the back button of our device we can finish our trip. After that the system is loading and presenting all information about our trip. All information of the current trip is presented in 3 different tabs.

In the info tab (Fig. 14) are presented some general statistics about the trip like the total distance, the total duration and the average speed. The total duration is the actual duration of the trip. Except the above statistics, there is also presented the percentage of the penalties of dangerous driving events. The presented percentages

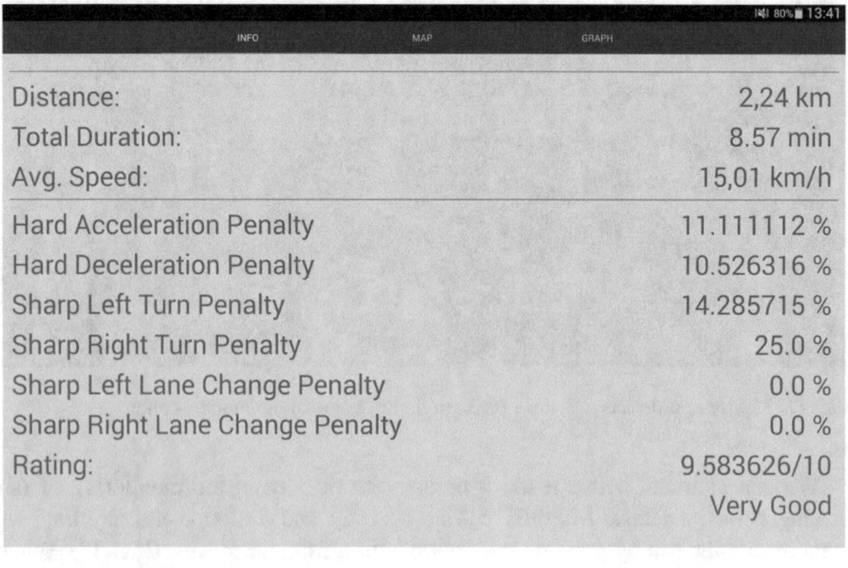

	80% 13:41
INFO MAP GRAPH	
Distance:	2,24 km
Total Duration:	8.57 min
Avg. Speed:	15,01 km/h
Hard Acceleration Penalty	11.111112 %
Hard Deceleration Penalty	10.526316 %
Sharp Left Turn Penalty	14.285715 %
Sharp Right Turn Penalty	25.0 %
Sharp Left Lane Change Penalty	0.0 %
Sharp Right Lane Change Penalty	0.0 %
Rating:	9.583626/10
	Very Good

Fig. 14 Info tab screen of our application

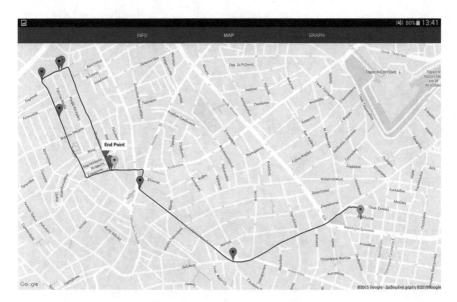

Fig. 15 Map tab screen of our application

of penalties are: Hard Acceleration Penalty, Hard Deceleration Penalty, Sharp Left Turn Penalty, Sharp Right Turn Penalty, Sharp Right Left Lane Change Penalty and Sharp Left Lane Change Penalty. Also is presented the rating (Score—up to 10) and the behavior (Very Bad to Excellent) of the current trip.

The second tab is the map tab (Fig. 15). In the map tab is presented the route of our trip based on Google maps. In the map we can see the starting point of our trip, which is represented by a blue marker and the end point of our trip, which is represented by a light green marker. Also we can see at which point of our route, we are commit a dangerous driving event. All dangerous driving points are represented with a red marker. When we tap in a dangerous driving point it is presented the name of the current event.

The last tab is the graph tab. In the graph tab are presented 3 line charts, the acceleration line chart, the deceleration line chart and the turn line chart (Figs. 16 and 17). We can see one of the 3 charts at the time. If we want to change graph chart we tap on the radio button of the chart we want under the displayed chart.

The x-axis represents the time in milliseconds and the y-axis represents the values of the detection method we have already chosen. These values can be can acceleration values (acceleration detection method—m/s^2) or orientation values (sensor fusion method—rad/s).

The charts are scrollable in the x-axis, which means that by scrolling horizontally we can see the acceleration or orientation values in relation with the time. The displayed x-axed duration is 60 s and as we scroll we can see the next 60 s.

Except the acceleration, deceleration and turn line charts we can see the thresholds lines for each chart. The thresholds line is represented with a red direct

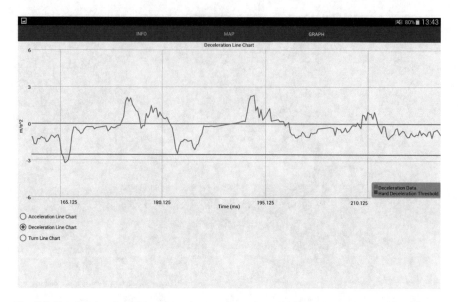

Fig. 16 Deceleration line chart of graph tab screen of our application

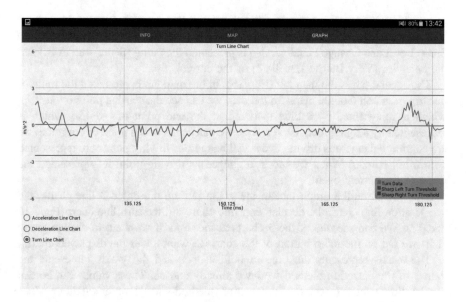

Fig. 17 Turn line chart of graph tab screen of our application

line with the threshold value for each event. If there is a value that has exceed the threshold line it means than we have commit a dangerous driving event from the current chart (acceleration, deceleration, turn) at this particular time.

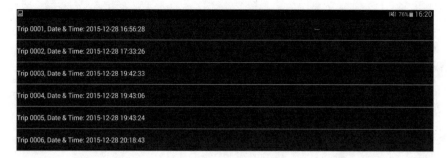

Fig. 18 "My Trip" screen. We can see the list of our trips

4.5 My Trips

In the main menu we can choose the "My Trips" option whenever we want to browse our previous trips (Fig. 18). All the recorded trips are presented in a scrollable list. Each trip is represented with the name "Trip" and an auto increment number, followed by a timestamp, which declares the date and the time that the trip began.

4.6 Settings

In the main menu, the "Settings" option offers the user the chance to set up his preferences (Fig. 19). One of the main settings that the user can change is the standard monitor, which declares the preferred monitoring style. The preferred

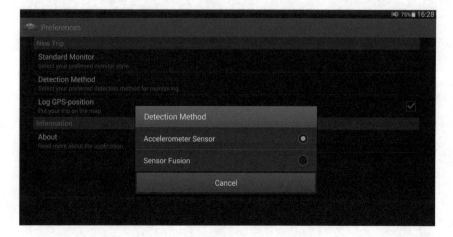

Fig. 19 Select detection method from the settings of our application

monitoring style can be the Basic Monitor and the Map Monitor. The other main setting is the choice of the applicable detection method, which includes the choice of the sensor data, which the system will evaluate in order to detect safe or dangerous driving events and classify the user's driving behavior. The detection method can be one of the accelerometer sensor or the sensor fusion method. Also, we can enable the GPS sensor, if it's not enabled and read more about the application.

5 E-Platform of Auto UBI System

In the context of this research, we have developed a web-based cloud platform that can be used by a usage-based insurance company. The purpose of the platform is to check and evaluate the trip data of the company's drivers (clients).

Via the particular platform an employee of the insurance company can have access to the list and data of all clients by entering the administrator's username and password (Fig. 20).

When a company's employee logs in successfully to the system, he can browse the table with drivers' data (Fig. 21). In this table are presented all data for every driver. The presented data of the driver are: Full name, e-mail, Address, Phone Number and the license plate number of his vehicle.

As we can see in the right side of every row (or driver) of the table, there is a "Select" button. When we want to browse the past trips of a particular driver, we

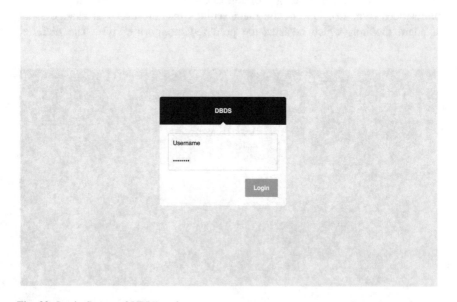

Fig. 20 Login Screen of UBI Portal

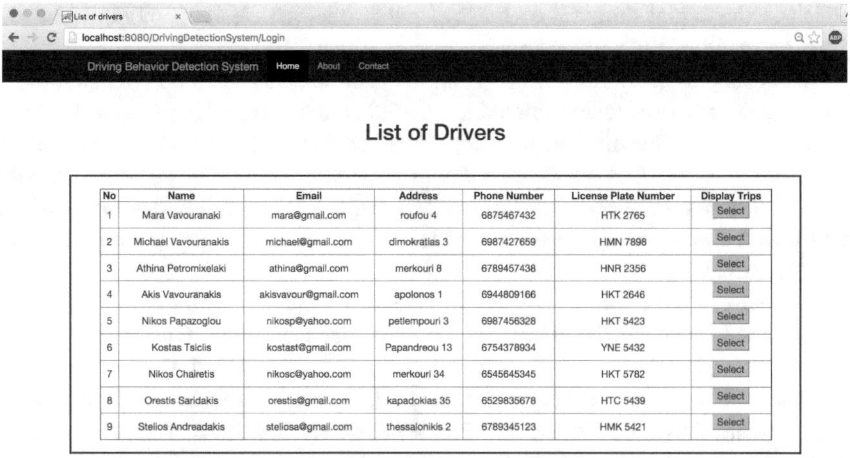

List of Drivers

No	Name	Email	Address	Phone Number	License Plate Number	Display Trips
1	Mara Vavouranaki	mara@gmail.com	roufou 4	6875467432	HTK 2765	Select
2	Michael Vavouranakis	michael@gmail.com	dimokratias 3	6987427659	HMN 7898	Select
3	Athina Petromixelaki	athina@gmail.com	merkouri 8	6789457438	HNR 2356	Select
4	Akis Vavouranakis	akisvavour@gmail.com	apolonos 1	6944809166	HKT 2646	Select
5	Nikos Papazoglou	nikosp@yahoo.com	petlempouri 3	6987456328	HKT 5423	Select
6	Kostas Tsiclis	kostast@gmail.com	Papandreou 13	6754378934	YNE 5432	Select
7	Nikos Chairetis	nikosc@yahoo.com	merkouri 34	6545645345	HKT 5782	Select
8	Orestis Saridakis	orestis@gmail.com	kapadokias 35	6529835678	HTC 5439	Select
9	Stelios Andreadakis	steliosa@gmail.com	thessalonikis 2	6789345123	HMK 5421	Select

Fig. 21 List of all registered drivers in the program of UBI company

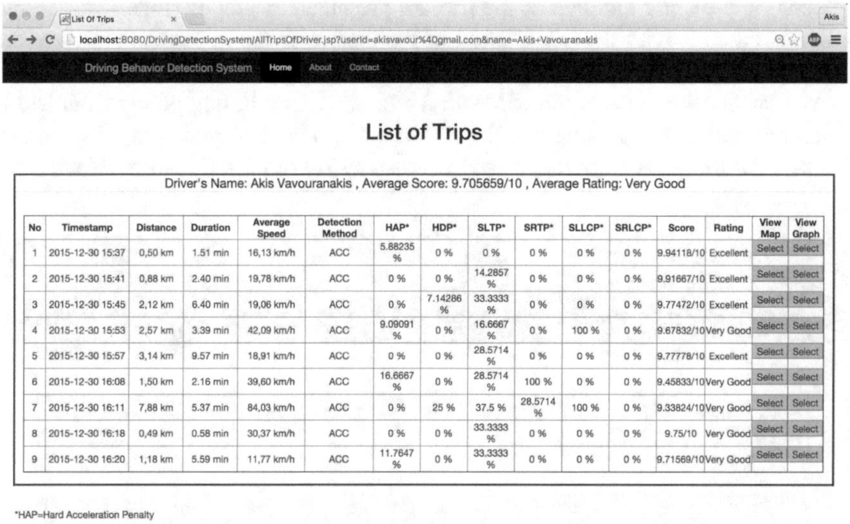

List of Trips

Driver's Name: Akis Vavouranakis , Average Score: 9.705659/10 , Average Rating: Very Good

No	Timestamp	Distance	Duration	Average Speed	Detection Method	HAP*	HDP*	SLTP*	SRTP*	SLLCP*	SRLCP*	Score	Rating	View Map	View Graph
1	2015-12-30 15:37	0,50 km	1.51 min	16,13 km/h	ACC	5.88235 %	0 %	0 %	0 %	0 %	0 %	9.94118/10	Excellent	Select	Select
2	2015-12-30 15:41	0,88 km	2.40 min	19,78 km/h	ACC	0 %	0 %	14.2857 %	0 %	0 %	0 %	9.91667/10	Excellent	Select	Select
3	2015-12-30 15:45	2,12 km	6.40 min	19,06 km/h	ACC	0 %	7.14286 %	33.3333 %	0 %	0 %	0 %	9.77472/10	Excellent	Select	Select
4	2015-12-30 15:53	2,57 km	3.39 min	42,09 km/h	ACC	9.09091 %	0 %	16.6667 %	0 %	100 %	0 %	9.67832/10	Very Good	Select	Select
5	2015-12-30 15:57	3,14 km	9.57 min	18,91 km/h	ACC	0 %	0 %	28.5714 %	0 %	0 %	0 %	9.77778/10	Excellent	Select	Select
6	2015-12-30 16:08	1,50 km	2.16 min	39,60 km/h	ACC	16.6667 %	0 %	28.5714 %	100 %	0 %	0 %	9.45833/10	Very Good	Select	Select
7	2015-12-30 16:11	7,88 km	5.37 min	84,03 km/h	ACC	0 %	25 %	37.5 %	28.5714 %	100 %	0 %	9.33824/10	Very Good	Select	Select
8	2015-12-30 16:18	0,49 km	0.58 min	30,37 km/h	ACC	0 %	0 %	33.3333 %	0 %	0 %	0 %	9.75/10	Very Good	Select	Select
9	2015-12-30 16:20	1,18 km	5.59 min	11,77 km/h	ACC	11.7647 %	0 %	33.3333 %	0 %	0 %	0 %	9.71569/10	Very Good	Select	Select

*HAP=Hard Acceleration Penalty

Fig. 22 List of trips (and data of trips) of a particular driver

click the "Select" button in the row of the particular driver. For example we click to browse the list of trips of a driver named "Akis Vavouranakis" (Fig. 22).

In the list of trips, we can see the driving history of every driver. In the presented table, every row of the table represents a trip. Lots of data are presented for every trip. The presented general data are: The ***timestamp*** of the trip, which is the date and the starting time of the trip, the ***distance*** of the trip in km, the ***duration*** of the trip, which is the actual duration of the trip (when the vehicle is in motion) and the *average speed* of the vehicle (km/h).

Also it is presented the selected from the user detection method (Accelerometer sensor data or Sensor fusion orientation data) and the percentage of penalties of all dangerous driving events. The percentages of penalties of the driving events that are presented are: ***Hard Acceleration Penalty*** (HAP), ***Hard Deceleration Penalty*** (HDP), ***Sharp Left Turn Penalty*** (SLTP), ***Sharp Right Turn Penalty*** (SRTP), ***Sharp Left Lane Change Penalty*** (SLLCP) and the ***Sharp Right Lane Change Penalty*** (SRLCP).

The *score* and the *rating* of every trip are also presented. The score of every trip is calculated based on the already above mentioned penalties. The max score for every trip is 10. Based on the score the rating (driver's behavior) of every trip is calculated. The rating can be: Very Bad, Bad, Good, Very Good or Excellent.

In the line above the array and its elements, we can see some average data of the driver. First of all, the name of the driver is displayed, which is followed by the *average score* and the *average rating* of all his trips.

As we can see, in the data table (which contains the list of trips of a particular user), in the last two columns of each row (trip) there are two "select" buttons. By selecting the first one, the route of the trip is displayed in a map (Fig. 23) and by selecting the second the data of the trip are presented in graphs (Figs. 24, 25, and 26). Also a table with the information of the particular trip is presented before the map.

We can see the route of our trip with a blue line. The starting point of our trip is displayed with a blue marker and the end of our trip is displayed with a light green marker. All dangerous driving events are displayed with a red marker. If we tap a

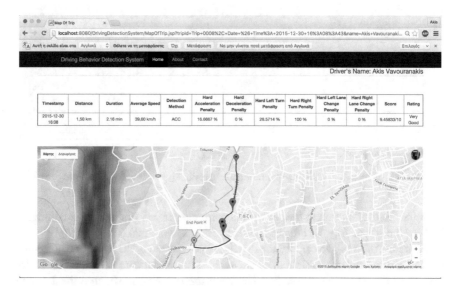

Fig. 23 Map screen of our portal with the route and data of a particular trip

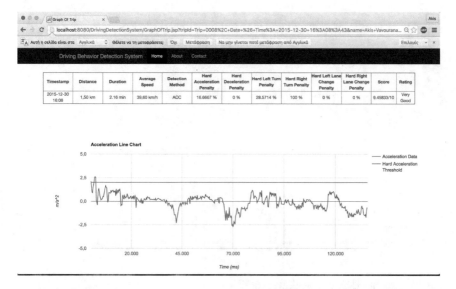

Fig. 24 Acceleration line chart with data of a particular trip

Fig. 25 Deceleration line chart of a particular trip

red marker, the name of the dangerous driving event is displayed. The map is based on Google maps and we can zoom In/Out on it.

By selecting the "View Graph" select button, 3 line charts are presented, the acceleration line chart, the deceleration line chart and the turn line chart. The x-axis represents the time in milliseconds and the y-axis represent the values of the detection

Fig. 26 Turn line chart of a particular trip

method we have already chosen. These values can be acceleration values (acceleration detection method—m/s^2) or orientation values (sensor fusion method—rad/s).

Except of the acceleration, we can see the thresholds lines for each chart for the deceleration and turn line charts. The thresholds line represented with a red direct line with the threshold value for each event. If there is a value that has exceed the threshold line it means than we have commit a dangerous driving event of the current chart (acceleration, deceleration, turn) at this particular time. Also a table with the information of the particular trip is presented before the graphs.

6 Conclusions

This work shows that the user (driver) can benefit from a Usage-Based system; the system rewards the user for being a safe driver and also provides him with feedback on his driving habits, making him a better and safer driver. The device measures various elements of user's driving behavior, safe or hard breaking, safe or sharp accelerating, safe or sharp turns and safe or sharp lane changes. Also, measures the start time of a trip, the duration of the trip and the total travelled distance. Also, for every trip, the user gets a score and a rating, depending on how well he drives. The insurance company has access to all these data (trips data, routes and graphs) of its drivers via the e-platform.

Also, in our work we demonstrated that the user's driving behavior could be estimated based either on acceleration data of the accelerometer sensor or based on orientation data of sensor fusion method. Sensor fusion method combines data from the accelerometer, geomagnetic field sensor and the gyroscope. The driver can choose the detection method from the settings of the application's main menu. We propose to drivers to use the accelerometer-based method as detection method. The reason is not that the sensor fusion method is not accurate or reliable, in total, but rather that sensor fusion is not the best option for the detection of sharp turns or sharp lane changes. When just the accelerometer sensor is used, the algorithm detects sharp turns or sharp lane changes faster than when we use the orientation data of sensor fusion method. On the other hand, when we use the sensor fusion as detection method the system detects faster the safe/hard acceleration or deceleration of the vehicle than when we use only the accelerometer data.

Regardless of the detection method that is used, safety comes first and the system takes into account that sometimes hard deceleration or rapid accelerations or other dangerous driving events are necessary in order to avoid a collision. Hence, our system works to identify a pattern in driving habits so the occasional hard brakes do not have a significant impact (if any) on the potential rating. Thus, the driver's discount depends mostly on his average rating over his trips. So, better rating means highest discount for his insurance.

By using the application and the e-platform of our UBI information system, the drivers and the insurance company have many benefits. Some of them are:

- Social and environmental benefits from more responsible and less unnecessary driving.
- Commercial benefits to the insurance company from better alignment of insurance with actual risk. Improved customer segmentation.
- Potential cost-savings for responsible customers.
- Technology that powers UBI enables other vehicle-to-infrastructure solutions including drive-through payments, emergency road assistance, etc.
- More choice for consumers on type of car insurance available to buy.
- Social benefits from accessibility to affordable insurance for young drivers— rather than paying for irresponsible peers, with this type of insurance young drivers pay for how they drive.
- Higher-risk drivers pay more per use. Thus, they have highest incentive to change driving patterns or get off the roads and leave roads safe.
- For telematics usage-based insurance: Continuous tracking of vehicle location enhances both personal security and vehicle security. The GPS technology could be used to trace the vehicle whereabouts following an accident, breakdown or theft.
- The same GPS technology can often be used to provide other (non insurance) benefits to consumers, e.g. satellite navigation.
- Gamification of the data encourages good driver behavior by comparing with other drivers.

Our system can serve some useful purposes but it has some limitations and drawbacks too. Some of them are:

- The system cannot detect the backward movement of the vehicle. When the user drives backwards, the detected driving events will be wrong. These wrong driving events will affect the score and the rating of the user.
- A limitation of our system has to do with the calibration process of the device. Before we start driving the application will tell us to follow some instructions for the calibration of the device. This process has to be repeated for each new trip and the whole process takes about 5 s to complete.
- Another limitation is when we start the calibration process the car is on a slope. If the car is on a slope during the calibration procedure the reading data will be affected and this will lead to poor results.
- The device has to be in a fixed position during the trip. After the calibration and while monitoring you cannot move the device from its fixed position. If the device moved while monitoring the sensor's output data it will be wrong and the evaluation of the data will not be accurate. In this case the user has to repeat the calibration procedure.
- Some of the low cost Android devices have low quality sensors or processing power. So the readings of the sensor's data are not so accurate and the application slows down due to low processing power.
- All Android devices have in-built the most used sensors like accelerometer and gyroscope. The magnetometer sensor on the other side is not included in all smartphone devices. So, without the magnetometer sensor the user cannot use the Sensor Fusion detection method.

References

1. Usage based insurance from Wikipedia (2016). https://en.wikipedia.org/wiki/Usage-based_insurance Accessed 10 May 2016
2. National Association of Insurance Commissioners (2016). Usage-Based Insurance and Telematics. http://www.naic.org/cipr_topics/topic_usage_based_insurance.html. Accessed 20 May 2016
3. Paefgen, J., Staake, T., Thiesse, F.: Resolving the misalignment between consumer privacy concerns and ubiquitous IS design: the case of usage-based insurance. In: International Conference on Information Systems (ICIS) (2012)
4. Truong, A.: A new take on auto insurance by the mile(2015).Fast Company. http://www.fastcompany.com/3033107/fast-feed/a-new-take-on-auto-insurance-pay-by-the-mile. Accessed 14 Dec 2015
5. Insurance Journal (2016). Allstate's Usage-Based Auto Insurance Expands to New York, New Jersey. http://www.insurancejournal.com/news/east/2012/10/23/267659.html. Accessed 12 Feb 2016
6. Paefgen, J., Staake, T., Thiesse, F.: Resolving the misalignment between consumer privacy concerns and ubiquitous IS design: the case of usage-based insurance. In: International Conference on Information Systems (ICIS) (2012)
7. Handel, P., Skog,I., Wahlstrom, J., Bonawide, F., Welsh, R., Ohlsson, J., Ohlsson, M.: Insurance telematics: opportunities and challenges with the smartphone solution. Intell. Transp. Syst. Mag. IEEE **6**(4), 57–70 (2014)

8. Iqbal, L.: A privacy preserving gps-based pay-as-you-drive insurance scheme. International Global Navigation Systems Society (2006)
9. Castignani, G., Derrmann, T., Frank, R., Engel, T.: Driver behavior profiling using smartphones: a low-cost platform for driver monitoring. IEEE Intell. Transp. Syst. Mag. 7(1) (2015)
10. Kalra, N., Bansal, D.: Analyzing driver behavior using smartphone sensors: a survey. Int. J. Electron. Electr. Eng. 7(7), 697–702 (2014). ISSN 0974-2174
11. Stoichkov, R.: Android smartphone application for driving style recognition. Department of Electrical Engineering and Information Technology Institute for Media Technology, July 20
12. Atan2 (2016). From Wikipedia. https://en.wikipedia.org/wiki/Atan2 Accessed 12 Jan 2016
13. Lawitzki, P.: Application of dynamic binaural signals in acoustic games. Master's thesis, Hochschule der Medien Stuttgart (2012)
14. Lawitzki, P.: Android sensor fusion tutorial. September 2012. http://plaw.info/2012/03/android-sensor-fusion-tutorial/. Accessed 10 Apr 2016
15. Singh, P., Juneja, N., Kapoor, S.: Using mobile phone sensors to detect driving behavior. In: Proceedings of the 3rd ACM Symposium on Computing for Development. ACM (2013)
16. Fazeen, M., Gozick, B., Dantu, R., Bhukhiya, M., Gonzalez, M.C.: Safe -driving using mobile phones. In: IEEE Transactions on Intelligent Transportation Systems (2012)

Collaborative Media Delivery in 5G Mobile Cloud Networks

Piotr Krawiec, Jordi Mongay Batalla, Joachim Bruneau-Queyreix,
Daniel Negru, George Mastorakis
and Constandinos X. Mavromoustakis

Abstract Majority of global traffic transferred through the Internet nowadays, as
well as in the near future, is related with video delivery service. For this reason each
solution focusing on improvement of media distribution is in the high interest of
both research community and industry. Particularly, new solutions are required in
mobile systems, since it is expected that mobile video traffic growth will be twice
faster than wired streaming. In this chapter, a new approach for media delivery in
5G mobile systems is proposed. It assumes collaboration of small base stations and
users' terminals, which create ad hoc peer-to-peer cloud streaming platform. The
proposed system provides high quality of content delivery due to
network-awareness assured by introducing resource allocation mechanisms into 5G
Radio Access Network infrastructure. The solution has been implemented in a proof
of concept and the performance evaluation tests verified that the system offers
significant improvement, in terms of quality and availability of the content as well
as lower delays in media delivery, comparing to traditional VoD service.

P. Krawiec (✉) · J.M. Batalla
National Institute of Telecommunications, Warsaw, Poland
e-mail: P.Krawiec@itl.waw.pl

J.M. Batalla
e-mail: J.Mongay@itl.waw.pl

P. Krawiec · J.M. Batalla
Warsaw University of Technology, Warsaw, Poland

J. Bruneau-Queyreix · D. Negru
University of Bourdeaux, Bordeaux, France
e-mail: jbruneau@labri.fr

D. Negru
e-mail: dnegru@labri.fr

G. Mastorakis
Technological Educational Institute of Crete, Heraklion, Greece
e-mail: gmastorakis@staff.teicrete.gr

C.X. Mavromoustakis
University of Nicosia, Nicosia, Cyprus
e-mail: mavromoustakis.c@unic.ac.cy

© Springer International Publishing Switzerland 2017
C.X. Mavromoustakis et al. (eds.), *Advances in Mobile Cloud Computing
and Big Data in the 5G Era*, Studies in Big Data 22,
DOI 10.1007/978-3-319-45145-9_14

1 Introduction

Video streaming is one of the most demanded Internet services and it is always increasing the data volume. Video will growth at a rate of more than 30 % until 2019 [1]. Then, it will represent more than 80 % of all consumer Internet traffic. Moreover, the number of devices playing video and the quality of such devices is growing, so the desired quality of video streaming will increase in the next years from High Definition (1280 × 720 pixels) and Full High Definition (1920 × 1080) until Ultra-High Definition (4K: 3840 × 2160 and 8K: 7680 × 4320). While Ultra-High Definition is reserved mainly for wired communications, High Definition and Full High Definition will be served by both wired and wireless networks. Furthermore, mobile video traffic will grow two times faster than fixed IP traffic from 2014 to 2019 [1]. Therefore, the management of video streaming service, with its ever-increasing the capacity requirements, will result crucial for the next-generation mobile networks.

Future 5G mobile systems aim to increase the overall capacity of the network for reaching users' expectations related to current (e.g. video streaming) and new applications (e.g. augmented reality) and to meet ever-emerging bandwidth demands. Although 5G system is currently under development and there is still a degree of uncertainty regarding the final 5G specification, it is expected that the capacity growth will be achieved thanks to several features. One of them is an improvement of network efficiency. Enabling technologies in this area are massive MIMO (Multiple-Input Multiple-Output) techniques which allow for boosting spectral efficiency, jointly with usage of short-range higher frequency bands, particularly mm-wave.

In addition to the increase of peak rate in the wireless space, Radio Access Networks (RAN) in 5G technology introduce new mechanisms directed to obtain an effective usage of wireless resources. Such mechanisms aim to manage highly dense HetNets (Heterogeneous Networks) for increasing the efficiency of the radio spectrum. HetNets are based on dense multiple antennas outdoor and indoor, which form groups of ultra-small (pico/femto) cells allowing for intensive spectrum spatial reuse. A HetNet will incorporate different Radio Access Technologies (RAT), both in licensed (e.g. LTE) and unlicensed (e.g. WiFi) bands. Moreover, spatial diversity in 5G will be also achieved by widespread exploitation of device-to-device (D2D) communication, realized between two adjacent terminals with limited, or even without involvement of network infrastructure.

In this chapter, we present a solution for increasing the capacity of the systems in Multimedia distribution by taking advantage of the mechanisms and algorithms deployed in 5G RAN. Concretely, we propose to take advantage of an existence of large number of local, small (femto-) base stations, jointly with D2D communication capabilities, to establish a Mobile Media Cloud (MMC). MMC is an ad hoc configuration, which exploits a combined peer-to-peer (P2P) content distribution with multi-source, multi-destination congestion control algorithms. It bases on Dynamic Adaptive Streaming over HTTP (DASH) mechanism and assumes that the ensemble of femto-stations and wireless terminals become a distributed local cache, which provides collaborative delivery of stored media segments to the users.

Proposed solution corresponds to the concept presented in [2], which also exploits femto-base stations to store media content and terminals as caching helpers. However, the architecture proposed in [2], in contrast to MMC, is not designed to, and does not take into account features of adaptive media delivery mode. In turn, the authors in [3] focused on an optimization of multimedia delivery in 5G mobile system by integration of DASH-based adaptive delivery scheme with RAN infrastructure. Nevertheless, they did not consider collaboration of base stations and wireless terminals for cost-effective P2P media streaming. On the other hand, some propositions of collaborative systems for media delivery in wireless networks have been already presented in literature [4–6], but they are generally created as overlay systems and thus they suffered high response delays. MMC assumes that streaming management is incorporated into the RAN, in order to decrease latency in media delivery and maximize the overall users' Quality of Experience (QoE).

This chapter contains description of proposed Mobile Media Cloud for collaborative, adaptive media delivery in 5G systems. Specifically, in Sects. 2 and 3 we will describe more technical details of MMC general architecture and architecture of MMC node (i.e. MMC Peer), respectively. Next section discusses implementation of MMC Peer middleware and presents results of the tests performed on the implemented prototype. At last, the chapter is concluded in Sect. 5.

2 Architecture for Adaptive Mobile Cloud Streaming

The most popular model of Internet video consumption is on-demand streaming. It assumes that different users request the same content, very often over a similar period, nevertheless in fully asynchronous way. This model applies to traditional Video on Demand (VoD) services, as well as social networking scenarios, when user-generated content, due to rapidly spreading recommendations, becomes very popular in a given time. Such pull-based streaming model results in a large number of unicast connections, initiated by users, which saturate network resource pool between streaming servers and users' terminals. The well-known solution to prevent this issue are dedicated Content Delivery Networks (CDNs), which optimize content placement by replication and bringing the content closer to the end users. However, CDN is characterized by high costs and often its deployment has no economic justification. As an alternative, peer-to-peer technology is investigated for content replication. P2P systems are low cost, because they rely on existing infrastructure. On the other hand, typical P2P solution cannot provide high QoE for the users due to lack of control and management.

Taking into account features of future mobile architecture such as deployment of ultra-dense small cells and the possibility of direct communication between users devices, we propose a Mobile Media Cloud concept to create low-cost media delivery system. MMC will be founded on existent 5G RAN infrastructure, i.e. femto-stations and wireless terminals, which create ad hoc mobile cloud adjusted to P2P-based streaming. To ensure better QoE for the end users, MMC assumes network-assisted management. Multi-source, multi-destination congestion control algorithms, which are responsible for selecting the best MMC Peers for delivery given content, are incorporated into 5G RAN, and bases on actual network context. For this purpose, we propose to use cloud resources available at RAN proposed for 5G.

Cloud-based Radio Access Network (C-RAN, also known as Centralized RAN) is considered to be an architecture which responds well to the challenges of 5G access domain. The basic idea of C-RAN is to detach the Baseband Units (BBU), which carry out signal processing, from the remote sites and shift them to centralized locations. The base stations host Remote Radio Head (RRH) modules only, that perform conversion between digital baseband signals and analog signals transmitted/received by antennas. To decrease deployment and operational costs, BBUs are aggregated, creating a BBU pool, and run in dedicated datacenter or using cloud service.

The architecture of MMC is presented in Fig. 1. It includes MMC Peers, which are created by base stations with MMC middleware running on them. MMC Peers act as distributed storage for content replication, decreasing distance between content source and receivers. Two models for base stations operating as MMC Peer can be considered, depending on where caching functionality is realized. The first

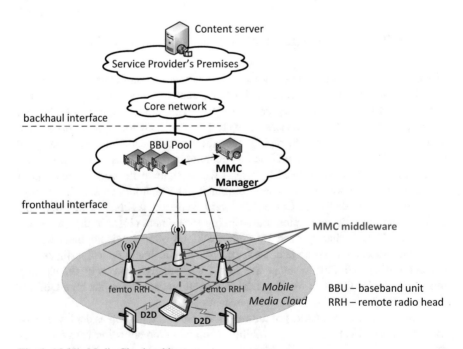

Fig. 1 Mobile Media Cloud architecture

one assumes that storing is performed at the BBU level, by using storing capabilities provided by BBU pool. In the second model, RRH modules are enhanced with disk storage. In that case, the bandwidth at fronthaul links can be saved since some content requests will be handled locally. On the other hand, it requires an additional logic at RRH to recognize content request and to decide if it can be served using cached content, or should be forwarded to BBU.

Please note that terminals are assumed as important element of 5G system. They move from simply traffic consumers toward intermediate nodes in 5G network topology, which actively interact with other nodes in media delivery process. For this reason, we can accept that MMC Peers will also be launched at users' terminals, especially those which are characterized by adequate caching capabilities, as laptops or tablets.

The MMC Peer M&C (management and control) plane, formed by MMC middleware, is responsible for establishment and management of a MMC ad hoc system. It handles media requests and controls streaming process from other peers. For optimal allocation of the available resources at RAN, the proposed system includes an additional entity called MMC Manager. The entity executes all necessary operations and determines all data required for a decision what and how to perform the streaming and adaptation functionalities, e.g., at which bitrate and from which peers.

To overcome a negative impact of variable network conditions on streaming process, MMC uses the Dynamic Adaptive Streaming over HTTP (DASH) platform. DASH is open standard and it is called to be the reference into the market due to its simplicity and flexibility. It is a stream-switching adaptive protocol that is based on fragmentation of the original video content in different segments that are encoded in multiple bitrates (called representations). The end user consecutively requests new video segments, selecting appropriate representation, in order to adapt the video streaming bitrate to the current network situation. During media content consumption process, replicas of segments transferred from content server (located at Service Provider's premises, or CDN) through the MMC are cached in MMC Peers. Decision which segments should be stored and on which peers is taken according with instructions received from MMC Manager, which can implement different caching strategies [7], taking into account capabilities of peers (e.g. storage size, upload capacity of peers' interfaces), information about content popularity obtained from Service Provider etc. Afterwards, the further requests for given content are handled locally by using distributed MMC storage.

2.1 Mobile Media Cloud Manager

A central point of proposed Mobile Media Cloud is MMC Manager since it is responsible for working out which resources should be used for handling given user's request, to achieve the best (in terms of efficiency) resource exploitation, addressing at the same time QoE requirements. More specifically, MMC Manager coordinates the optimal exploitation of the resources which are available at each MMC Peer by managing, in a scalable way, the upload bandwidth of each peer participating in the MMC.

In order to ensure both high resource utilization and quality guarantees, we propose to take advantage of new resource reservation and provision schemes which base on the bandwidth auto-scaling algorithms, such as presented in [8, 9]. Resource prediction engine of MMC Manager predicts the upcoming demands on resources and plans the needed bandwidth capacity for media delivery based on models described in [8, 9]. Then, MMC Manager makes decision, if requested content should be delivered using the MMC, and manages the arrangement of the content among MMC Peers.

MMC Manager controls the P2P delivery mechanism process based on flow control and multi-source multi-destination congestion algorithms that maximize the utilization of the upload bandwidth of each participating peer in MMC configuration. In response to MMC Peer query, MMC Manager selects the optimal set of the peers that should contribute to the content delivery. The selection is performed taking into account current information about state and context of network and associated terminals. To this end, MMC Manager interacts with RAN for requesting specific monitoring data for each MMC Peer. Based on the collected values of metrics depicting the current and upcoming ability of the each peer in content provision, MMC Manager decides on the best MMC Peers for content delivery at the moment.

To ensure MMC Manager's response times at a low level and easy access to the status of RAN resources, we propose to deploy the module within the C-RAN, using virtualized pool of processing resources available at the BBU pool (as it is depicted in Fig. 1). The novel Mobile Edge Computing (MEC) technology [10], which is designed for modern and future mobile systems, offers such capabilities. MEC provides an open Application Programming Interface (API) to authorized third parties, offering them access to cloud-computing capabilities at the RAN, as well as to crucial information related with the current network context.

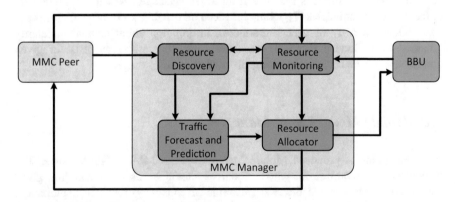

Fig. 2 Functional architecture of the MMC Manager

The MMC Manager functional architecture is presented in Fig. 2 and includes the following components:

- *Resource Discovery*—it gathers information related with active peers in MMC domain, such as supported RATs, their storage capacity etc., and provides the functionality required to enable cooperation between peers to construct a distributed cloud based on a P2P topology. In case of wireless terminals acting as MMC Peers, this component decides which devices can be considered as neighbors, taking into account not only their physical proximity, but also device capabilities (for example available wireless interfaces which can be used to establish direct connections). Neighboring terminals can collaborate in P2P-based media delivery process thanks to D2D link established between them.
- *Resource Monitoring*—it collects monitoring information related with the state and context of network and MMC Peers. This component implements the MEC API to obtain from BBUs low-level monitoring data, both overall (femto-) cell statistics as well as individual channel state information.
- *Traffic Forecast and Prediction*—it implements resource prediction engine, which takes advantage of well-established workload and traffic prediction, on one hand, as well as of forecasting algorithms, on the other hand, in order to provide timely small-term and mid-term estimation of the required resources needed to accommodate an anticipated demand.
- *Resource Allocator*—it calculates the optimal bandwidth allocation for the P2P media delivery between MMC Peers and selects the best peers for handling a given request. It is also responsible for coordination of the resources allocation and optimization processes performed at the 5G infrastructure provider's level. Based on information provided by this component (through MEC API), BBUs can make use of advanced mechanisms for optimization of radio resources assignment, such as Coordinated Multi-Point (CoMP), enabling better resource utilization due to reduction of inter-cell interferences.

3 Internal Architecture of Mobile Media Cloud Peer

The MMC middleware provides mechanisms and functionalities required for establishing MMC Peer based on infrastructure of small base-stations and wireless terminals. MMC Peer is a media-centric node, providing QoS/QoE-enabled multimedia services delivery to a wide set of user terminals in 5G environment. A set of the peers can have an ad hoc organization as a cloud (MMC). When used in an MMC configuration, MMC Peer transmits media content to other Peers and is responsible for adapting streaming process to current network environment conditions and related users' context.

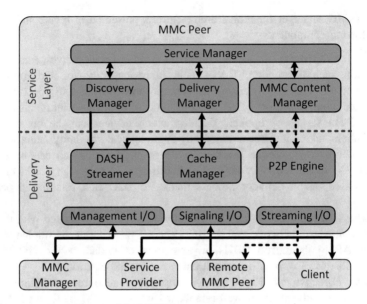

Fig. 3 Functional architecture of the MMC Peer

The main MMC Peer component's services are: creation of the MMC, the P2P content delivery from remote Peer to end user's client application and content caching. The internal architecture of MMC Peer is presented in Fig. 3.

The MMC Peer functional architecture consists of two layers. The bottom is Delivery layer, which implements low-level protocol/data and signaling related functionalities. It provides common abstractions, by masking the heterogeneity and the distribution of different RAT supported by given peer. It is composed of:

- *Signaling I/O*—is responsible for exchanging signaling information with remote peer, related with DASH adaptive streaming and content upload;
- *Streaming I/O*—is responsible for receiving content requests from the remote end-point and, in response, streaming data to her/him;
- *Management I/O*—is responsible for requesting/receiving of cloud configuration/service Admission Control information from the MMC Manager;
- *DASH Streamer*—implements the HTTP streaming functions;
- *P2P Engine*—is responsible for P2P streaming between peers in the MMC configuration;
- *Cache Manager*—manage local and remote access to local cache.

The upper-layer is Service layer. Relying on the Delivery layer, it enables the allowed user terminals to perform the discovery and delivery in Client/server or cloud mode for high level services at the MMC Peer. The main components are:

- *Service Manager*—is responsible for service Admission Control, i.e. controlling local and remote users' requests and managing access to media;
- *Discovery Manager*—is responsible for providing means to query other MMC Peers or Service Provider about available services/content that match a given search criteria;
- *Delivery Manager*—it controls overall streaming delivery process, handles HTTP requests arrived from remote end-point and launches the eventual adaptation action;
- *MMC Content Manager*—is responsible for MMC Peer to MMC Peer communications to retrieve content availability.

Figure 4 presents the general scenario for media streaming in MMC, when a user wants to retrieve a content that is fully available in the MMC. MMC Peers exploit functionalities provided by MMC Manager to find out which Peer should be prioritize to stream the content to the end user. In the meantime, RAN (through MEC API) sends monitoring reports to the MMC Manager on the traffic observed on each Peer. Thanks to them, the MMC Manager can take a proper decision about the appropriated Peer to be used in the MMC.

MMC streaming process covers two phases. The first phase is related with obtaining dynamic DASH MPD (Media Profile Description) file of a desired content. This phase starts when user's client application sends request to its local MMC Peer (Peer1 on the sequence diagram in Fig. 4) for the content ID and its

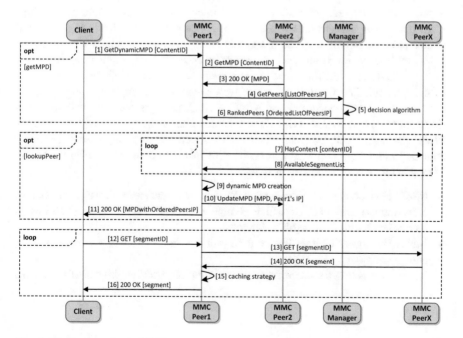

Fig. 4 Media streaming in MMC

associated MPD file. Next, the local Peer will request the MPD file on the originally hosting Peer (Peer2 in Fig. 4). The MPD file contains the list of Peer's IP addresses that previously cached the content. Then the local Peer requests the MMC Manager to evaluate quality of each Peer referenced in the MPD file (which contains required media) in order for the local Peer to best select Peers for the upcoming streaming session. The local Peer can then probe each remote Peer (PeerX on the sequence diagram) to determine which part (i.e. segments) of the content they previously cached. Once all Peers are probed, the local Peer dynamically re-create the MPD file pointing at all the pre-selected remote Peers and deliver the latter file to the client. The local Peer also updates the original MPD file with its IP address in order for future streaming sessions to include the local Peer in the choice of selected content sources.

The second phase refers to actual streaming process. This phase consists in sending by the client requests to the local Peer (Peer1 in Fig. 4) for segments identified inside the dynamic MPD. The local Peer requests the content segments from remote Peers (PeerX in Fig. 4) on behalf of the client, caches the segments if needed, and finally deliveries the content segments to the DASH client.

4 Prototype of MMC Peer Middleware

The MMC Peer composes the atom of the Mobile Media Cloud and has a major role in the content delivery process. The MMC Peer represents the penultimate equipment that relays content to end users. Components of MMC Peer's Service layer realize the functionalities of the M&C plane—they are responsible for creating the MMC ad hoc system from a set of connected base stations and/or terminals.

To validate and test the proposed approach, we implemented functionalities of MMC Peer middleware in the form of web services. Our implementation emphasizes on the interconnectivity of MMC Peers, the resource retrieval upon the demands from end user as well as best resource allocation strategy with the help of MMC Manager as an external service. Developed software allows for streaming content to end user in a P2P approach, with instruction set received from MMC Manager for peer selection, as follows:

- MMC Peer creates an ephemeral MPD file upon end user demand for content consumption and decides which remote Peers should be used according to recommendation from MMC Manager;
- MMC Peer holds content from online caching process as well as contents from end user upload;
- MMC Peer retrieve content from multiple remote Peers on behalf of the client.

4.1 Middleware Implementation

This section focuses on the design and implementation of MMC Peer middleware, in particular Service layer, which is responsible for performing MMC M&C tasks. The middleware has been developed as RESTful web service using the Java EE framework with Java 8 programming language (Java EE has been chosen for its large-scale, scalable and reliable API and platform). Jersey RESTful Web Services framework [11] was used as implementation of JAX-RS API (Java API for RESTful Services), which runs on top of lightweight Grizzly HTTP server [12]. Subsections below describe MMC Peer's internal components implementation and interfaces with external modules.

4.1.1 Discovery Manager

The Discovery Manager handles HTTP GET requests to serve end user (i.e. client) application with DASH manifest (MPD) files. Two methods have been defined:

- *getDynamicMPD()* method issued by end users and then retrieve the content manifest files (on a remote MMC Peer or an original content server located at Service Provider's CDN) which contain all meta data required to stream contents back to the end user. The query's response is a dynamically generated MPD providing any DASH client with a DASH segment list referring to remote MMC Peers ready to stream the required content. The latter MPD points at the local MMC Peer which will be queried (cf. Delivery Manager subsection below) to properly retrieve and cache the content;
- *getMPD()* method issued by a remote MMC Peer wishing to access a MPD resource. Such method exists to return the original MPD to a remote MMC Peer which had some of its users requesting a specific content with the *getDynamicMPD()* method. When a *getMPD()* method is issued by a remote MMC Peer, the local Peer updates the requested MPD with the remote Peer's IP. The updated MPD is now referring to a content that has been consumed and probably cached by a new Peer. Therefore, the MPD now references a new potential streaming MMC Peer.

Any content is identified by a unique *contentID*. The Discovery Manager is made available via the following URLs:

- getDynamicMPD():
  HTTP1.1 GET api/app/content/DMPD/{IP}/{contentID}/playlist.mpd
- getMPD():
  HTTP1.1 GET api/app/content/MPD/{IP}/{contentID}/playlist.mpd

where *{IP}* parameter denotes IP of the remote MMC Peer hosting the original content and its MPD and *{contentID}* denotes unique content ID.

4.1.2 Delivery Manager

The Delivery Manager handles DASH segment requests from remote MMC Peers. Basically, the Discovery Manager serves end users with dynamically generated MPD, then the Delivery Manager of local MMC Peer (upon user's DASH player's request) fetches DASH segments and caches them according to some predefined online caching strategy. Squid cache manager [13] is used to perform caching. In case of user-generated content (UGC) scenario, the Delivery Manager is also responsible for triggering all necessary actions when end users upload content on their EHG device. The Delivery Manager is made available via the following URLs:

- consumeSegment():
  HTTP1.1 GET api/app/content/consume/{mode}/{IP}/{contentID}/{segName}
  where *{mode}* parameter denotes caching strategy and *{segName}* denotes segment name.
- postContent():
  HTTP1.1 POST
  Parameters passed as *@formDataParam*:

  – *uploadedInputStream*: uploaded content input stream;
  – *fileDetail*: meta data about uploaded content.

4.1.3 MMC Content Manager

The MMC Content Manager allows MMC Peers to communicate through a RESTful API to exchange information about content availability. MMC Peers can send a *hasCachedVideo()* request to probe remote Peers about a specific content. In return, remote Peers respond with a list of available, previously cached DASH segments. The MMC Content Manager is made available via the following URLs:

- hasVideo():
  HTTP1.1 GET api/app/content/hasVideo/{IP}/{contentID}.

The MMC Content Manager is also composed of interconnectivity capability offered by the Squid proxy, which enables the module to retrieve content from remote Peer via a local proxy on the local Peer or via a remote proxy on a remote Peer. In developed prototype, the Squid proxy is configured with MMC Peer siblings, which are used for content retrieval when necessary, and under the MMC Discovery and Delivery Managers Agreement.

4.1.4   Service Manager

The MMC Content Manager, Discovery Manager, Delivery Manager are all made available through the Service Manager which is in charge of all service access control.

4.1.5   Interfaces and Interactions with Other Components

For proper interactions between the components of the network architecture, there is a need of interfaces between them. MMC Peer middleware provides interface with MMC Manager placed in C-RAN cloud resource pool. On every end user's request for content retrieval *getDMPD()*, the local MMC Peer contacts the MMC Manager with the function *GetPeers()* in order to list the available Peers listed in the MPD file provided by the MMC Peer holder. The request format is as follows:

   {Ordered list of ContentPeer_ID} GetPeers (DestinationPeer; {List of ContentPeer_ID}; RequiredResources)

   where *DestinationPeer* is the consuming MMC Peer (i.e. local Peer); the *List of ContentPeer_ID* are the potential remote Peers and the *RequiredResources* is the required content.

4.2   Test Results

The final version of MMC Peer middleware implementation was checked by functional and non-functional requirements validation. It consists of the external and internal functionality verifications to confirm if these functionalities worked and performed well in a standalone manner. To that matter, validation scenarios including mock MMC Manager and mock content server were addressed, hence interaction with remote MMC Peers and internal functionalities were achieved offline without any support from other modules.

The testbed setup is presented in Fig. 5. In order to evaluate the efficiency of MMC system, we deployed a representative number of MMC Peers (see Table 1). Each Peer has a limited cache size. A DASH-based Video-on-Demand service is running on top of the architecture (represented by the content server which emulates CDN domain). Each MMC Peer periodically computes a summary report based on its local activity log files, and sends the report to a monitoring service which monitors the overall systems.

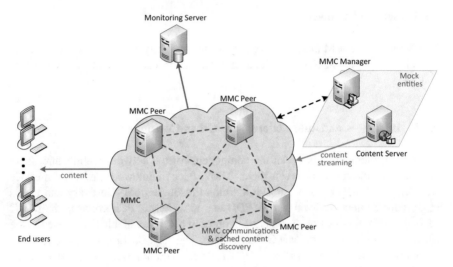

Fig. 5 Test-bed set up

Table 1 Parameters of MMC test-bed

Parameter	Value
Total number of peers for the test	10 MMC Peers
Cache size	500 MB
Catalog size (number of available videos)	20
Segment numbers for each video	15 segments
Segment length	2 s

4.2.1 Conformance Tests

The MMC Peer middleware control modules are responsible for the MMC M&C plane functionalities, namely:

- creating the MMC ad hoc system from a set of peer-to-peer connected nodes;
- handling content requests from the users, following up with appropriate behavior and sending back the appropriate responses;
- asking the MMC Manager to order a list of potential MMC Peers to be involved in the upcoming streaming session in order to optimize the state and the availability of the MMC;
- collaborating with Service Provider entities (i.e. content server) to obtain media content requested by the user, if this content (or part of it) is not stored on any of MMC Peers belonging to given MMC;
- exchanging information with MMC Manager about eventual content to be uploaded on the MMC Peer.

Regarding the data plane, MMC Peer middleware functionalities are:

- handling HTTP stream from other entity (MMC Peers, content server); basically, the MMC Peer behaves as a proxy for the end users;
- replicating content on local MMC Peer according to pre-defined content caching policy;
- streaming content to local end users or MMC Peers.

The correctness of middleware functionalities presented above was confirmed by controlling the log messages generated by involved entities. All these features were functional and the prototype passed all the conformance tests provided.

4.2.2 Performance Tests

Upon the success of the conformance tests, performance tests could be driven and could exhibit the real asset the MMC provides to existing CDNs for content delivery. The "the closer the data, the better the service" paradigm was evaluated in terms of CDN offload. Since content could come from several entities (the original content server, remote MMC Peers, both remote MMC Peers and content server), the content server load was evaluated in terms of percentage of content requests it handles. On the other hand, the evaluation of the MMC load was performed as the average percentage of content requests which MMC Peers can actually handle. This benchmark platform included a connected set of MMC Peers (that forms the MMC). The MMC Manager was mock (but functional) entity since this benchmark purpose was the evaluation of MMC Peer middleware only. Several end users were placed behind MMC Peers (they can use the same MMC Peer to access video contents). A Monitoring Server gathered MMC Peers activity logs.

During the tests, we observe the metrics presented in Table 2; they reflect the performance of the MMC to deliver content to end users.

Table 2 Definition of assessment metrics for MMC Peer middleware performance evaluation

Name of metrics	Definition and unit
Data hit ratio	(Amount of content served by caching)/(total amount of data traffic)
Request hit ratio	(Number of requests fulfilled by MMC Peers)/(total number of user requests)
Data sources (from content server)	Percentage of data (byte) coming from the content server
Data sources (from MMC)	Percentage of data (byte) coming from the remote Peers
Delay to content server	Request delay (ms) if the content is served by the origin server
Delay to MMC Peers	Request delay (ms) if the content is served by the sibling caches
Number of overhead requests	Number of the signalling messages
Bandwidth of overhead traffic	Bandwidth of the signalling messages

The test scenario assumed a set of video streaming sessions from content server or local MMC Peer to end users or to remote MMC Peers (in the meantime MMC Peers replicate content according to a caching policy). During these sessions Monitoring Server periodically retrieves MMC Peers activity logs and produces comprehensive graphs about the performance metrics. The reports related with the state of MMC Peers' caches are generated by the log analysis tools Calamaris [14], which analyzes the access log of the Squid cache proxy.

At the Server Provider side (represented by a mock content server instance), DASH-based video segments and their associated MPDs are hosted by an Apache HTTP server. At client side, user requests are generated randomly: when one video playback is over, another is started just after with a randomly selected video.

Observed Results

This section describes the evaluation results for the MMC Peer middleware performances. For each metric described in Table 2 related to the evaluation of the MMC, the monitoring tool depicts min/max/average values among all MMC Peers, which are presented in Figs. 6, 7, 8 and 9.

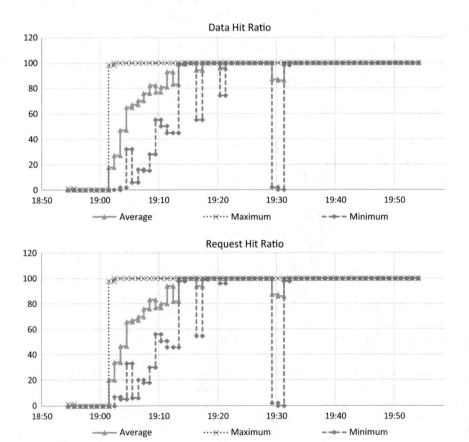

Fig. 6 Caching evaluation: data/request hit ratio

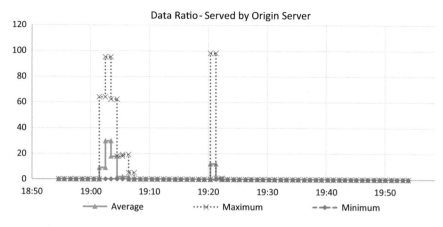

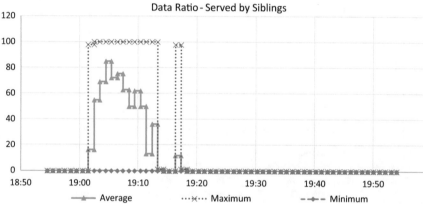

Fig. 7 Caching evaluation: data sources (from origin server or sibling caches)

The hit ratio is the most important metric for MMC Peer middleware performance. In Fig. 6, we can observe that the hit ratio equals to zero at the beginning, meaning that all requests go towards the content server. Then, the hit ratio increases quickly thanks to the caching enabled at the MMC Peers. The content requested by the end users can be directly served by the local MMC Peers or by the remote Peers, thanks to the MMC peer-to-peer communication to discover previously cached content. At the end of tests, the entire catalog is stored in the MMC and we obtains (practically) the hit ratio = 100 %.

The quick increase of the hit ratio is confirmed by the data source figure (Fig. 7). We can observe that the data source for the user requests is the original content server at the beginning. After the content is injected into the MMC system, it is directly forwarded from one MMC Peer to another to serves end users' requests. Therefore, we can see an important percentage of "sibling MMC Peer cache" data sources.

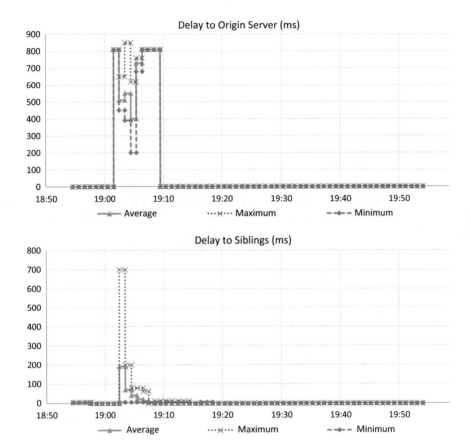

Fig. 8 Caching evaluation: request delay

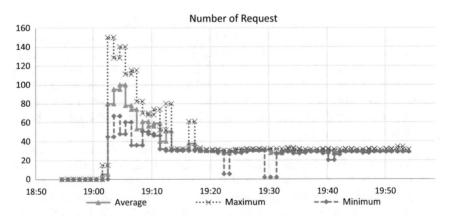

Fig. 9 MMC overhead evaluation (MMC communications and content discovery requests)

One advantage of the MMC delivery system is to reduce the request delay and increase the QoS/QoE for the end users. In Fig. 8, we measure the average delay for each request from the content server or from the sibling MMC Peer cache. We can observe that the delay to the content server is between 400 and 800 ms, depending on its load. With the sibling MMC Peer caches, the request delay is clearly reduced: 200 ms at the beginning and near 0 ms later. The 200 ms delay at the beginning is due to the fact that one MMC Peer cached a specific content and was highly requested for the delivery of this suddenly popular content. Therefore, this Peer was overloaded with requests and needed to delay some requests. This delay decreased quickly since other MMC Peers cached the content and could quickly offload the overloaded Peer.

Finally, we evaluate the overhead traffic generated by the establishment and functioning of the MMC, in our case, signaling messages used for the dynamic MPD creation. Results are shown in Fig. 9. The number of requests is important only at the beginning, for the MMC communications and content discovery. The average request size is 1 Kbits and the observed average request number per MMC Peer is 35 after the MMC bootstrap phase. Therefore, the average overhead bandwidth observed for each MMC Peer is 1 Kbits $\times$ 35 requests/8 = 4.375 kB. We can thus conclude that the overhead traffic for MMC functioning is not significant.

To summarize, the test-bed and results described in this section allowed us to validate the concept of MMC collaborative caching under real-world scenarios (MPEG-DASH Streaming).

5 Summary

This chapter presents a novel solution which aims at improving media distribution in future 5G mobile systems. The proposed streaming approach assumes collaboration of numerous base stations of dense femto-cell structure foreseen for 5G, which jointly with associated wireless terminals create an ad hoc Mobile Media Cloud system. MMC offers an innovative and low-cost way to move media close to end users. As a result, media consumers benefit from better quality and higher availability of the content, as well as lower delays in media delivery.

MMC is composed of MMC Peers, which are formed by MMC middleware running on them. The MMC Peer is responsible for the MMC creation as well as several content management services. Pushing and moving content as it is being consumed within the MMC is also of the MMC Peer's responsibilities. By collaborating with a remote MMC Manager peer ranking service, the MMC Peer is able to stream DASH content in a collaborative P2P manner in order to achieve cost-effective VoD services. By incorporation of MMC Manager directly into 5G RAN infrastructure, the proposed approach can perform optimized peer selection process with low latency and high awareness of the current status of radio resources.

The prototype implementation of the MMC Peer middleware was developed as Java-based web services. The performance tests demonstrated the effectiveness of presented solution in terms of CDN offload at the expense of small additional traffic related with MMC functioning.

Acknowledgements This work was undertaken under the Pollux IDSECOM project supported by the National Research Fund Luxembourg (FNR) and the National Centre for Research and Development (NCBiR) in Poland.

References

1. Cisco White Paper: Cisco Visual Networking Index: Forecast and Methodology, 2014–2019 (2015)
2. Golrezaei, N., Molisch, A.F., Dimakis, A.G., Caire, G.: Femtocaching and device-to-device collaboration: a new architecture for wireless video distribution. IEEE Commun. Mag. **51**(4), 142–149 (2013). doi:10.1109/MCOM.2013.6495773
3. Fajardo, J.O., Taboada, I., Liberal, F.: Improving content delivery efficiency through multi-layer mobile edge adaptation. IEEE Netw. **29**(6), 40–46 (2015). doi:10.1109/MNET.2015.7340423
4. Toledano, E., Sawada, D., Lippman, A., Holtzman, H., Casalegno, F.: CoCam: a collaborative content sharing framework based on opportunistic P2P networking. In: 2013 IEEE 10th Consumer Communications and Networking Conference (CCNC), Las Vegas, NV, 2013, pp. 158–163. doi:10.1109/CCNC.2013.6488440
5. Zhong, M. et al.: ColStream: collaborative streaming of on-demand videos for mobile devices. In: Proceedings of IEEE 15th International Symposium on a World of Wireless, Mobile and Multimedia Networks (WoWMoM), Sydney, June 2014
6. Seenivasan, T.V., Claypool, M.: Cstream: neighborhood bandwidth aggregation for better video streaming. Multimed. Tools Appl. **70**(1), 379–408 (2011)
7. Bruneau-Queyreix, J., Négru, D., Batalla, J.M., Borcoci, E.: Home-boxes: context-aware distributed middleware assisting content delivery solutions. In: Proceedings of the Posters and Demos Session (Middleware Posters and Demos'14), pp. 37–38. ACM, New York, NY, USA (2014)
8. Resource Usage Prediction Algorithms for Optimal Selection of Multimedia Content Delivery Methods
9. Efficient Entertainment Services Provision over a Novel Network Architecture
10. Patel, M. et al.: Mobile-Edge Computing Introductory Technical White Paper (2014)
11. Jersey framework [Online]. https://jersey.java.net. Accessed Oct 2015
12. Grizzly framework [Online]. https://grizzly.java.net. Accessed Oct 2015
13. Squid caching proxy [Online]. http://www.squid-cache.org/. Accessed Oct 2015
14. Calamaris log parser [Online]. http://cord.de/calamaris-english. Accessed Oct 2015

Online Music Application
with Recommendation System

Iulia Paraicu and Ciprian Dobre

Abstract Unlike regular DVD stores that allow the customer to choose from a relatively small number of products, online music platforms such as Spotify or YouTube offer large numbers of songs to their users, making the online selection process quite different from the conventional one. The goal of any recommendation system is to solve this issue by making suggestions that fit the user's preferences. The InVibe project offers a free web platform for music listening that uses its custom recommendation system to help users explore the amount of music in a natural and exciting manner. The paper will focus on the collaborative filtering algorithms used to build the recommender system, the implementation of the web application and the overall architecture designed to integrate the recommender module with the web platform.

Keywords Recommendation algorithm · Online application · Music streaming · Radio station · Web technology

1 Introduction

Today the Internet offers its users countless possibilities when it comes to online music players. However, most of them fail in attracting and maintaining the potential users. That is mostly because the big majority of music applications leave the exploring process solely in the hands of the user. The problem is that the volume of audio content available on such a platform is far too big to be properly explored by the user alone, so usually he gets confused by the endless listening possibilities. Other issues that may keep people away from the online music

I. Paraicu · C. Dobre (✉)
Faculty of Automatic Control and Computers, Department of Computer Science,
University Politehnica of Bucharest, Bucharest, Romania
e-mail: ciprian.dobre@cs.pub.ro

I. Paraicu
e-mail: iulia.paraicu@cti.pub.ro

© Springer International Publishing Switzerland 2017
C.X. Mavromoustakis et al. (eds.), *Advances in Mobile Cloud Computing
and Big Data in the 5G Era*, Studies in Big Data 22,
DOI 10.1007/978-3-319-45145-9_15

listening platforms are pricing, geographical restrictions, lack of diversity, low audio quality, online adverts that block the content, bad user experience and bad user interface.

We present InVibe, an online music platform that integrates available services and custom modules to create a complete solution for issues such as the listed above.

Firstly, like all big competitors in the online music environment, InVibe integrates a recommender system that learns the preferences of the user from her listening history and generates individual suggestions daily. Also, the recommendation system is used to create radio stations. A radio station is a playlist based on one track, chosen by the user, and filled automatically with other tracks related to the first one, so if the user feels like listening to that one track, he will probably be in the mood to listen to the other ones as well. These facilities make InVibe not only a platform for music listening, but also a music discovery system, designed to maintain the users by keeping them up-to-date and excited about the provided content.

Secondly, InVibe uses the API provided by YouTube to get its music. This way, InVibe does not have trouble with content rights. The main advantage is that music content costs nothing, so the platform can be used at no costs. Also, the application is available in all countries.

Last but not least, the application offers to music lovers a clean environment, without any other type of content, or adverts that interrupt the process of music listening.

This chapter focuses on presenting the details behind InVibe. Following this introductory section, the structure of this book chapter is organized as follows: Sect. 2 presents an overview of the related status for online music playing applications. Section 3 demonstrates a study of the recommendation algorithm, followed by a presentation of technology details for building the application in Sect. 4. Finally, Sect. 5 concludes the chapter. Due to the fact that it covers most of the problems that online music listeners face with, there are strong reasons to believe that InVibe can become a big competitor on the market.

2 Related Work

There are several examples of (online) platforms that provide music content and suggestions to users. Edison Research released their Infinite Dial Report [3], describing today's trends in music, streaming, radio and digital music. A relevant statistic in the report shows the sources that the young people use, to stay up-to-date with music. The conclusion of the report is that YouTube is considered by the young people the best source for music discovery. Multiple conclusions can be drawn from the report and statistics being presented [3]:

- The best source for music discovery is considered to be an online platform that uses a recommender system.
- Half of the sources used in the report are online applications.
- The intelligence behind these online applications has overcome the traditional radio and television, but also the suggestions that come from friends or family.
- The market of online music discovery is not dominated by only one provider, meaning that there has not been discovered a solution that fully satisfies the large majority of users yet.

We analyzed some of the web applications mentioned in the study, with their advantages and disadvantages.

YouTube[1] is a video-sharing website created in 2005 and bought by Google in late 2006. Although it was not its principal purpose, YouTube has become the most used online application for music listening and is the best music discovery source for young people. The platform has some strong points that place itself in front of the competitors. Unlike other music applications, YouTube is free, therefore, accessible to everybody. But its biggest asset is that it also represents a strongly social platform. Its users can like/dislike, share, comment or even upload content. In this way, YouTube keeps the users engaged more than any other platform.

On the other side, the social part and the 70 % videos that are not related to music can be seen as obstacles for the users that only come for the music. Also, the user interface is not optimized for a music player type of application. Therefore, simple operations like creating, populating or accessing a playlist take more time than on a music purpose platform. That is why people nowadays tend to choose applications like Spotify and Pandora when it comes to only listening to music.

Pandora[2] is an online music streaming service and automated music recommendation system that lets the user listen to music by creating radio stations. The recommender system of the application is based on the Music Genome Project that classifies songs taking into consideration more than 400 attributes. Most of these attributes values come from analyzing directly the audio signals of the music track.

Pandora probably has the best recommender system in terms of accuracy, because it has the only recommendation system that classifies the songs based on complex signal processing.

The main problem with Pandora is that it contains only 1 million songs, far less than other similar services. For example, Spotify and Beats index around 20 million songs. That is because the audio signal analysis is very expensive in terms of cost, time and other resources. Also, the free version of Pandora Radio is highly limited and full of advertisement.

Spotify[3] is an online music service with recommendations system launched in late 2008. It includes more than 20 million songs and it is available in both paid and free version. Unlike Pandora, Spotify has a lot of songs to offer to its users

---

[1]https://www.youtube.com/.

[2]http://www.pandora.com/.

[3]https://www.spotify.com.

organized in various forms: play-lists, radio-stations, suggestions and tops. That is because its recommender system is based on collaborative filtering, meaning that it calculates recommendations based on collective knowledge. According to this type of systems, the similarity between two songs is proportional to the number of the individual playlists that contains both songs. The application is also well known for a modern, dynamic and user-friendly interface.

However, Spotify is not available in some countries (including Romania) due to content rights policies. Also, its free version is full of advertisement that interrupts the online streaming.

The InVibe project comes as a result of studying these platforms, as a solution that combines from these applications their strong points, while minimizing their weaknesses.

3   The Recommender System

A recommender/recommendation system is an application module that aims to predict the preferences of the people that use the application. Such systems have become extremely popular in recent years, being integrated with lots of web applications such as movies, music, books and e-commerce platforms.

Unlike the stores from the real world, which let the customer to choose from a relatively small number of items, web platforms contain extremely large numbers of items that simply cannot be selected by the user in a totally informed manner. The purpose of recommender systems is to present to the user mostly the items that he is likely to be interested in, in order to help him decide faster.

Giving suggestions based on personal tastes, not only helps the user make faster decisions, but also changes the life-cycle of the recommended items. Physical organizations can only provide what is popular at the moment, most of the time due to good marketing strategies, but online organizations have everything available and can raise the popularity of truly valuable items.

There have been made various steps into improving the efficiency of the recommendation systems since they first appeared [1]. In 2006 Netflix, an online DVD rental company, has announced a prize of $1 million to the first team to improve the system by 10 % [2]. The prize was given in 2009 to BellKor's Pragmatic Chaos, team that managed to create an algorithm that was 10.6 % more accurate than the one at Netflix [4].

Today, machine learning algorithms for recommender systems are constantly being developed and improved, because online platforms such as Amazon, Spotify or YouTube are trying to create an experience that feels more and more natural to their users.

Recommender systems try to solve the problem of giving accurate suggestions using one of the following approaches:

- Content-based. This type of system tags all items according to some features and subsequently applies classification algorithms to organize and suggest them to the users. This approach can be unrealistic in some situations because a system does not always possess sufficient classification information about its products, or this kind of information is really expensive to obtain.
- Collaborative filtering (short CF) recommends items based on similarity measures between users and/or items. The basic idea is that users who shared the same interests in the past e.g. have viewed or bought the same books, are likely to have similar tastes in the future.

3.1 Collaborative Filtering

We define the problem of collaborative filtering in the following way. The problem can be represented by the triple (U, I, R), where:

- U is the user identifier and takes values between $1, \ldots, N$
- I is the item identifier and takes values between $1, \ldots, M$
- R represents the rating given by the user U to the item I

The CF algorithms learn from a given training set, formed by a large number of (U, I, R) triplets, than have to calculate an estimated rating, R, for each entry of a test set. The root mean squared error (RMSE) and the mean absolute error (MAE) are used to compare the performances of the algorithms.

n—the number of entries in the test set

$$MAE = \frac{\sum_1^n (R_i' - R_i)}{n}$$

$$RMSE = \sqrt{\frac{\sum_1^n (R_i' - R_i)^2}{n}}$$

In the context of recommender systems, the items, I, can be represented by songs (like in our case), movies, products and more. In the following sections, the algorithms described operate on movies, but the principles remain the same for other types of items, such as songs.

3.1.1 The Utility Matrix

The utility matrix is an $N \times M$ matrix containing the ratings that users gave to the items.

Figure 1 illustrates a very simple example in which users A, B, C, D have rated items HP1, HP2, HP3, TW, SW1, SW2, SW3 (Harry Potter, Twilight and StarWars

	HP1	HP2	HP3	TW	SW1	SW2	SW3
A	4			5	1		
B	5	5	4				
C				2	4	5	
D		3					3

Fig. 1 Utility matrix

	HP1	HP2	HP3	TW	SW1	SW2	SW3
A	2/3			5/3	−7/3		
B	1/3	1/3	−2/3				
C				−5/3	1/3	4/3	
D		0					0

Fig. 2 Normalized utility matrix

movie series). A value in the matrix corresponds to the rating given by the user in the same row to the item in the same column. If the user has not rated an item, the corresponding matrix position is empty.

Some users may have similar tastes, but different habits in rating items. For example, some users will rate with 1 a movie that they consider bad while other will consider 3 an appropriate low mark. For this reason, the matrix should be normalized before applying any algorithm. The normalization is done by subtracting from each rating the average rating of the corresponding user, which turns the good ratings into positive numbers and the bad ones into negative numbers (Fig. 2).

3.1.2 K-Nearest Neighbor

KNN (K-Nearest Neighbor) identifies the users that have the most similar interests with the current user, called neighbors and predicts the users rating based on the ones given by the neighbors.

The first challenge is to find a way to calculate the similarity between users. One method is to compute the cosine distance between the vectors describing two users. The vectors $A = [\frac{2}{3}; 0; 0; \frac{5}{3}; -\frac{7}{3}; 0; 0]$ and $B = [\frac{1}{3}; \frac{1}{3}; -\frac{2}{3}; 0; 0; 0; 0]$ describe the users A and B. The cosine of two vectors can be found by using the Euclidean dot product formula:

$$A \cdot B = \|A\|\|B\| \cos \theta$$

$$similarity = \cos \theta = \frac{A \cdot B}{\|A\|\|B\|} = \frac{\sum_1^n (A_i \times AB_i)}{\sqrt{\sum_1^n (A_i)^2} \times \sqrt{\sum_1^n (B_i)^2}}$$

The similarity is the result of the cosine angle, which means it will take values from the interval $[-1; 1]$. If the cosine angle is close to 1, it means that the users are similar, while values approaching—1 correspond to users that have opposite tastes.

After identifying the nearest neighbors, the algorithm predicts user ratings using the formula:

$$R' = \frac{\sum_1^k (similarity(u, u_i) \cdot rating(u_i))}{\sum_1^k similarity(u, u_i)}$$

3.1.3 QR Decomposition

There are two ways to look at the QR Decomposition. First is the standard definition: $A = QR$, where Q and R must be found. This can be done easily following the straightforward decomposition algorithm. However, in the context of recommender systems the utility matrix, A, contains lots of empty entries, therefore, cannot be decomposed by the standard method. Moreover, if the missing values were known, there would be no need to apply the decomposition because all the missing ratings would be known. It is clear that the definition QR Decomposition method cannot be used to solve the recommendations problem.

The second way to look at the QR Decomposition is the one proposed by Simon Funk in 2006, at Netflix Prize [2]. He started from the idea that the utility matrix does not contain random numbers, but values somehow related which are described by some generalities. For example, a movie can be approximately described by some basic attributes like gender, director, what stars play in it and so on. Also, every user's preferences can be described by whether he likes or not a gender, a director, some stars and so on. If these are true, then we need less than $M \times N$ numbers to explain all the ratings. Assuming that there are k attributes of this kind, the user's preferences can be represented by a vector of length k. The movie's characteristics can also be described by a similar vector.

The rating given by a user to an item will be equal to the product of the users vector and the movies vector:

$$ratingMatrix[user][item]$$
$$= \sum_1^k userFeature[i][user] \cdot itemFeature[i][item]$$

All the user vectors compose a user-feature matrix, Q, and all the items vector compose an item-feature matrix, R (Fig. 3).

Simon Funk used to say that, in machine learning, the reasoning works in both ways, therefore if meaningful generalities can help you represent your data with fewer numbers, finding a way to represent your data in fewer numbers can often help you find meaningful generalities. Meaning that if we find two matrices Q and R_t that multiplied provide values very close to the one already known in A

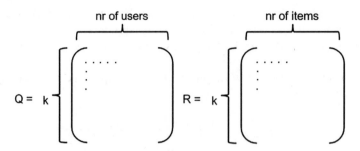

Fig. 3 Q—user-feature matrix and R—item-feature matrix

(the utility matrix), then the rest of the values from QRt correctly approximate the missing entries of A.

In our case:

$$A \approx Q \times R_t$$

$$\begin{bmatrix} \frac{2}{3} & 0 & 0 & \frac{5}{3} & -\frac{7}{3} & 0 & 0 \\ \frac{1}{3} & \frac{1}{3} & -\frac{2}{3} & 0 & 0 & 0 & 0 \\ 0 & 0 & 0 & -\frac{5}{3} & \frac{1}{3} & \frac{4}{3} & 0 \\ 0 & 0 & 0 & 0 & 0 & 0 & 0 \end{bmatrix}$$

$$= \begin{bmatrix} q_{11} & q_{12} \\ q_{21} & q_{22} \\ q_{31} & q_{32} \\ q_{41} & q_{42} \end{bmatrix} \times \begin{bmatrix} r_{11} & r_{12} & r_{13} & r_{14} & r_{15} & r_{16} & r_{17} \\ r_{21} & r_{22} & r_{23} & r_{24} & r_{25} & r_{26} & r_{27} \end{bmatrix}$$

The algorithm to find the Q and R matrices is described in the following paragraphs. Firstly, the matrices are initialized with random values, then each iteration consists in updating the values from the matrices, so the $Q \times R_t$ product will be closer to the A matrix.

To obtain the update rules, the squared error equation is written, after that the gradient of the current values must be find out. Therefore, the following equation should be differentiated with respect to q and r variables separately:

$$e^2 = \left(R^{'} - R\right)^2$$

$$\frac{\partial}{\partial q} e^2 = -2\left(R^{'} - R\right)r = -2er$$

$$\frac{\partial}{\partial r} e^2 = -2\left(R^{'} - R\right)q = -2eq$$

The update rules can now be formulated, considering $\propto$ as the learning rate:

$$q' = q + 2\propto er$$

$$r' = r + 2\propto eq$$

The algorithm should be run more than once with different starting values for Q and R to avoid being stuck in a local minimum and increase the chances of finding the global minimum.

3.2 Results

A data set provided by Movie Lens (Grouplens 2016)[4] was used to test the previously described algorithms. It contains real records of ratings given by people to movies. The set contains 100 K entries of the form $\langle user_{id}, movie_{id}, rating \rangle$ that are split into a training set (80 %) and a test set (20 %). The data was already prepared for cross-validation, by being organized in five pairs of files $\langle training\_set, test\_set \rangle$, which joined contain the same 100 k entries, but all five test sets are disjoint. There is also additional information about users and movies that can be used for a content-based approach.

KNN versus QRD

The algorithms were compared based on accuracy, time and error distribution. When it comes to time, the two algorithms do quite different, depending on the situation. That is because KNN is lazy learning method, which defers the generalization of the training data until a request is made to the system, while QR Decomposition is an eager learning method, which generalize the training set before receiving queries. Because of this, a system that needs to calculate in real time a relatively small number of ratings per time unit should use the KNN algorithm, whereas a system that computes large numbers of ratings per time unit but can base its prediction on slightly old versions of the utility matrix should definitely use the QR Decomposition approach.

After running tests with the two previously described algorithms, similar performances have been noticed in terms of accuracy. However, the QR decomposition had slightly better results (Table 1).

Table 1 Experimental results

	MAE	RMSE
QR decomposition	0.741	0.938
Nearest neighbors	0.749	0.955
Arithmetic mean	0.735	0.932

---

[4]http://grouplens.org/datasets/movielens/

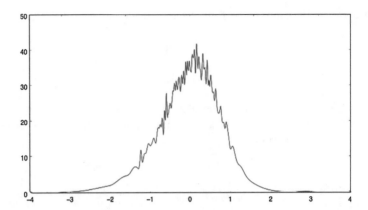

Fig. 4 The density of the errors for the KNN algorithm

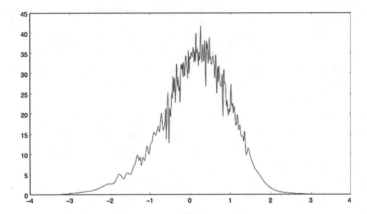

Fig. 5 The density of the errors for the QR algorithm

The arithmetic mean of all pairs of results generated by the two methods has better accuracy then each algorithm alone. This could have meant that the error distribution is different for the two methods, and they can be combined in a convenient way. That is why the distribution of the error is studied further on for both methods.

In the graphs in Figs. 4 and 5, the x-axis represents the value of the error, and the y-axis represents the number of examples. The graphs show the density of processed examples on each error level. Firstly, it can be noticed that 80 % of the examples have errors lower than 25 % for both algorithms, which means that the methods can be trusted when it comes to predicting the user's preferences. On the other hand, the graphs look very similar, revealing that the distribution error is similar for both methods.

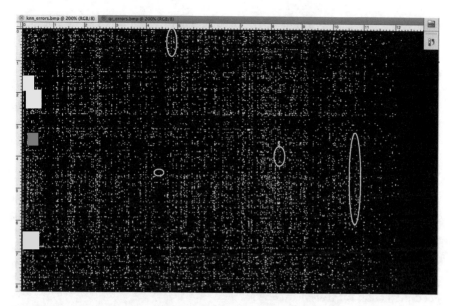

Fig. 6 Section of the utility matrix after the KNN algorithm was applied

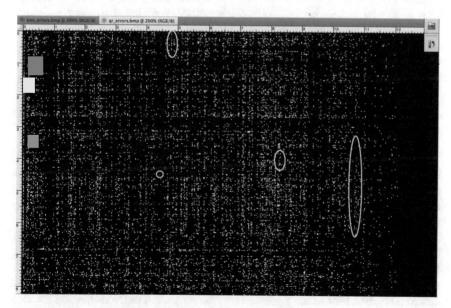

Fig. 7 Section of the utility matrix after the QR algorithm was applied

To understand better how the errors are spread in the utility matrix, we analyzed them graphically.

In Figs. 6 and 7, each pixel represents an entry from the utility matrix. The nonblack pixels represent all the entries from the test set. The red pixels indicate a positive error $(predicted\_rating > real\_rating)$, while the blue ones show a negative error $(predicted_{rating} < real\_rating)$. The intensity of the color grows proportionally with the value of the error, so white pixels represent perfect predictions, and brightly colored ones indicate maximum values for the error.

As it can be seen, the images look very similar and the maximum points of error (circled in yellow) are the same in both cases, meaning that the results of the two approaches have the same distribution error.

3.2.1   Discussion

Before comparing the error distributions for the two algorithms, we thought of combining the methods with a trained neural network in order to obtain better accuracy. However, because the errors are distributed quite similarly, we concluded that such an approach would not generate significant or consistent improvements. On the positive side, the tests have proven that even a simple recommender system relying on textbook-algorithms with minimal tuning and improvements can achieve good prediction for real-life datasets.

Due to the fact that the algorithms have close performances when it comes to precision, and also because the data sets can become really large, we concluded that time parameter is the most important when making a choice. However, time is directly influenced by the nature of the problem.

In our particular case of recommender system, there are two situations where the application needs to provide automatically calculated suggestions:

- When providing its users with a top of ten personal recommendations (that will happen once every week).
- When the user wants to create a radio-station (a playlist based on a selected song).

For the first scenario, we decided to use the QR Decomposition algorithm. That is because the system will need to generate a lot of suggestion at once, so we preferred the eager learning method that firstly calculates the utility matrix decomposition, then can provide any rating in almost no time by multiplying one row from the Q matrix with one column from the R one.

A radio station must provide a number of songs related to one track (chosen by the user) that the user is supposed to enjoy if he is in the mood for the first song. Finding items similar to a specific item is exactly what the K-Nearest Neighbour algorithm does. This time the implementation is a bit different from the one described earlier. Firstly, the algorithm runs on songs, not on users, base on the idea that if two different songs are rated high by the same group of users then they are similar. Secondly, in this case, the algorithm stops right after finding the neighbors, because it is no need to calculate anything else.

As the system grows in terms of users and listened to tracks, one should consider applying parallel computing techniques for the implementations. This will allow processing bigger data sets in less time.

4 An Integration into an Online Application

The web application and the recommender system are completely separated modules. That is way their integration can be presented without knowing how the web platform works.

The recommender system is composed of five python scripts that are connected. It communicates with the web platform only through the database by the following scenario:

1. The users feedback, collected from the player each time the user rates a song, is sent to the web server and then stored in the database.
2. The database communicates with the recommendations system through the MySQL-Python connector installed on the same machine with the RS, so the input querier script, from the RS, can directly query the database to get the stored ratings.
3. The build um script uses the ratings to generate the initial utility matrix.
4. The system applies QR Decomposition on the utility matrix to generate the suggestions for each user and K-nearest Neighbor on the same matrix to build the radio stations.
5. Both suggestions and playlists for radio stations are written to the database by the output writer script.
6. When suggestions page or radio-station page need to be shown, the web server reads the data generated by the RS from the database and sends it to the client application running in the browser.

The main advantage of this architecture is that the web platform and the recommendation system can be developed and tested completely separated, which minimize the risk of integration problems. Also, the platforms can be written in distinct languages and are able to run on different machines, which have the resources and the configurations appropriate for each system (Fig. 8).

4.1 Used Technologies

Besides the standard languages understood by browsers (HTML, CSS and Java-Script), there are a lot of decisions to be made when it comes to choosing the web technologies appropriate for an application. This section describes the tools and libraries used to create the InVibe web platform.

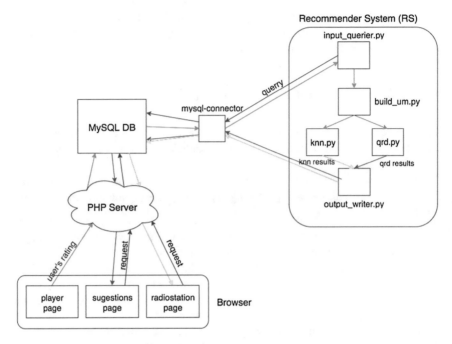

Fig. 8 The recommender system architecture and the integration with the web application

MySQL is the most popular open-source relational database management system. MySQL offers a high degree of connectivity, speed, security and scalability, and is an appropriate choice for most of the web applications.

PHP is an open-source, general propose scripting language that runs on top of a web server such as Apache. PHP scripts can be embedded into HTML code, are executed on the server, and then the result is returned to the browser. The result is usually pure HTML that will be directly interpreted by the browser, a file or a JSON, which contains information that will be used by a JavaScript script through AJAX.

Laravel is an open PHP framework designed over an MVC architecture. Laravel has an expressive syntax and provides a high-level modularization. The framework has a lot of facilities that aim to eliminate the routine procedures that the developer does, such as routing, authentication, sessions and queries. Laravel also integrates other tools and libraries to make the developer experience more pleasant.

One of those tools used in developing the web application is Vagrant. Vagrant lets the developer work on a Linux virtual machine through the ssh tool. The Vagrant virtual machine can easily be configured to contain folders that are synchronized with ones from the local machine in real time. Laravel contains a prepackaged Vagrant box (homestead) which has installed all the software necessary for the applications server, so the developer can program without having to install PHP, a web server, MySQL or any other server software on the local machine. The application can also be easily configured to be tested directly from the local machines browser.

The Laravel framework helps the developer to communicate with the database in an efficient way through the Eloquent ORM library. The models are created only by extending the Eloquent/Model class and specifying the corresponding database table. Querying the database with Eloquent will directly return model instances that match the query. Also, to insert or update database information the developer only has to apply the save method provided by the library on the model object.

Another useful tool is Laravel Elixir. It run on top of Gulp and is used to transform LESS and SASS code into CSS in real time.

JQuery was used to simplify the client scripting of the HTML. JQuery offers a much simpler way to manipulate the HTML elements, create animations and use AJAX techniques. Its Syntax is also really compact, so it makes the applications faster to develop. 28.

AJAX was used to create dynamic pages by changing small amounts of data between server and client. In other words, the web page can change its contents without having to reload.

SASS, is a scripting language that is interpreted to CSS. SASS provides the developer with some powerful tools such as variables, nesting and inheritance, that make the process of creating stylesheets faster and more organized than with plain CSS code.

By using all these technologies, the development of the web platform was faster, and the main effort was put into structuring and customizing the application rather than into doing routine tasks.

4.2   The Application

4.2.1   The Database

The database structure is presented in Fig. 9.

The users table stores information about the user such as name, email and password. The songs table contains the YouTube id of the song, the title (artist—song name), a URL to a thumbnail image resource and the YouTube video id. The songs-users table has entries that represent many-to-many relations between users and songs. The table is used to recreate the playlist of each user.

The ratings table stores every rating given by any user to any song. The radio stations table contains entries described by their own id and the id of the base song from that radio station.

The radio stations-songs table contains many-to-many relations between radio stations and songs. Each song may be in more than one radio station and each radio station has at least one song. Finally, the recommendations table stores the suggestions calculated by the recommender system, each containing the id of the user, the id of the song and the rank of the recommendation.

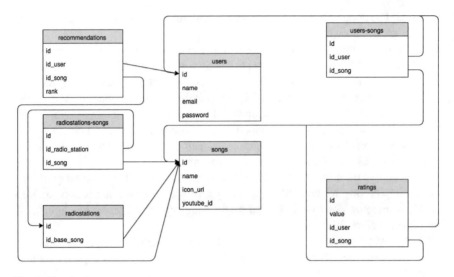

Fig. 9 The database structure

4.2.2 The MVC Structure of the Application

MVC (Model-View-Controller) is the most popular architectural pattern when it comes to web applications. According to it, the application should be divided into three components (model, view and controller) that have different roles and communicate respecting specific rules, like in Fig. 10 Laravel is already structured according to the MVC pattern, as most of the PHP frameworks.

This section presents the general concepts of MVC and analyses how they are applied to the InVibe application.

The model (Fig. 11) contains entities mapped to the tables of the database. Each table is represented by an object, and its attributes are the tables columns. The model can only communicate with the controller, which requests these data objects.

In Laravel, the model data objects are created automatically just by extending the Model class and specifying the database table name. The object may eventually be sent to the client application in JSON format, for this some attributes can be hidden, like in the example in Fig. 11.

After defining the model, it can be used anywhere in the controllers.

The view (Fig. 12) acts as a template for the model data received from the controller and formats it in a way accessible to the user. Views completely isolate the HTML code from the rest of the application.

Laravel introduced the concept of blade templates, which provide an inheritance mechanism that works with sections.

The child inherits the HTML structure of the parent through the annotation extends. The child is able to insert sections of code in the inherited structure with the annotation section. For example, the code between *@section(content)* and *@endsection* in the child view (Fig. 13), will be inserted in place of *@yield(content)*

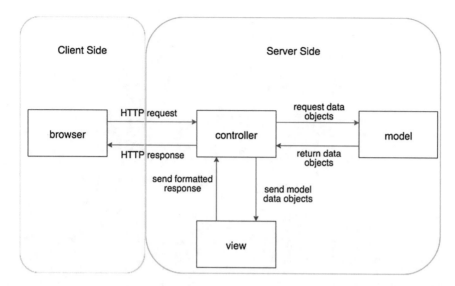

Fig. 10 The MVC structure

```
class User extends Model implements AuthenticatableContract, CanResetPasswordContract
{
    use Authenticatable, CanResetPassword;

    /**
     * The database table used by the model.
     *
     * @var string
     */
    protected $table = 'users';

    /**
     * The attributes that are mass assignable.
     *
     * @var array
     */
    protected $fillable = ['name', 'email', 'password'];

    /**
     * The attributes excluded from the model's JSON form.
     *
     * @var array
     */
    protected $hidden = ['password', 'remember_token'];
}
```

Fig. 11 Laravel Model class

in the parent view (Fig. 12). The child can also append code to a section already defined in the parent like shown in the example above.

The diagram in Fig. 14 represents how the views are structured in blade templates in the application InVibe. Each template on the second level is used by a different controller to format data.

```
<!-- Stored in resources/views/layouts/master.blade.php -->

<html>
    <head>
        <title>App Name - @yield('title')</title>
    </head>
    <body>
        @section('sidebar')
            This is the master sidebar.
        @show

        <div class="container">
            @yield('content')
        </div>
    </body>
</html>
```

Fig. 12 Parent blade template

```
<!-- Stored in resources/views/layouts/child.blade.php -->

@extends('layouts.master')

@section('title', 'Page Title')

@section('sidebar')
    @parent

    <p>This is appended to the master sidebar.</p>
@endsection

@section('content')
    <p>This is my body content.</p>
@endsection
```

Fig. 13 Child blade template

Finally, the controller is the central component in an MVC architecture. It receives the HTTP request from the client side, sends a data object request to the model, takes the data received from the model and sends it to the view, then receives the formatted data and finally sends it back to the client as a HTTP response.

In Laravel, each HTTP request is taken over by a particular method in a controller, as specified in the file routes.php (Figs. 15 and 16).

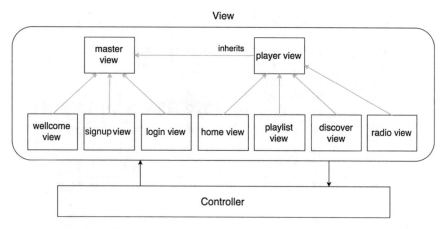

Fig. 14 Blade templates structure

```
Route::get('/', 'WelcomeController@show');
```

Fig. 15 Section of routes.php file

Fig. 16 Laravel controller

```
class WelcomeController extends Controller
{
    public function show()
    {
        if (Auth::check()) {
            return redirect('home');
        }

        return view('welcome');
    }
}
```

The controller above returns to the browser the welcome page formatted by the welcome.blade.php view when the user makes a request at invibe.app/. This controller does not require information from the database, so it does not access any model.

The InVibe application has different controller for each web page. They communicate with the browser through simple HTTP requests, for displaying the page and through AJAX requests to create dynamic elements in the page.

The diagram in Fig. 17 represents the structure of the web pages in the browser and the controllers which generate them. When accessing invibe.app the welcome page is shown. It contains few details about the application, the sign-up and the

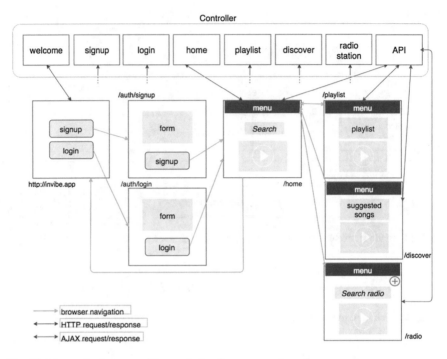

Fig. 17 The applications controllers and who they manage the web pages

log-in button. If the user is already authenticated, he will be redirected to the home page.

The user enters the login page or the sign-up one, in case he does not have an account, then he completes the required form to access the home page of the application.

The home page has a menu from which the user can navigate to the playlist, discover, or radio station sections. It also contains a search input, a video player and a rating bar.

The playlist page is similar to the home page. In addition, it contains a list of song added by the user to the personal playlist. The songs are added and removed dynamically to the playlist, through AJAX.

The user can see the daily recommendations in the discover section and give them feedback according to his preferences.

The radio station page provides a dynamic search input to the user, from which a new radio station is shown with the list of the contained songs.

All the controllers described above have only one method each. The method, named *show*, is called when the user accesses the page URL and returns the HTML page to the browser.

The dynamic actions of the application are realized through AJAX requests. All of these requests are directed to the *APIController*, which handles them. The API between the client and the server contains the following requests:

- Search Song—post request to the *api/searchSong* endpoint. The request parameter is the search text input, and the response is a list of songs with YouTube id, name and thumbnail URL.
- Choose Song—post request to the *api/chooseSong* endpoint. The request parameter is the YouTube id of the chosen song, and the response contains the songs rating and a flag that indicates whether the song is in the user's playlist or not.
- Rate Song—post request to the *api/rateSong* endpoint. The request parameters are the YouTube id, the song name, the song thumbnail URL and the value of the rating. The song is added to the database if it does not exist yet or just updated with the new rating value.
- Add/Remove Song—post request to the *api/addSong* endpoint. The request parameters are the YouTube id, the song name, the song thumbnail URL and the *inPlaylist* flag. The song is added or removed from the playlist database table based on the flag's value.
- Load Playlist—get request to the *api/loadPlaylist* endpoint. The request response returns the list of songs from the user's playlist with id, YouTube id and name.
- Load Recommendations—get request to the *api/loadRecom* endpoint. The request response returns the list of songs generated by the recommender system and stored in the recommendation database table.
- Load Radiostation—post request to the *api/loadRadio* endpoint. The request parameter is the radio station base song YouTube id, and the response is the list of songs from the corresponding radio station, calculated by the recommendation system.

5 Concluding Remarks

The online music market is certainly a still growing one, being full of diversity and trying to please every music lover on the Internet. Working at the InVibe project, we had the opportunity to study this market and understand what most of the users expect from such an application. That is why we ended up creating a modern web platform, with an intelligent recommendation system integrated, that is available everywhere and costs the user nothing.

To create the recommender system, we studied multiple collaborative filtering and content-based approaches, ending up to use the QR Decomposition and the K-Nearest Neighbour algorithms. We have chosen these methods because they have good performances, but also because they are the best-suited ones for the application requirements. The algorithms were implemented in Python. For the web application, we used the server-side scripting language PHP, the Laravel MVC framework and a MySQL database. These techniques are easy to learn and fast to implement, providing the developer with powerful tools that require a minimum effort.

The recommendations system and the web platform are completely separate modules, for this reason even changing entirely one of them would not affect the functionalities of the other one.

Acknowledgements The authors would like to thank the anonymous reviewers for their constructive comments and feedback on the manuscript.

References

1. Adomavicius, G., Tuzhilin, A.: Toward the next generation of recommender systems: A survey of the state-of-the-art and possible extensions. IEEE transactions on knowledge and data engineering, **17**(6), 734–749 (2005)
2. Bennett, J., Lanning, S.: The netflix prize. In: Proceedings of KDD cup and workshop, vol. 2007, p. 35 (2007)
3. Edison Research: The Infinite Dial (2015). Report http://www.edisonresearch.com/the-infinite-dial-2015/. Accessed 09 May 2016
4. Koren, Y.: The bellkor solution to the netflix grand prize. Netflix prize documentation, 81 (2009)

Author Biographies

Iulia Paraicu finished her Bachelor studies within the University POLITEHNICA of Bucharest. Her main research areas are ccollaborative systems, artificial intelligence and algorithms, and Web technologies. Full contact details of author: Iulia Paraicu, University POLITEHNICA of Bucharest, 313, Splaiul Independentei, Office EG403, sector 6, 060042, Bucharest, Romania, E-mail: iulia.paraicu@cti.pub.ro.

Ciprian Dobre completed his Ph.D. at the Computer Science Department, University Politehnica of Bucharest, Romania, where he is currently working as a full-time Associate Professor. His main research interests are in the areas of modeling and simulation, monitoring and control of large scale distributed systems, vehicular ad-hoc networks, context-aware mobile wireless applications and pervasive services. He has participated as a team member in more than 10 national projects the last four years and he was member of the project teams for 5 international projects. He is currently involved in various organizing committees or committees for scientific conferences. He has developed MONARC 2, a simulator for LSDS used to evaluate the computational models of the LHC experiments at CERN and other complex experiments. He collaborated with Caltech in developing MonALISA, a monitoring framework used in production in more than 300 Grids and network infrastructures around the world. He is the developer of LISA, a lightweight monitoring framework that is used for the controlling part in projects like EVO or the establishment of world-wide records for data transferring at SuperComputing Bandwidth Challenge events (from 2006 to 2010). His research activities were awarded with the Innovations in Networking Award for Experimental Applications in 2008 by the Corporation for Education Network Initiatives (CENIC) and a Ph.D. scholarship by Oracle between 2006 and 2008.